Selected Titles in This Series

175 **R. L. Dobrushin, R. A. Minlos, M. A. Shubin, and A. M. Vershik, Editors,** Contemporary Mathematical Physics (F. A. Berezin Memorial Volume)

174 **A. A. Bolibruch, A. S. Merkur′ev, and N. Yu. Netsvetaev, Editors,** Mathematics in St. Petersburg

173 **V. Kharlamov, A. Korchagin, G. Polotovskiĭ, and O. Viro, Editors,** Topology of Real Algebraic Varieties and Related Topics

172 **K. Nomizu, Editor,** Selected Papers on Number Theory and Algebraic Geometry

171 **L. A. Bunimovich, B. M. Gurevich, and Ya. B. Pesin, Editors,** Sinai's Moscow Seminar on Dynamical Systems

170 **S. P. Novikov, Editor,** Topics in Topology and Mathematical Physics

169 **S. G. Gindikin and E. B. Vinberg, Editors,** Lie Groups and Lie Algebras: E. B. Dynkin's Seminar

168 **V. V. Kozlov, Editor,** Dynamical Systems in Classical Mechanics

167 **V. V. Lychagin, Editor,** The Interplay between Differential Geometry and Differential Equations

166 **O. A. Ladyzhenskaya, Editor,** Proceedings of the St. Petersburg Mathematical Society, Volume III

165 **Yu. Ilyashenko and S. Yakovenko, Editors,** Concerning the Hilbert 16th Problem

164 **N. N. Uraltseva, Editor,** Nonlinear Evolution Equations

163 **L. A. Bokut′, M. Hazewinkel, and Yu. G. Reshetnyak, Editors,** Third Siberian School "Algebra and Analysis"

162 **S. G. Gindikin, Editor,** Applied Problems of Radon Transform

161 **K. Nomizu, Editor,** Selected Papers on Analysis, Probability, and Statistics

160 **K. Nomizu, Editor,** Selected Papers on Number Theory, Algebraic Geometry, and Differential Geometry

159 **O. A. Ladyzhenskaya, Editor,** Proceedings of the St. Petersburg Mathematical Society, Volume II

158 **A. K. Kelmans, Editor,** Selected Topics in Discrete Mathematics: Proceedings of the Moscow Discrete Mathematics Seminar 1972–1990

157 **M. Sh. Birman, Editor,** Wave Propagation. Scattering Theory

156 **V. N. Gerasimov, N. G. Nesterenko, and A. I. Valitskas,** Three Papers on Algebras and Their Representations

155 **O. A. Ladyzhenskaya and A. M. Vershik, Editors,** Proceedings of the St. Petersburg Mathematical Society, Volume I

154 **V. A. Artamonov et al.,** Selected Papers in K-Theory

153 **S. G. Gindikin, Editor,** Singularity Theory and Some Problems of Functional Analysis

152 **H. Draškovičová et al.,** Ordered Sets and Lattices II

151 **I. A. Aleksandrov, L. A. Bokut′, and Yu. G. Reshetnyak, Editors,** Second Siberian Winter School "Algebra and Analysis"

150 **S. G. Gindikin, Editor,** Spectral Theory of Operators

149 **V. S. Afraĭmovich et al.,** Thirteen Papers in Algebra, Functional Analysis, Topology, and Probability, Translated from the Russian

148 **A. D. Aleksandrov, O. V. Belegradek, L. A. Bokut′, and Yu. L. Ershov, Editors,** First Siberian Winter School "Algebra and Analysis"

147 **I. G. Bashmakova et al.,** Nine Papers from the International Congress of Mathematicians, 1986

146 **L. A. Aĭzenberg et al.,** Fifteen Papers in Complex Analysis

145 **S. G. Dalalyan et al.,** Eight Papers Translated from the Russian

(Continued in the back of this publication)

Contemporary Mathematical Physics

F. A. Berezin Memorial Volume

Felix Alexandrovich Berezin
(1931–1980)

American Mathematical Society

TRANSLATIONS

Series 2 • Volume 175

Advances in the Mathematical Sciences – 31

(*Formerly Advances in Soviet Mathematics*)

Contemporary Mathematical Physics

F. A. Berezin Memorial Volume

R. L. Dobrushin
R. A. Minlos
M. A. Shubin
A. M. Vershik
Editors

American Mathematical Society
Providence, Rhode Island

Translation edited by A. B. Sossinsky

1991 *Mathematics Subject Classification.* Primary 22-XX, 47-XX, 81-XX.

ABSTRACT. The content of the volume is connected in various ways to the scientific heritage of F. A. Berezin (1931–1980), an outstanding Moscow mathematician who tragically died in an accident. F. A. Berezin discovered new notions and ideas in mathematical physics, representation theory, analysis, geometry and other areas of mathematics. In particular, he introduced the notion of Grassmannian analysis ("supermathematics") and a new notion of deformation quantization. The topics of papers in this volume include quantization, invariant integration, spectral theory, representation theory, etc.

The book is useful for graduate students and researchers working in various areas of mathematics and mathematical physics.

Library of Congress Card Number 91-640741
ISBN 0-8218-0426-X
ISSN 0065-9290

---

10 9 8 7 6 5 4 3 2 1   01 00 99 98 97 96

Contents

Foreword

Over fifteen years have passed since Felix Alexandrovich Berezin died. Many things have changed since then, and life today is very different from what it was at that time. Berezin was among those who most ardently strived for changes; unfortunately, he lived neither to see the day when his ideas and work had become popular and widely accepted, nor to work in a period when scientific contacts with friends and colleagues dispersed in the world at large had become routinely possible.

The editors of this collection, who knew F. A. Berezin very closely, have called on all those who have worked with him at one time or another, studied under his supervision, developed his ideas, were his friends or simply wanted to be one of the authors of a volume devoted to the memory of this outstanding scientist. The idea of such a publication attracted many mathematicians and physicists. The result of their work is collected in two volumes of the present series, which represent a wide spectrum of topics in mathematical physics—F. A. Berezin's main area of research—and related branches of mathematics.

In discussing F. A. Berezin's scientific heritage, one should begin by mentioning his two most significant achievements: quantization theory and supermathematics. But his name lives on in representation theory, statistical physics, operator theory, and the geometry of homogeneous spaces. His papers are remarkable by their breadth of insight and their conceptual character. We are now witnesses of the powerful development of Berezin's quantization and Berezin's Grassmannian analysis (supermathematics). But to an observer from the outside, the influence that Berezin exerted on mathematicians and physicists may perhaps not be too obvious, more precisely, the role that he played in **making physics popular among mathematicians and mathematics among physicists.**

In this volume, the first of two, we have assembled papers in group representation theory and several mathematical papers in supermathematics and spectral analysis.

It was not easy to organize the interaction of such a group of mathematicians, dispersed in many cities and countries; all the organizational and technical tasks involved in the preparation of this volume were successfully carried out by N. D. Vvedenskaya, to whom the editors express their gratitude.

We hope that this volume will express our deep respect for the memory of F. A. Berezin and will promote his work and ideas.

R. Dobrushin,[1] R. Minlos, M. Shubin, A. Vershik

[1]The editors want to announce that Roland L'vovich Dobrushin died November 12, 1995. He was a friend of F. A. Berezin and one of the organizers of this collection.

Amer. Math. Soc. Transl.
(2) Vol. **175**, 1996

Felix Alexandrovich Berezin (A Brief Scientific Biography)

R. A. Minlos

1. What he achieved in science (*an overview of F. A. Berezin's work, the modern understanding of mathematical physics*)[1].

In his relatively brief life (he died in an accident before reaching the age of 50), F. A. Berezin succeeded in doing a great deal in mathematics and mathematical physics. Not only did he leave a deep trace in several branches of mathematics that existed before him (group representation theory, the spectral theory of operators, quantum mechanics, statistical physics, constructive quantum field theory), but he also initiated several new concepts, methods, and theories: a general approach to the quantization problem, the construction of the second quantization formalism in terms of functional integrals, which later became the so-called "calculus of symbols" (a forerunner of the theory of pseudo-differential operators), and finally (this was his most important and long nurtured achievement) the theory of supersymmetry and supermanifolds, i.e., what mathematicians now usually call supermathematics.

Further we shall discuss all these topics in more detail. Here I would only like to stress that perhaps the most valuable and important characteristic of Berezin's mathematical life was not his concrete achievements, but the overall stubborn direction of his research, whose main backbone was mathematical physics. He was one of the very few people who transformed mathematical physics into what it has become today. In fact, until the end of the fifties, the expression "mathematical physics", at least in Russia, was mainly associated with the study of special types of differential equations arising in physical theories (the wave equation, the heat equation, etc.). Berezin was one of the first to notice that, as the old saying goes, the old barrels are not too ancient for the young wine, and the term "mathematical physics" should be applied to a much wider class of mathematical problems, namely to all theories and structures in mathematics that arise in attempts to clearly understand the fundamental physical theories (quantum physics, kinetics, statistical physics, gravitation). Today mathematical physics, precisely in this understanding, has developed tremendously and has attracted many mathematicians (and even some physicists), while some 35–40 years ago, at the very outset of Berezin's

[1]In the preparation of this essay, I have used materials made available to me a few years ago by A. A. Kirillov, D. A. Leites, V. N. Sushko, M. A. Shubin. Several facts I learned from N. D. Vvedenskaya. I am grateful to all of them.

scientific life, nearly all the physicists regarded this activity with poorly disguised sneers, while the mathematicians did not disguise that they couldn't care less. One needed a great deal of courage and determination, being aware of this total lack of understanding and secretly suffering from it, to persevere in working in the chosen direction.

Thus, in a few large strokes, we can sketch the main inner motivations of the mathematical work of F. A. Berezin.

2. Early years (*family, school, university, the graduate studies that never took place*).

Alik (Felix Alexandrovich) Berezin was born on April 25, 1931, in Moscow to a typical intellectual family: his father was an economist, his mother was a doctor. Alik's parents separated early, and he was brought up by his mother and her parents. In 1948, after graduating from high school, he entered the first year of the Mechanics and Mathematics Department of Moscow State University.

His interest in mathematics arose long before that: from the 8th grade he began to participate in school mathematical olympiads, very absorbing mathematical problem-solving competitions organized by young enthusiasts (mostly graduate and undergraduate students) every spring at the Mechanics and Mathematics Department. These same enthusiasts conducted weekly mathematical "circles", something like math seminars for beginners, where elegant theorems and even fragments of mathematical theories, accessible to high school students, were presented, difficult problems were discussed and solved on the spot by the participants. Alik Berezin took part in the work of such a circle, headed by E. B. Dynkin, then still a graduate student.

In his first years at university he also participated in Dynkin's seminar (for undergraduates), which in fact was the continuation of the circle for high school students. This seminar had two main topics (algebra and probability), corresponding to E. B. Dynkin's two main research interests at the time. Berezin was more interested in algebra and received a strong introduction to the subject, which was to serve him well in all his subsequent work. In the math circle, and later in Dynkin's seminar, Berezin made a very early acquaintance with several budding mathematicians, who were then his fellow students at the department (R. Dobrushin, S. Kamennomostskaya, F. Karpelevich, R. Minlos, I. Shapiro-Pyatetski, N. Vvedenskaya, A. Yushkevich). These acquaintances, many of whom grew to be life-long friends, were to play an essential role in his life. A few years later, while still an undergraduate, Alik Berezin began to participate in the famous Gelfand seminar, and for a long period of time fell under the influence of Izrael Gelfand. In this seminar he wrote his first important research paper on group representation theory (see below).

In 1953 Berezin graduated from the Mechanics and Mathematics Department of Moscow University. Although by that time he had established himself as a talented young research mathematician (he was apparently the strongest student in his graduating class), he was not recommended for graduate work: in the last years of Stalin's life, antisemitism had become a state policy, and Berezin, whose mother was Jewish, was automatically denied this privilege (by that time practically all ethnic Jews could not even become undergraduate students at Moscow University). For three years, Berezin taught mathematics in one of the Moscow high schools, continued to attend the Gelfand seminar and to do research in representation theory.

In 1956, with the advent of the Khrushchev liberalization, the atmosphere at the Department of Mechanics and Mathematics changed somewhat for the better, so that I. G. Petrovski, then the Rector of the University, succeeded in giving a job to Berezin at the Chair of Theory of Functions and Functional Analysis at the insistence of I. M. Gelfand. Berezin was only 25, and he was to work at that chair until the end of his life.

3. The first period of work at the university (*the chair in the fifties and sixties, how mathematical physics started, works of the first period, Berezin as a teacher*).

The first years of his work occurred in a period of absolutely exceptional intellectual and spiritual revival that characterized the Mechanics and Mathematics Department at the end of the fifties and the sixties. This was especially obvious at the chair where Alik worked. Until the mid-fifties the chair, headed for many years by the marvelous and childishly pure D. E. Menshov, had mostly consisted of specialists in the theory of functions of a real or complex variable (D. E. Menshov, N. K. Bari, A. I. Markushevich). During the subsequent years this group was also augmented by several qualified experts (P. L. Ulyanov, B. V. Shabat, A. G. Vitushkin, A. A. Gonchar, E. P. Dolzhenko and their pupils), but the most intensive development of the chair took place along the lines of functional analysis, the direction headed by I. M. Gelfand. Thus, R. A. Minlos was hired together with Berezin, and shortly afterward G. E. Shilov came followed by his pupil A. G. Kostuchenko. A few years later a large group of I. M. Gelfand's and G. E. Shilov's brilliant pupils were working there (A. A. Kirillov, V. P. Palamodov, E. A. Gorin, and others). At the beginning of the sixties Professor B. M. Levitan was invited to the chair, and for several years S. V. Fomin and V. M. Tikhomirov worked there.

Thus, thanks to the efforts of I. M. Gelfand and G. E. Shilov and to the support of I. G. Petrovski, the chair acquired a first class group of analysts; no other university in the world could boast of a group at the same level. This team, which was occasionally supplemented by good specialists as the years went by, continued to exist with almost the same members until the early nineties, when the progressive disintegration of the department (which began in the late fifties under the deanship of P. M. Ogibalov) passed from the hidden phase to the overt one. Of course the chair suffered several losses during this long period: the death of G. E. Shilov in 1975, Berezin's death in 1980. And in fact I. M. Gelfand lost interest in affairs of the chair in the late sixties. But in the fifties and sixties, the intense scientific life at the chair, the appearance of young and talented undergraduate and graduate students, created an exhilarating and beneficial background for research.

In 1957 Berezin defended his *kandidat's* (PhD) dissertation, which incorporated his paper on Laplace operators on semi-simple Lie groups [**1**]. This paper contained the following remarkable result: the description of all irreducible infinite-dimensional representations of complex semi-simple Lie groups in Banach spaces. In modern language Berezin's theorem may be stated as follows: any irreducible representation of the group G is isomorphic to a subfactor of an elementary representation (i.e., a representation induced by a one-dimensional character of a Borel subgroup). The depth of this fact can be seen from the circumstance that the next step in this direction was made only 20 years later. Namely, when D. P. Zhelobenko and M. Duflo obtained the explicit classification of all irreducible representations by indicating which subfactors of the elementary representations are equivalent to each other.

In 1956 Berezin, following I. M. Gelfand's advice, began a deep study of quantum field theory, and this was the starting point of his work in mathematical physics.

In the first period of this work, from the second half of the fifties to the mid-sixties, Berezin thought a lot about spectral theory, in particular, about scattering theory for the Schrödinger operator. There are only a few conclusive results of his in this direction, in several papers where various particular cases are considered (see [**2–5**]), but the observations, considerations, and ideas that arose from these studies had a significant influence on other mathematicians and physicists who were in contact with him and, in the long run, led to the understanding of the spectral and scattering picture for the quantum problem for N particles that we have today (see [**5a**]).

At the same time as Berezin, several other young mathematicians in Russia started studying related problems (L. D. Faddeev, V. P. Maslov, R. A. Minlos, G. M. Zhislin), thus initiating the movement of mathematicians towards mathematical physics that we mentioned above. Members of this circle often talked together and rightly regarded Alik Berezin as their leader. The cooperation between Berezin and L. D. Faddeev was especially fruitful; Alik's influence on the latter was apparently very strong. Later, in the mid-fifties, the research interests of these people diverged, and their spirit of comradeship waned somewhat, but memories of that period live on in some of us.

At the very beginning of the sixties, Berezin wrote his paper on the second quantization formalism, later presented in his monograph *The Method of Second Quantization* [**6**]. This formalism, long used by the physicists, is based on the representation of linear operators acting in the so-called Fock space in the form of functions (usually polynomials) in certain special generators of the algebra of all such operators, the so-called *creation* and *annihilation* operators. Berezin gave this calculus a very elegant form by assigning to each such polynomial a polynomial functional on the algebra of functions (in the case of a symmetric Fock space) or on the Grassmann algebra (in the case of an anti-symmetric Fock space), so that for operations with operators (multiplication, dualization, transformations arising from canonical changes of variables, etc.) the corresponding functionals undergo transformations that are very commonplace for mathematicians: derivation, multiplication, change of variables, continual integration. This method was applied by Berezin and his pupils to the study of some one-dimensional models of quantum field theory: the Tirring model (both for the massless case and the case of positive mass), the nonlinear twice quantized Schrödinger equation (see [**7–9**]). It should be noted that these papers had a significant influence on the development of contemporary constructive field theory. The paper on second quantization constituted the main contents of Berezin's doctoral dissertation, which he successfully defended at the Mechanics and Mathematics Department in 1965.

Berezin's study of second quantization had several important scientific consequences.

First of all, it stimulated renewed interest in the old problem of representing the so-called commutational (and anti-commutational) relations (in this connection, see V. Ya. Golodets' survey in *Uspekhi* [**10**]).

Another topic that partially arose from the study of second quantization and was developed by Berezin for many years is the general understanding of the quantization procedure. Although Berezin studied these questions from the mid-sixties, his perception is best expressed in a cycle of articles that appeared in 1973-76 (see [**11–13**]). According to the main idea of these papers, quantization has the following precise mathematical meaning: the algebra of quantum observables is a deformation of the algebra of classical observables, and the deformation parameter is Planck's constant, while the direction of deformation (the first derivative with respect to the parameter at zero) is the Poisson bracket. In the case of a flat phase space this point of view is equivalent to the ordinary one. In the other cases it leads to a new meaningful theory. In particular, in his articles in *Izvestia* [**11, 12**] Berezin considered the case when the phase portrait is a homogeneous symmetric domain in complex space. He discovered a new interesting effect: the set of possible values of Planck's constant is discrete and bounded from above.

Even earlier, in the second half of the sixties, in connection with his work on second quantization, Berezin published the paper [**14**] in which he studied the representation of operators in Hilbert space by using various systems of generators in the algebra of such operators (pq-symbols, qp-symbols, Weyl symbols, the Wick symbol ordinarily used in second quantization). Note that in many aspects this paper is close to the theory of pseudo-differential operators that arose at the time and now plays an important role in mathematical physics. Thus in the work of Berezin many crucial ideas of this theory appeared independently, although, unfortunately, the significance of Berezin's work in this direction was not understood at the time.

An example of Berezin's concrete activity in this field, which led to his discovery of beautiful and important mathematical objects, is his approach to the study of the Feynman inequality:

$$\operatorname{Sp} e^{-t\widehat{H}} < (2\pi)^{-n} \int_{\mathbb{R}^n \times \mathbb{R}^n} e^{-tH(p,q)}\, dp\, dq, \tag{1}$$

where $H(p,q) = p^2 + V(q)$, $V(q)$ being the potential, while $\widehat{H} = -\Delta + V(q)$ is the corresponding quantum Hamiltonian, i.e., the Schrödinger operator acting in $L_2(\mathbb{R}^n)$. Berezin wanted to understand for which more general Hamiltonians $\widehat{H}$ this inequality remains valid. It became clear that the answer depended upon the chosen quantization, i.e., on the correspondence between H and $\widehat{H}$. It finally turned out that for any operator $\widehat{H}$ the following inequalities hold

$$(2\pi)^{-n} \int_{\mathbb{R}^n \times \mathbb{R}^n} e^{-tH_W(p,q)}\, dp\, dq \leqslant \operatorname{Sp}(e^{-t\widehat{H}}) < (2\pi)^{-n} \int_{\mathbb{R}^n \times \mathbb{R}^n} e^{-tH_{aW}(p,q)}\, dp\, dq, \tag{2}$$

where $H_W(p,q)$ is the Wick symbol of the operator $\widehat{H}$, while $H_{aW}(p,q)$ is the so-called anti-Wick symbol of this operator, first introduced by Berezin in connection with inequality (2). In the paper [**15**], it was proved that the exponent e^{-t} in (2) can be replaced by any downward convex function. Later inequality (2) and its generalization just described were carried over to the case when, instead of H_W and H_{aW}, one considers the covariant and contravariant symbols introduced by Berezin in [**16**], which are defined by using an overcomplete system of vectors in Hilbert space. The abstract scheme for the introduction of these symbols in [**16**] was later used by Berezin for the construction of quantization on Kaehler manifolds. Moreover, already in the papers [**15, 16**], Berezin used inequalities similar to (2) in order to obtain various spectral asymptotics for the operator $\widehat{H}$ for sufficiently large values of the spectral parameter, as well as semiclassical asymptotics. In particular the paper [**15**] contains the first rigorous proof of the semiclassical asymptotics of the distribution function for the eigenvalues of sufficiently general Hamiltonians.

These are the main topics of Alik Berezin's research in the fifties and sixties. We shall return to our survey of his further achievements below. But in order to assess the role of the research in mathematical physics already described, we must also discuss Berezin's pedagogical activities, understood in the wide sense. He would patiently try to develop in the physicists with whom he was in contact a taste and a feel for mathematical thinking, for the elegance of abstract deductions, and would show how to apply them to specific problems. Berezin had perfectly mastered the language and the rather loose ("galloping", so to speak) physical style of thinking, easily conversing with physicists in their own manner, thus giving a good lesson to his colleagues and pupils. He lectured in mathematics to physicists with great pleasure. A great deal more patience and work was required to interest mathematicians in physics, to overcome their deeply rooted attitude to physics as a science beyond the limits of the understandable. For more than 20 years Berezin directed a seminar in mathematical physics and functional analysis at the Mathematics and Mechanics Department of Moscow University, sometimes by himself, sometimes together with someone else. This seminar was well known among the younger physicists and mathematicians: several first-rate scientists grew up in it, and it was the place where many outstanding papers were written. At different times he also conducted seminars in representation theory and functional analysis, and lectured in quantum mechanics, statistical physics, quantum field theory, and path integrals. His courses in statistical physics and in quantum mechanics were published in rotaprint form. Just before his death, he had started to revise the latter; this job was completed by M. A. Shubin on the basis of the notes that Berezin had prepared (see the book [**17**]).

Berezin was very fond of discussing things with his pupils, colleagues, and friends, and he had a lot of joint papers: his coauthors in different years were I. M. Gelfand, V. L. Golo, R. I. Karpelevich, G. I. Kats, D. A. Leites, M. S. Marinov, A. M. Perelomov, G. P. Pokhil, V. S. Retakh, Ya. G. Sinai, L. D. Faddeev, I. I. Shapiro-Pyatetski, M. A. Shubin, and V. M. Finkelberg.

4. The last period (*the flourishing of mathematical physics, everyone drifts to his own corner, supermathematics, some other topics*).

During the sixties the scope of ideas that interested mathematical physicists was constantly widening, and by the early seventies became too wide to be grasped by one person. This period (from the mid-sixties to the mid-seventies) was truly the heroic period in the history of mathematical physics, not only in Russia but worldwide: advances in the theory of phase transfer, in the general theory of Gibbs fields, the so-called "Markovian revolution" in constructive quantum field theory, new methods in the study of one-dimensional integrable systems, the renormalization group and the Wilson program for the study of critical phenomena, the appearance of supermathematics (which will be discussed below), are only some of the most striking topics of the time. Of course, such a drastic expansion of mathematical physics and the increase of its proponents (which could have hardly seemed possible in the fifties) led to a natural differentiation of their research interests; mathematical physicists progressively split up into several weakly related groups, each united around its own *maestro*. Berezin became one of these leaders and during this period worked in the much narrower circle of his nearest collaborators and pupils. This relative isolation was caused, besides external reasons, by a deep inner motivation: by that time the general approach for the construction of supermathematics became clear to Berezin, and its implementation occupied him for most of the seventies. This approach involved a sort of "slight madness", a psychological barrier that was most difficult to overcome. This explains the small circle of people to whom Berezin was willing to disclose his plans.

We have reached the point in our exposition when we must describe in more detail this last and most significant period of Berezin's scientific carrier. The field also had other sources, but Berezin came to supermathematics, as in many other cases, from his work in second quantization. The formal calculus in the Grassmann algebra, which was developed by Berezin in connection with the second quantization formalism in antisymmetric Fock space, led him to the thought that "there exists a nontrivial analog of analysis in which the role of functions is played by elements of the Grassmann algebra" ([**6, 18–20**]), i.e., a calculus in which anticommuting variables play their role together with commuting variables. He unceasingly advertised this idea and carefully collected examples and construction to support it.

The first construction, i.e., the Berezin integral in anticommuting variables, still remains the most impressive in the new theory, the most complicated and most difficult to really understand, although its formulation is quite simple (see [**6**]). This construction is closely connected to another one, also discovered by Berezin and now bearing his name, the *Berezinian*. In [**19**] Berezin developed the key case (when all the variables are odd), and in 1971 in a letter to G. I. Kats wrote out a hypothetic general formula for the Berezinian, later established by his graduate student V. F. Pakhomov.

The end result of the cooperation of Berezin and Kats was their joint paper [**20**]. Its results are close to those of Milnor, Moore, and Quillen in the sixties, however Berezin and Kats treat the Hopf algebras as formal Lie supergroups and indicate the relationships between formal Lie supergroups and Lie superalgebras, generalizing the exponential map and Lie theory. This paper first sets the problem of constructing Lie superalgebras globally, and not only as formal objects. Two years later this problem was solved.

Finally, the last crucial new object of the theory, the notion of supermanifold, was defined by D. A. Leites [**21**] on the basis of an idea proposed by Berezin [**22**]. The construction of supermanifolds is effected along the lines of algebraic geometry (by studying the manifold by means of the local algebra of smooth functions on it) with the only difference being that in the case of supermanifolds one must use superalgebras (see Berezin's survey [**23**]).

In the mid-seventies, Berezin's pioneering ideas began to spread, and supersymmetry groups, i.e., Lie supergroups of transformations of "superspace-time", began to appear in the work of physicists. Thanks to the work of Yu. A. Golfand and E. P. Lichtman, D. V. Volkov and V. A. Akulov, G. Wess and B. Zumino, V. I. Ogievetski, and many others, it became clear that supermanifolds provide an adequate language for the formulation of unified field theory. This is related to the following fundamental assumption about the structure of space-time: space-time is a supermanifold each point of which is an ordinary space-time, while the transformation group is the supergroup extending the Poincaré group via odd generators.

In the last year of his life Berezin began writing a book on supermathematics, which he was not destined to finish. The book was completed by V. P. Palamodov, using the notes and rough copies left by Berezin (see [**24**]).

As we approach the end of our survey of Berezin's mathematical achievements, we would like to touch upon two other topics in mathematical physics that Berezin addressed from time to time.

One of his hopes (as was the case for many others) was to construct a noncontradictory quantum field theory. Without exaggeration, it can be said that almost all of his work (on the N particle problem, quantization, superanalysis) he regarded as stepping stones to this difficult problem. He had some ideas and considerations, for example he long believed that the renormalization procedure in quantum field theory can be correctly understood in the framework of the theory of extensions: the original Hamiltonian is well defined only as a symmetric operator on an appropriate subset of Fock space, while the true Hamiltonian can be obtained as its self-adjoint extension. This idea was nicely illustrated in his joint paper with L. D. Faddeev on δ-like interaction of two quantum particles [**4**]. The same idea was the basis of his own paper on the so-called Lie model [**25**]. Here Berezin made use of Heisenberg's idea that this model should be studied in a space with indefinite metric and constructed the Hamiltonian of the Lie model as the extension of the symmetric operator to a space with indefinite metric. Many people believe that this approach may turn out to be useful in contemporary quantum theories of gauge fields, which necessarily require the introduction of an indefinite metric.

In the sixties Berezin addressed statistical physics fairly often. In 1965 his joint paper with Ya. G. Sinai on the existence of a phase transfer in ferro-magnetic lattice structures with finite interaction was published [**26**]. In subsequent years Berezin repeatedly attempted to find explicit solutions for the three-dimensional Ising model, again using the techniques of second quantization (of which he was very fond and apparently regarded as a universal approach) for the purpose. Some results obtained in this direction were published in [**27, 28**]. Unfortunately, the significant achievements in statistical physics of the end of the sixties and the early seventies and the related advances in constructive quantum field theory remained practically unnoticed by him. In the seventies he never returned to this topic.

Such was, in its main traits, the scientific path of F. A. Berezin.

5. F. A. Berezin's social and political status and position at the Department (*general traits of the scientific life at the Mechanics and Mathematics Department; Berezin as an alien body for the powers that be, the party rule in the Department, the opera story, the letter to the rector R. V. Khokhlov*).

It is difficult to assess the scientific career of Berezin, as well as that of any other important and honest scientist working in Russia at the time, outside the scientific, social, and political context within which they worked, and without taking into consideration their own social and political position.

Above we had the occasion to mention the remarkable scientific atmosphere that prevailed in the Mechanics and Mathematics Department of Moscow University at the end of the fifties and in the sixties. This atmosphere, despite the subsequent "tightening of the screws" (see below) did not entirely disappear until the beginning of the nineties. At the Mechanics and Mathematics Department every year several dozen scientific seminars in various topics in mathematics and mechanics regularly functioned and about as many optional lecture courses were held. The goals and the levels of these seminars and courses could be very different, but most of them were aimed at giving additional material to the undergraduate and graduate students. To clarify the situation it is useful to understand how the traditional educational system works at the Mechanics and Mathematics Department: there is a syllabus, consisting of ten to twelve compulsory courses of lectures that usually last two (or even three) semesters. As a rule, these courses are supplemented by exercise classes, where smaller groups of students solve problems that illustrate the subject of the lecture course. On the other hand, the seminars and brief courses mentioned above are entirely optional and are chosen by the students themselves or in accord with the suggestions of their scientific advisor (only the minimal total number of such courses and seminars is fixed by the syllabus).

Beside these educational seminars, the Department traditionally had several research seminars of the highest level, which would bring together mature mathematicians and where the latest achievements in the given field were discussed. Of course these research seminars were regularly attended by many graduate students and the most advanced undergraduates, and were an excellent school for them. We have already mentioned two such seminars: the famous I. M. Gelfand seminar and the one in mathematical physics and functional analysis directed by Berezin at the end of the fifties and early sixties jointly with R. A. Minlos. Another well-known seminar in mathematical physics, working at the Mechanics and Mathematics Department since 1962 and still in existence today, headed by R. L. Dobrushin, V. A. Malyshev, R. A. Minlos, and Ya. G. Sinai, was traditionally devoted to statistical physics. Mathematical physics (its geometrical aspects and the theory of integrable systems) was also the topic of the well-known seminar headed by S. P. Novikov that has been functioning at the Department for many years. All these seminars, like several others at the Department, were known world-wide, and many scientists (from Russia or from abroad) made it a point to visit them and/or give a talk there. One recalls the unique spirit of free and serious discussion at all the talks prevailing at these seminars: each participant would try to understand the speaker completely, the talk could be interrupted at any moment by a question, a clarifying remark, or by a whole flow of improvised comments by one of the participants. There was not even a hint of subordination, any participant that had something to say on the question under discussion could come to the blackboard (sometimes even during the talk) and be heard. An attitude of respect and consideration for all participants was the rule. These spontaneous and exciting discussions, often spiced with clever jokes, were truly a first for the intellect and were perhaps the most valuable ingredients of the seminars. They often led to a new understanding of the problem under question, sometimes unexpected even for the speaker, new ideas and questions would arise and later develop into serious research work. This was very important for the younger participants, teaching them the proper attitude to creative research and scientific intercourse, and often delighted foreign visitors, bored by the stiff etiquette at their own seminars. One Italian colleague, who lived in Moscow for a long time and regularly went to the seminar in statistical physics (and sometimes, out of pure curiosity, would go to the political meetings that took place at the Department from time to time), joked that these seminars reminded him of political meetings at the University of Rome, while, conversely, our political meetings reminded him of dreadfully boring scientific seminars in Rome.

Among the other important aspects of Moscow mathematical life were the sessions of the Moscow Mathematical Society, especially in the seventies and eighties, when I. M. Gelfand became its President and succeeded in remarkably invigorating its work. Each session of the Society was a carefully prepared survey of some new and interesting mathematical topic. The survey would usually be delivered by the leading expert in the field. The sessions were very widely attended, by the youngest mathematicians as well as by the more experienced researchers.

Listing the outstanding conditions for research at the Mechanics and Mathematics Department, one must mention the rich university library, in particular its mathematical part, which until recent times was systematically supplied with all kinds of mathematical publications appearing in Russia and most of the leading journals from Europe, America and Japan.

However, despite the excellent working conditions at the department, Berezin's life at the Mechanics and Mathematics Department did not proceed very smoothly. We have already mentioned the discrimination to which he was subjected upon graduation from university. The Khrushchev "thaw" gave him the opportunity to return to the University to stay. However, in his work Berezin was often faced with various external obstructions, a sort of pre-planned injustice rooted in the system itself. The discrimination and obstructions, which increased noticeably in the seventies, often made life miserable for him. To try to elucidate the mechanism of covert pressure applied to Berezin, I should perhaps explain the traditional distribution of power at the Mechanics and Mathematics Department that prevailed until the end of the Soviet regime. An important part of the power belonged to the so-called "party bureau", the executive group of the Department's communist party organization. The party bureau consisted mostly of creatively unproductive functionaries who had found in the party a haven and a justification of their own worthlessness. These people were usually spiteful (and, as a rule, antisemitic) and directed their spite at the really active researchers in the department (especially if the latter were Jews). Of course, the spitefulness of the party bureau was partially balanced by certain positive rules and traditions, as well as by the influence and administrative prerogatives of the Scientific Council (and sometimes the Dean's office), which mainly consisted of real research scientists. In the period of Khrushchev liberalism and the first years of the post-Khrushchev era, when the party bosses were still in a state of relative indecision, the influence of what may be called the "scientific party" grew noticeably. This was especially so during the deanship of N. V. Efimov, a remarkable and noble personality. At the end of the sixties, when P. M. Ogibalov (a party functionary from way back, known in the Stalin years for his active participation in various "party cleanings" and "denunciations") became the Dean, the party leaders united with the Dean's office and a dismal atmosphere pervaded the Department for years to come. Specifically, the party bureau according to the existing traditions could (and did) direct the life and work of any employee of the Department by means of the following prohibitions:

1. forbid a raise or an appointment to a better position;
2. forbid a trip abroad (both in the case of a private or a scientific invitation);
3. forbid graduate studies to a pupil of a researcher who displeases the party;
4. forbid lecturing in a compulsory course;
5. forbid one's reelection in a permanent position for the next five-year period.

The last veto was only applied in exceptional cases (one of these led to the untimely death of G. E. Shilov).

All these prohibitions (except the last one) were applied to Berezin consistently at the Department. Perhaps at this point I should recall an amusing and typical episode in which I took part together with Berezin. One of the standard pretexts for various prohibitions was that the employee concerned did not have a "social workload" or did not perform it adequately. Here "social workload" meant an unimportant and necessarily nonrenumerated activity, usually quite dull and/or meaningless; for example, the organization of so-called "political informations" at which the person responsible would retell to students (or to his own colleagues) the contents of the latest Soviet newspapers, or the so-called "civil defense sessions", where year after year one would be told what to do if an H-bomb drops on your head, or the like. It was imperative that each employee have a specific "social workload" of this type. Of course, no normal human being could take such farcical activity seriously and usually only made the motions of carrying it out (or even quietly avoided doing it altogether). The party bureau was usually aware of this and looked at this deceit through its fingers, but at any moment could demand an explanation, keeping a person under stress and control, reminding him of its pervasive existence. It is typical that no serious socially useful activity, e.g., membership in an editorial board, a position in the administration of the Moscow Mathematical Society, the organization of "mathematical circles" for high school students, was acceptable as one's "social workload", unless of course this was specially allowed by the party bureau (I recall hearing the following phrase several times: "what kind of a social workload is that if he enjoys doing it?").

To come back to my story, in the early sixties after Berezin and I had worked at the Chair of the Theory of Functions and Functional Analysis for five years in the position of junior researchers, the question of our appointments to the positions of senior researchers arose. The party bureau did not agree to this, on the pretext of the absence of any "social workload". Negotiations on this topic with the party bureau were conducted by G. E. Shilov, who, as an active music lover, was in constant contact with the opera studio functioning at the university. He decided to help us by using the studio, which was about to stage an opera whose original libretto was in the Bielorussian language; he proposed that Berezin and I perform the translation. We worked hard at this for several weeks, completing a rather good Russian version (alas, this was to be our only joint work [**29**]) and then spent a long time with G. E. Shilov to make the text fit the music. The opera studio was satisfied with the result, the opera ran with success (our names were on the posters), but we did not get the expected appointments, because the party bureau refused to regard all this activity as an acceptable social workload. We were both appointed senior researchers only two years later, when part of the members of the party bureau were replaced. This was Berezin's last advancement to the end of his life.

There were serious problems with some of Berezin pupils as well, who were not allowed to do graduate studies. This was the case with D. A. Leites, his favorite pupil, who was a key figure in the construction of supermathematics. Concerning Berezin's trips abroad, they ceased entirely after 1975, despite an endless stream of invitations from Europe and America (one of the drawers of his desk was filled to overflowing by these invitations, as we discovered after his death). The trips that he was forbidden to go on were important to him not only professionally, but also psychologically: during those years the recognition that he so badly needed was becoming a reality.

In the mid-seventies, Berezin wrote a letter to the new rector of Moscow University, the physicist R. V. Khokhlov, in which he described the general situation then prevailing at the Mechanics and Mathematics Department: discrimination against Jews at the entrance examinations to the University and to enter graduate studies; the related exclusion of many honest teachers actively working in research from all important affairs of the Department, such as entrance and final examinations, and lectures in the obligatory courses; the almost total prohibition of trips abroad for an overwhelming majority of teachers; the specially organized unmotivated "failures" at thesis defenses for *kandidat*'s (=PhD) and doctoral degrees for ethnical Jews; and many other aspects. It is known that R. V. Khokhlov had the intention of taking decisive measures to make life at the Mechanics and Mathematics Department healthier (his sudden death as the result of a mountain climbing accident put an end to that), and apparently Berezin's letter played a significant role in Khokhlov's unrealized plans.

After R. V. Khokhlov's death, the contents of the letter reached the party bosses, only increasing their hostility toward Berezin (the first act of reprisal was the sudden and anonymous cancellation of his already approved trip to Czekoslovakia on a private invitation).

Despite all this harassment and all the humiliations, Berezin always retained his freedom-loving and independent personality, observing the vileness that surrounded him with disgust and sorrow. Being a pessimist by nature, in the last years of his life he became increasingly gloomy, unable to see any ray of light in our dismal life of those years.

In the summer of 1980 F. A. Berezin drowned during a trip in Kolyma. His body was found and brought back to Moscow. An urn with his cinders reposes in his grave at the Vostryakov cemetery in Moscow. It is a pity that he did not live to see the present days, nor to experience the world-wide recognition of his scientific work. He would have rejoiced in the one and in the other.

Moscow, February 1994

References

1. F. A. Berezin, *Laplace operators on semi-simple Lie groups*, Trudy Moskov. Mat. Obshch. **6** (1957), 371–463; **12** (1963), 453–466.
2. ———, *Asymptotics of eigenfunctions of the multiparticle Schrödinger equation*, Dokl. Akad. Nauk SSSR **163** (1965), no. 4, 795–798.
3. ———, *Trace formula for the multiparticle Schrödinger equation*, Dokl. Akad. Nauk SSSR **157** (1964), no. 5, 1069–1072.
4. ———, *Remark on the Schrödinger equation with singular potential*, Dokl. Akad. Nauk SSSR **137** (1961), no. 5, 1011–1014.
5. F. A. Berezin, G. N. Pokhil, and V. M. Finkelberg, *The Schrödinger equation for systems of one-dimensional particles with point-like interaction*, Vestnik Moskov. Univ. **1** (1964), 21–28.

5a. F. A. Berezin, R. A. Minlos, and L. D. Faddeev, *Some mathematical questions in the quantum mechanics of systems with a large number of degrees of freedom*, Proc. 4th Soviet Math. Congress **2** (1964), Moscow, 532–541.

6. F. A. Berezin, *The method of second quantization*, "Nauka", Moscow, 1965; English transl., Academic Press, 1966.
7. F. A. Berezin, *On the Tirring model*, Zh. Èksper. Teoret. Fiz. **40** (1961), no. 3, 885–894.
8. F. A. Berezin and V. N. Syshko, *Relativistic two-dimensional model of a self-interacting fermion field with nonzero mass in the state of rest*, Zh. Èksper. Teoret. Fiz. **48** (1965), no. 5, 1293–1306.
9. F. A. Berezin, *About a model of quantum field theory*, Mat. Sb. **76** (1968), no. 1, 3–25; English transl. in Math. USSR-Sb..

10. V. Ya. Golodets, *Description of the representations of anti-commuting relations*, Uspekhi Mat. Nauk **24** (1969), no. 4, 3–64; English transl. in Russian Math. Surveys.
11. F. A. Berezin, *Quantization*, Izv. Akad. Nauk SSSR Ser. Mat. **38** (1974), no. 5, 1116–1175; English transl. in Math. USSR-Izv..
12. ———, *General concept of quantization*, Comm. Math. Phys. **40** (1975), 153–174.
13. ———, *Quantization on complex symmetric spaces*, Izv. Akad. Nauk SSSR Ser. Mat. **39** (1975), no. 2, 363–403; English transl., Math. USSR-Izv. **9** (1975), 341–379.
14. ———, *About a representation of operators by means of functionals*, Trudy Moskov. Mat. Obshch. **17** (1967), 117–196.
15. ———, *Wick and anti-Wick symbols of operators*, Mat. Sb. **86** (1971), no. 4, 578–610; English transl. in Math. USSR-Sb..
16. ———, *Covariant and contravariant symbols of operators*, Izv. Akad. Nauk SSSR Ser. Mat. **36** (1972), no. 5, 1134–1167; English transl. in Math. USSR-Izv..
17. F. A. Berezin and M. A. Shubin, *The Schrödinger equation*, Moscow State Univ., Moscow, 1983.
18. F. A. Berezin, *On canonical transformations in representations of second quantization*, Dokl. Akad. Nauk SSSR **150** (1963), no. 5, 959–962.
19. ———, *Automorphisms of the Grassmann algebra*, Mat. Zametki **1** (1967), no. 3, 269–276.
20. F. A. Berezin and G. I. Kats, *Lie groups with commuting and anticommuting parameters*, Mat. Sb. **82** (1970), no. 3, 343–359; English transl. in Math. USSR-Sb..
21. D. A. Leites, *Spectra of graded commutative rings*, Uspekhi Mat. Nauk **29** (1974), no. 2, 209–210; English transl. in Russian Math. Surveys.
22. F. A. Berezin and D. A. Leites, *Supermanifolds*, Dokl. Akad. Nauk SSSR **224** (1975), no. 3, 505–508; English transl. in Soviet Math. Dokl..
23. F. A. Berezin, *Mathematical foundations of supersymmetric field theories*, Yadernaya Fiz. **29** (1979), no. 6, 1670–1687.
24. ———, *Introduction to the algebra and analysis of anticommuting variables*, Moscow State Univ. Publ., Moscow, 1983.
25. ———, *About the Lie model*, Mat. Sb. **60** (1963), no. 4, 425–446; English transl. in Math. USSR-Sb..
26. F. A. Berezin and Ya. G. Sinai, *Existence of phase transfer of a lattice gas with attracting particles*, Trudy Moskov. Mat. Obshch. **17** (1967), 197–212.
27. F. A. Berezin, *The plane Ising model*, Uspekhi Mat. Nauk (1969), no. 3, 2–22; English transl. in Russian Math. Surveys.
28. ———, *The number of closed nonselfintersecting contours on a plane lattice*, Mat. Sb. **85** (1971), no. 1, 49–64; English transl. in Math. USSR-Sb..
29. F. A. Berezin and R. A. Minlos, *"The thorny rose", opera libretto (translated from the Bielorussian)*, Moscow univ. opera studio, 1962.

Amer. Math. Soc. Transl.
(2) Vol. **175**, 1996

On the Functional Equation Related to the Quantum Three-Body Problem

V. M. Buchstaber and A. M. Perelomov

In memory of F. A. Berezin

ABSTRACT. In the present paper we give the general solution of the functional equation

$$(f(x)+g(y)+h(z))^2 = F(x)+G(y)+H(z), \qquad x+y+z=0,$$

which is related to the exact factorized ground-state wave function for the quantum one-dimensional problem of three different particles with pairwise interaction.

Functional equations connecting several functions and admitting a general analytic solution have recently attracted the attention of many mathematicians as well as physicists (for recent results, see, for example, [**BC, BP, BK, BFV**]).

In modern mathematical physics such equations arise in connection with the integrable systems of classical and quantum mechanics (for a survey, see for example [**P, OP**]).

In the present paper (a previous version of which appeared as [**BP**]) we investigate one such equation. Namely, we investigate the functional equation connecting six unknown functions

$$(f(x)+g(y)+h(z))^2 = F(x)+G(y)+H(z), \qquad x+y+z=0, \tag{1}$$

which generalizes the well-known Frobenius–Stickelberger equation [**FS**] and is related to the exact factorized ground-state wave function for the quantum one-dimensional problem of three different particles with pairwise interaction. We present the general solution of this equation.

1. Let us recall first that an analogous (but simpler) equation for the special case of three identical particles was considered earlier by B. Sutherland [**S**] and F. Calogero [**Cal1**]. Namely, in the paper [**S**], the one-dimensional many-body

1991 *Mathematics Subject Classification.* Primary 39B22; Secondary 81U10.

problem of n identical particles with pairwise interaction was considered for the case in which the exact ground-state wave function $\Psi_0(x_1, \dots, x_n)$ is factorized

$$\Psi_0(x_1, \dots, x_n) = \prod_{j<k} \psi(x_j - x_k). \tag{2}$$

It was shown that the logarithmic derivative of $\psi(x)$

$$f(x) = \psi'(x)/\psi(x) \tag{3}$$

must satisfy the functional equation

$$\begin{gathered} f(x)f(y) + f(y)f(z) + f(z)f(x) = F(x) + F(y) + F(z), \\ x + y + z = 0, \end{gathered} \tag{4}$$

where $f(x)$ (respectively $F(x)$) is an odd (resp. even) function

$$f(-x) = -f(x), \qquad F(-x) = F(x). \tag{5}$$

In [**S**], a partial solution of equations (4), (5) was also found.

The general solution of equations (4), (5) was found in [**Cal1**] (see also [**OP**] for a survey of this and related problems). This solution has the form

$$f(x) = \alpha\zeta(x; g_2, g_3) + \beta x, \tag{6}$$

where $\zeta(x)$ is the Weierstrass zeta-function (see, for instance, [**WW**]).

In the present paper we consider only the three-body problem, but in the general case, i.e., when all three particles are different from each other.

In this case the ground-state wave function has the form

$$\Psi_0(x_1, x_2, x_3) = \psi_1(x_2 - x_3)\psi_2(x_3 - x_1)\psi_3(x_1 - x_2) \tag{7}$$

and satisfies the Schrödinger equation

$$-\Delta\psi_0 + U\psi_0 = E_0\Psi_0, \tag{8}$$

$$U = u_1(x_2 - x_3) + u_2(x_3 - x_1) + u_3(x_1 - x_2). \tag{9}$$

Substituting Ψ_0 from (7) into (8), we obtain

$$\begin{aligned} \Psi_0^{-1}\Delta\Psi_0 = U - E_0 = 3(f_1^2(x_2 - x_3) &+ f_2^2(x_3 - x_1) + f_3^2(x_1 - x_2)) \\ &- (f_1(x_2 - x_3) + f_2(x_3 - x_1) + f_3(x_1 - x_2))^2 \\ &+ 2(f_1'(x_2 - x_3) + f_2'(x_3 - x_1) + f_3'(x_1 - x_2)), \\ f_j = \psi_j'/\psi_j. & \end{aligned} \tag{10}$$

Hence, for the potential energy $U(x_1, x_2, x_3)$ to have the form of pairwise interactions (9), the three functions

$$f(x) = f_1(x), \quad g(y) = f_2(y), \quad h(z) = f_3(z) \tag{11}$$

must satisfy the functional equation

$$(f(x) + g(y) + h(z))^2 = F(x) + G(y) + H(z), \qquad x + y + z = 0. \tag{12}$$

The following expression for the potential energies results from (10)–(12).

$$
\begin{aligned}
u_1(x) &= 3f^2(x) + 2f'(x) - F(x) + \varepsilon_1, \\
u_2(x) &= 3g^2(x) + 2g'(x) - G(x) + \varepsilon_2, \\
u_3(x) &= 3h^2(x) + 2h'(x) - H(x) + \varepsilon_3, \\
&\varepsilon_1 + \varepsilon_2 + \varepsilon_3 = E_0 .
\end{aligned} \tag{13}
$$

2. Let us consider the meromorphic solutions of the equation

$$(f(x) + g(y) + h(z))^2 = F(x) + G(y) + H(z) \tag{14}$$

satisfying the condition $x + y + z = 0$.

Let us call the solution of equation (14) *nondegenerate*, if the functions $f(x)$, $g(x)$, and $h(x)$ have a pole in a finite domain of the complex x-plane.

The main result of this paper is the following

THEOREM. *The general nondegenerate solution of equation* (14) *in the class of meromorphic functions has the form*

$$f(x) = \alpha\zeta(x - a_1; g_2, g_3) + \beta x + \gamma_1, \tag{15}$$
$$g(x) = \alpha\zeta(x - a_2; g_2, g_3) + \beta x + \gamma_2, \tag{16}$$
$$h(x) = \alpha\zeta(x - a_3; g_2, g_3) + \beta x + \gamma_3, \tag{17}$$
$$F(x) = \alpha^2\wp(x - a_1; g_2, g_3) + 2\gamma\alpha\zeta(x - a_1; g_2, g_3) + \gamma^2/3, \tag{18}$$
$$G(x) = \alpha^2\wp(x - a_2; g_2, g_3) + 2\gamma\alpha\zeta(x - a_2; g_2, g_3) + \gamma^2/3, \tag{19}$$
$$H(x) = \alpha^2\wp(x - a_3; g_2, g_3) + 2\gamma\alpha\zeta(x - a_3; g_2, g_3) + \gamma^2/3, \tag{20}$$

whcrc

$$a_1 + a_2 + a_3 = 0, \qquad \gamma_1 + \gamma_2 + \gamma_3 = \gamma . \tag{21}$$

PROOF. The proof of the theorem is divided into several steps.

Let us begin with

LEMMA 1. *The functions* $(f(x), g(y), h(z))$ *satisfy equation* (14) *for the corresponding functions* $(F(x), G(y), H(z))$ *if and only if the equation*

$$\det \begin{pmatrix} f''(x) & g''(y) & h''(z) \\ f'(x) & g'(y) & h'(z) \\ 1 & 1 & 1 \end{pmatrix} = 0 \tag{22}$$

can be solved under the condition $x + y + z = 0$.

PROOF. Let us apply the operator

$$\partial_- \cdot \frac{\partial}{\partial y} \cdot \frac{\partial}{\partial x}, \quad \text{where } \partial_- = \frac{\partial}{\partial x} - \frac{\partial}{\partial y}$$

to equation (1). This gives:

$$\frac{\partial}{\partial x}: \quad 2(f'(x)-h'(z))(f(x)+g(y)+h(z)) = F'(x)-H'(z), \tag{23}$$

$$\frac{\partial}{\partial y}\frac{\partial}{\partial x}: \quad 2h''(z)(f(x)+g(y)+h(z)) + 2(f'(x)-h'(z))(g'(y)-h'(z)) = H''(z), \tag{24}$$

$$\partial_-\frac{\partial}{\partial y}\frac{\partial}{\partial x}: \quad h''(z)(f'(x)-g'(y)) + f''(x)(g'(y)-h'(z)) + g''(y)(h'(z)-f'(x)) = 0. \tag{25}$$

Here we use the fact that ∂_- is a differential operator, and that $\partial_- h'(z) = \partial_- h''(z) = 0$.

Hence, if the functions $(f(x), g(y), h(z))$ satisfy equation (1), then these functions also satisfy equation (25), which can obviously be rewritten in the form (22).

Conversely, let the functions $(f(x), g(y), h(z))$ satisfy (22) and, consequently, (25). Equation (25) may be rewritten as

$$\partial_-[h''(z)(f(x)+g(y)+h(z)) + (f'(x)-h'(z))(g'(y)-h'(z))] = 0;$$

then there is a function $H_1(z)$ satisfying the following equation:

$$h''(z)(f(x)+g(y)+h(z)) + (f'(x)-h'(z))(g'(y)-h'(z)) = H_1(z). \tag{26}$$

Let us note that equation (26) is equivalent to the equation

$$\frac{\partial}{\partial y}[(f'(x)-h'(z))(f(x)+g(y)+h(z))] = H_1(z).$$

Therefore, there are functions $F_1(x)$ and $H_2(z)$ such that $H_2'(z) = H_1(z)$, and

$$(f'(x)-h'(z))(f(x)+g(y)+h(z)) = F_1(x) - H_2(z). \tag{27}$$

On the other hand, equation (27) is equivalent to

$$\frac{\partial}{\partial x}(f(x)+g(y)+h(z))^2 = 2(F_1(x)-H_2(z)),$$

i.e., there are functions $F(x)$, $G(y)$ and $H(z)$ such that $F'(x) = 2F_1(x)$, $H'(z) = 2H_2(z)$, and

$$(f(x)+g(y)+h(z))^2 = F(x)+G(y)+H(z).$$

Thus Lemma 1 is proved.

LEMMA 2. *Equation* (14) *is invariant under the following transformations*:

$$\begin{aligned}
f(x) &\to f_0 + a_1 x + a_2 f(a_3 x + \alpha_1),\\
F(x) &\to F_0 + a_4 x + a_2^2 F(a_3 x + \alpha_1) + 2a_2 c f(a_3 x + \alpha_1),\\
g(y) &\to g_0 + a_1 y + a_2 g(a_3 y + \alpha_2),\\
G(y) &\to G_0 + a_4 y + a_2^2 G(a_3 y + \alpha_2) + 2a_2 c g(a_3 y + \alpha_2),\\
h(z) &\to h_0 + a_1 z + a_2 h(a_3 z + \alpha_3),\\
H(z) &\to H_0 + a_4 z + a_2^2 H(a_3 z + \alpha_3) + 2a_2 c \cdot h(a_3 z + \alpha_3),
\end{aligned}$$

where a_k $(k = 1, \dots, 4)$ *and* c *are free parameters*

$$f_0 + g_0 + h_0 = c, \quad F_0 + G_0 + H_0 = c^2, \quad \alpha_1 + \alpha_2 + \alpha_3 = 0. \tag{28}$$

This Lemma is proved by a direct calculation.

COROLLARY 3. *For appropriate values of the parameters* α_1, α_2, α_3, *all the functions* $(f(x), g(y), h(z))$, $(F(x), G(y), H(z))$ *are regular at* $x = 0$, $y = 0$, $z = 0$, *respectively.*

The proof follows from the fact that the set of poles of a meromorphic function of one complex variable is discrete. Thus in what follows we may suppose that all the functions are regular at $x = 0$, $y = 0$, $z = 0$.

DEFINITION 4. Let us call the solution of equation (14) *totally degenerate*, if at least one of the functions $f(x)$, $g(x)$, and $h(x)$ is linear.

The next Lemma describes all the totally degenerate solutions of equation (14).

LEMMA 5. *Let* $(f(x), g(y), h(z))$, $(F(x), G(y), H(z))$ *be a totally degenerate solution of equation* (14).

Three cases are possible.

1. *All three functions* $f(x)$, $g(y)$, $h(z)$ *are linear. Then*

$$\begin{aligned} f(x) &= f_0 + f_1 x, & F(x) &= F_0 + F_1 x + (f_1 - g_1)(f_1 - h_1)x^2, \\ g(y) &= g_0 + g_1 y, & G(y) &= G_0 + G_1 y + (g_1 - f_1)(g_1 - h_1)y^2, \\ h(z) &= h_0 + h_1 z, & H(z) &= H_0 + H_1 z + (h_1 - g_1)(h_1 - f_1)z^2, \end{aligned}$$

where f_0, f_1, g_0, g_1, h_0, h_1 *are free parameters.*

Let $f_0 + g_0 + h_0 = c$. *Then*

$$F_0 + G_0 + H_0 = c^2, \quad F_1 = b + 2cf_1, \quad G_1 = b + 2cg_1, \quad H_1 = b + 2ch_1,$$

and b *is a free parameter.*

2. *Two of the functions* $f(x)$, $g(y)$, $h(z)$ *are linear. For example,*

$$g(y) = g_0 + g_1 y, \qquad h(z) = h_0 + h_1 z.$$

Then $f(x)$ *is an arbitrary function and*

$$\begin{gathered} g(y) = g_0 + ay, \quad h(z) = h_0 + az, \quad G(y) = G_0 + by, \quad H(z) = H_0 + bz, \\ F(x) = [g_0 + h_0 - ax + f(x)]^2 - (G_0 + H_0 - bx). \end{gathered}$$

Here g_0, h_0, a, b, G_0, H_0 *are free parameters.*

3. *Only one of the functions* $f(x)$, $g(y)$, $h(z)$ *is linear. For example,* $h(z) = h_0 + h_1 z$. *Then*

$$\begin{aligned} f(x) &= f_0 + ax + c_1 e^{\lambda x}, & F(x) &= F_0 + bx + c_1 e^{\lambda x}(2c + c_1 e^{\lambda x}), \\ g(y) &= g_0 + ay + c_2 e^{\lambda y}, & G(y) &= G_0 + by + c_2 e^{\lambda y}(2c + c_2 e^{\lambda y}), \\ h(z) &= h_0 + az, & H(z) &= H_0 + bz + 2c_1 c_2 e^{-\lambda z}. \end{aligned}$$

Here a, b, c, c_1, c_2, λ *are free parameters, and*

$$f_0 + g_0 + h_0 = c, \qquad F_0 + G_0 + H_0 = c^2.$$

PROOF. *Case* 1. It follows from (22) that $f(x)$, $g(y)$, $h(z)$ are arbitrary linear functions. The form of the functions $F(x)$, $G(y)$, $H(z)$ can be reconstructed directly from (14) by taking into account the identity $2xy = z^2 - x^2 - y^2$.

Case 2. We obtain from (22) $f''(x)(g_1 - h_1) = 0$.

If $f''(x) \neq 0$, then $g_1 = h_1$ and $f(x)$ is arbitrary. The form of the functions $F(x)$, $G(y)$, $H(z)$ can be reconstructed immediately.

Case 3. We get from (24):

$$2(f'(x) - h_1)(g'(y) - h_1) = H''(-x - y).$$

If $f'(x)$ and $g'(y)$ are not constants, then according to the classical Cauchy–Pexider result [**Cau**] (see also [**A**]) we obtain:

$$f'(x) - h_1 = \tilde{c}_1 e^{\lambda x}, \qquad g'(y) - h_1 = \tilde{c}_2 e^{\lambda x},$$

where $\tilde{c}_1$, $\tilde{c}_2$, and λ are free parameters. Therefore

$$f(x) = f_0 + h_1 x + c_1 e^{\lambda x}, \qquad g(y) = g_0 + h_1 y + c_2 e^{\lambda y},$$

where $c_k = \tilde{c}_k/\lambda$, $k = 1, 2$. The form of the functions $F(x)$, $G(y)$, $H(z)$ can be reconstructed easily. The Lemma is proved.

The functions $f(x)$, $g(x)$, $h(x)$ from equation (1), will be regarded as nondegenerate solutions of equation (1).

LEMMA 6. *For appropriate values of the parameters f_0, g_0, h_0, a_1, F_0, G_0 (see Lemma 2) we have*

$$f(0) = g(0) = h(0) = 0, \quad h'(0) = 0, \quad F(0) = G(0). \tag{29}$$

The proof is easy.

LEMMA 7. *For an appropriate choice of the parameters α_1 and α_2, we have the relation $f(x) \neq g(x)$.*

PROOF. Suppose on the contrary that

$$f(x + \alpha_1) - f(\alpha_1) \equiv g(x + \alpha_1) - g(\alpha_2) \tag{30}$$

for all α_1 and α_2 in any neighborhood of the point $x = 0$. Differentiating (30), we obtain

$$\frac{\partial f(x + \alpha_1)}{\partial x} = \frac{\partial f(x + \alpha_1)}{\partial \alpha_1} = f'(\alpha_1),$$

i.e.,

$$f(x + \alpha_1) = f'(\alpha_1)x + f(\alpha_1),$$

contradicting the assumption that the solution is nondegenerate. The Lemma is proved.

Hence it is sufficient to find all the nondegenerate solutions of equation (1) under the following additional conditions: $f(x) \neq g(x)$ and $f(0) = g(0) = h(0)$, $h'(0) = 0$, $F(0) = G(0) = 0$.

Interchanging x and y in equation (14), we obtain

$$(f(y) + g(x) + h(z))^2 = F(y) + G(x) + H(z). \tag{31}$$

Subtracting (31) from (14), we see that

$$[(f(x)-g(x))-(f(y)-g(y))][(f(x)+g(x))+(f(y)+g(y)+2h(z))] \\ =(F(x)-G(x))-(F(y)-G(y)).$$

The last equation can be rewritten as

$$\varphi(x+y)=\eta(x)+\eta(y)-\frac{\gamma(x)-\gamma(y)}{\xi(x)-\xi(y)}, \tag{32}$$

where $\varphi(x)=-2h(-x)$, $\eta(x)=f(x)+g(x)$, $\xi(x)=f(x)-g(x)$, $\gamma(x)=F(x)-G(x)$, and $\varphi(0)=\varphi'(0)=\eta(0)=\gamma(0)=\xi(0)=0$ and $\varphi''(x)\neq 0$.

DEFINITION 8. Let us call a solution $(\varphi,\eta,\xi,\gamma)$ of equation (32) *normalized,* if the following initial conditions are satisfied:

$$\xi'(0)=1, \quad \eta'(0)=0.$$

LEMMA 9. *The map*

$$(\varphi,\eta,\xi,\gamma)\to(\varphi,\eta+b_1\xi,b_2\xi,b_2(\gamma+b_1\xi^2)), \tag{33}$$

where b_1 and b_2 are parameters, $b_2\neq 0$, defines a group action. Each orbit of this group contains one and only one solution.

PROOF. The first statement may be checked by a direct computation. To prove the second statement, let us differentiate equation (32) with respect to y. At the point $y=0$ we have:

$$\varphi'(x)=\eta'(0)+\frac{\gamma'(0)}{\xi(x)}-\xi'(0)\frac{\gamma(x)}{\xi(x)^2}.$$

Assuming $\varphi(x)$ is regular at $x=0$ and $\varphi''(x)\neq 0$, it is easy to check that $\xi'(0)\neq 0$.

Applying the transformation (33) with $b_2=(\xi'(0))^{-1}$, $b_1=-\eta'(0)/\xi'(0)$ to the solution $(\varphi,\eta,\xi,\gamma)$, we obtain a normalized solution, and the Lemma is proved.

In what follows solutions are assumed to be normalized, unless the contrary is asserted. Let us now consider the functional equation

$$\begin{gathered}\varphi(x+y)=\varphi(x)+\varphi(y)+\tau(x)\tau(y)A(x+y),\\ \varphi(0)=\varphi'(0)=\tau(0)=\tau''(0)=0, \qquad \tau'(0)=1.\end{gathered} \tag{34}$$

LEMMA 10. *For any solution $(\varphi,\eta,\xi,\gamma)$ of equation (32), there is a unique solution (φ,τ,A) of equation (34) such that*

$$\xi(x)=\frac{\tau(x)}{\tau'(x)-b_3\tau(x)}, \tag{35}$$

$$\eta(x)=\varphi(x)-\varphi'(x)\xi(x), \tag{36}$$

$$\gamma(x)=-\varphi'(x)\xi(x)^2, \tag{37}$$

where $b_3=\xi''(0)$ is a free parameter.

PROOF. Let (φ, τ, A) be a solution of equation (34). Then acting on (34) by the operator $\partial_- = (\partial/\partial x - \partial/\partial y)$, we obtain

$$0 = \varphi'(x) - \varphi'(y) + (\tau'(x)\tau(y) - \tau(x)\tau'(y))\, A(x+y),$$

i.e.,

$$A(x+y) = -\frac{\varphi'(x) - \varphi'(y)}{\tau'(x)\tau(y) - \tau(x)\tau'(y)}. \tag{38}$$

Hence, we can transform the equation (34) to the equation

$$\varphi(x+y) = \varphi(x) + \varphi(y) + \tau(x)\tau(y)\,\frac{\varphi'(x) - \varphi'(y)}{\tau(x)\tau'(y) - \tau'(x)\tau(y)}. \tag{39}$$

On the other hand,

$$\begin{aligned}\frac{\tau(x)\tau(y)}{\tau(x)\tau'(y) - \tau'(x)\tau(y)} &= \frac{\tau(x)}{\tau'(x)}\,\frac{\tau(y)}{\tau'(y)}\,\frac{1}{(\tau(x)/\tau'(x) - b_3) - (\tau(y)/\tau'(y) - b_3)}\\ &= \frac{\xi(x)\xi(y)}{\xi(x) - \xi(y)},\end{aligned}$$

where the function $\xi(x)$ may be expressed in terms of $\tau(x)$ by the formula (35) with a free parameter b_3. Therefore,

$$\varphi(x+y) = \varphi(x) + \varphi(y) + \xi(x)\xi(y)\,\frac{\varphi'(x) - \varphi'(y)}{\xi(x) - \xi(y)}.$$

Substituting the expressions for $\eta(x)$ and $\gamma(x)$ from (36) and (37), we obtain a solution $(\varphi, \eta, \xi, \gamma)$ of equation (32).

Now let $(\varphi, \eta, \xi, \gamma)$ be a solution of equation (32). Substituting $y = 0$ in equation (32), we obtain

$$\varphi(x) = \eta(x) - \gamma(x)/\xi(x),$$

i.e., $\gamma(x) = \xi(x)\delta(x)$, where $\delta(x) = \eta(x) - \varphi(x)$, and our initial conditions $\varphi'(0) = \eta'(0) = 0 = \varphi(0) = \eta(0)$ are satisfied.

Hence, $\gamma'(0) = 0$, and from the formula for $\varphi'(x)$ obtained in the proof of Lemma 9, we obtain

$$\gamma(x) = -\varphi'(x)\xi^2(x), \qquad \eta(x) = \varphi(x) - \varphi'(x)\xi(x),$$

as asserted in (36) and (37). Let us note that formula (35) may be regarded as the differential equation for the function $\tau(x)$. Solving this equation with initial conditions $\tau(0) = 0$, $\tau'(0) = 1$, we obtain the function $\tau(x)$; if, moreover, we take $b_3 = \xi''(0)$, this function will satisfy the condition $\tau''(0) = 0$.

Now substituting the expressions for $\xi(x)$, $\eta(x)$, $\gamma(x)$ into equation (32), we obtain equation (39).

Let us apply the operator ∂_- to the equation (39); we obtain

$$\partial_-\left(\frac{\varphi'(x) - \varphi'(y)}{\tau(x)\tau'(y) - \tau'(x)\tau(y)}\right) \equiv 0.$$

Thus we have proved that the functions $\varphi(x)$ and $\tau(x)$ determine the function $A(x)$ given by expression (38). The Lemma is proved.

So we have shown how to construct all the solutions of equation (32) using the solutions of equation (34).

Now we describe the general analytical solution of equation (34).

LEMMA 11. *Let* (φ, τ, A) *be a solution of equation* (34). *(Let us recall that* $\varphi(0) = \varphi'(0) = \tau(0) = \tau''(0) = 0$ *and* $\tau'(0) = 1$.*) Then the function* $u(x) = \varphi'(x)$ *is a solution of the equation*

$$(u')^2 = c_3u^3 + 4c_2u^2 + 2c_1u + c_0^2, \qquad u(0) = 0,\ u'(0) = c_0, \tag{40}$$

and if $c_0 = 0$, *then* $c_1 \neq 0$.

The functions $\tau(x)$ *and* $A(x)$ *satisfy the following equations*:

$$\frac{\tau'(x)}{\tau(x)} = \frac{1}{2}\frac{u'(x) + c_0}{u(x)}, \tag{41}$$

$$\frac{A'(x)}{A(x)} = \frac{1}{2}\frac{u'(x) - c_0}{u(x)}. \tag{42}$$

If $c_0 = 0$, *then* $u(x) = \frac{1}{2}c_1\tau(x)^2$, *and* $A(x) = \frac{1}{2}c_1\tau(x)$.

PROOF. Let us consider the first three derivatives with respect to y of equation (34)

$$\varphi'(x+y) = \varphi'(y) + \tau(x)[\tau'(y)A(x+y) + \tau(y)A'(x+y)],$$
$$\varphi''(x+y) = \varphi''(y) + \tau(x)[\tau''(y)A(x+y) + 2\tau'(y)A'(x+y) + \tau(y)A''(x+y)],$$
$$\varphi'''(x+y) = \varphi'''(y) + \tau(x)[\tau'''(y)A(x+y) + 3\tau''(y)A'(x+y) + 3\tau'(y)A''(x+y) + \tau(y)A'''(x+y)].$$

Taking $y = 0$ and making use of the initial conditions for $\varphi(x)$ and $\tau(x)$, we get

$$\varphi'(x) = \tau(x)A(x), \tag{43}$$
$$\varphi''(x) = \varphi''(0) + 2\tau(x)A'(x), \tag{44}$$
$$\varphi'''(x) = \varphi'''(0) + \tau(x)[\tau'''(0)A(x) + 3A''(x)]. \tag{45}$$

Let $\varphi_k = \varphi^{(k)}(0)$ and $\tau_3 = \tau'''(0)$. From (43) and (44) we obtain

$$\frac{\varphi''(x) - \varphi_2}{\varphi'(x)} = 2\,\frac{A'(x)}{A(x)}; \tag{46}$$

from (45) and (43) it follows that

$$\frac{\varphi'''(x) - \varphi_3}{\varphi'(x)} = \frac{\tau_3A(x) + 3A''(x)}{A(x)}. \tag{47}$$

Making use of the identity

$$\frac{A''}{A} = \left(\frac{A'}{A}\right)' + \left(\frac{A'}{A}\right)^2$$

for the quantity $\varphi'(x) = u(x)$, we obtain the following equation (see equations (46), (47)):

$$\frac{u'' - \varphi_3}{u} = \tau_3 + 3\left(\frac{1}{2}\frac{u' - \varphi_2}{u}\right)' + \frac{3}{4}\left(\frac{u' - \varphi_2}{u}\right)^2.$$

This equation may be rewritten as follows:

$$\begin{gathered} 4(u'' - \varphi_3)u = 4\tau_3u^2 + 6[uu'' - u'(u' - \varphi_2)] + 3(u' - \varphi_2)^2, \\ 2uu'' - 3(u')^2 + 4\tau_3u^2 + 4\varphi_3u + 3\varphi_2^2 = 0. \end{gathered} \tag{48}$$

Let

$$\tau_3 = c_2, \quad \varphi_3 = c_1, \quad \varphi_2 = c_0 .$$

Equation (48) admits the integrating factor $u^{-4}u'$ and may be reduced to the following equation

$$(u^{-3}(u')^2)' = 4c_2(u^{-1})' + 2c_1(u^{-2})' + c_0^2(u^{-3})'. \tag{49}$$

Integrating (49) and multiplying the result by u^3, we obtain equation (40), where c_3 is the integration constant. Equation (42) follows from (46). Then from equation (43) we obtain:

$$u'(x) = \tau'(x)A(x) + \tau(x)A'(x).$$

It follows from (44) that

$$\tau(x)A'(x) = (u'(x) - c_0)/2.$$

Making use of this fact, we obtain

$$\tau'(x)A(x) = (u'(x) + c_0)/2.$$

Dividing this equation by equation (43), we come to equation (41). Note that if $c_0 = 0$, equations (41), (42), and conditions $\tau(0) = 0$, $\tau'(0) = 1$ imply

$$u(x) = \frac{c_1}{2}\,\tau(x)^2, \qquad A(x) = \frac{c_1}{2}\,\tau(x),$$

and it follows, in particular, that $c_1 \neq 0$ if $c_0 = 0$. The Lemma is proved.

Consider the Weierstrass function $\wp(x)$ with parameters g_2 and g_3. We have

$$\wp'(x)^2 = 4\wp(x)^3 - g_2\wp(x) - g_3 .$$

LEMMA 12. *The general solution of the equation* (40) *may be written in one of the following equivalent forms*:

$$u(x) = \frac{4}{c_3}\,(\wp(x+\alpha) - \wp(\alpha)), \tag{50}$$

$$u(x) = c_1\psi(x) + \frac{c_0^2 c_3}{2}\,\psi(x)^2 + c_0\psi'(x), \tag{51}$$

where

$$\psi(x) = \frac{1}{2}\,\frac{1}{\wp(x) - c_2/3}. \tag{52}$$

Here $\wp(x)$ *is the Weierstrass function with parameters*

$$g_2 = 3\left(\frac{2c_2}{3}\right)^2 - \frac{c_1c_3}{2}, \qquad g_3 = -\left(\frac{2c_2}{3}\right)^3 + \frac{c_1c_2c_3}{6} - \left(\frac{c_0c_3}{4}\right)^2, \tag{53}$$

and

$$\wp(\alpha) = c_2/3, \qquad \wp'(\alpha) = c_0c_3/4.$$

PROOF. Formula (50) gives:

$$(u'(x))^2 = \frac{16}{c_3^2}\,[4\wp(x+\alpha)^3 - g_2\wp(x+\alpha) - g_3].$$

On the other hand,

$$\begin{aligned}(u'(x))^2 &= c_3\left[\frac{4}{c_3}\,(\wp(x+\alpha)-\wp(\alpha))\right]^3 \\ &\quad + 4c_2\left[\frac{4}{c_3}\,(\wp(x+\alpha)-\wp(\alpha))\right]^2 + 2c_1\left[\frac{4}{c_3}\,(\wp(x+\alpha)-\wp(\alpha))\right] + c_0^2.\end{aligned}$$

Hence

$$\begin{aligned}&16\,[4\wp(x+\alpha)^3 - g_2\wp(x+\alpha) - g_3] \\ &\quad = 4^3[\wp(x+\alpha)-\wp(\alpha)]^3 + 4^3c_2[\wp(x+\alpha)-\wp(\alpha)]^2 \\ &\qquad + 8c_1c_3[\wp(x+\alpha)-\wp(\alpha)] + c_0^2c_3^2.\end{aligned}$$

Let us compare the coefficients of terms of the same degree in $\wp(x+\alpha)$. This shows that formula (50) with parameters g_2, g_3 follows from (53). To deduce (51) from (50), one makes use of the addition theorem for the $\wp$-function (cf., e.g., [**WW**]).

$$\wp(x+\alpha) - \wp(\alpha) = -(\wp(x) + 2\wp(\alpha)) + \frac{1}{4}\left(\frac{\wp'(x)-\wp'(\alpha)}{\wp(x)-\wp(\alpha)}\right)^2.$$

Therefore

$$\begin{aligned}&(\wp(x+\alpha)-\wp(\alpha))(\wp(x)-\wp(\alpha))^2 \\ &\quad = -(\wp(x)+2\wp(\alpha))(\wp(x)^2 - 2\wp(x)\wp(\alpha) + \wp(\alpha)^2) \\ &\qquad + \frac{1}{4}\,(4\wp(x)^3 - g_2\wp(x) - g_3 - 2\wp'(x)\wp'(\alpha) + \wp'(\alpha)^2) \\ &\quad = 3\wp(x)\wp(\alpha)^2 - 2\wp(\alpha)^3 - \frac{g_2}{4}\,\wp(x) - \frac{1}{4}\,g_3 - \frac{1}{2}\,\wp'(x)\wp'(\alpha) + \left(\frac{\wp'(\alpha)}{2}\right)^2 \\ &\quad = \left(3\wp(\alpha)^2 - \frac{1}{4}\,g_2\right)(\wp(x)-\wp(\alpha)) - \frac{1}{2}\,\wp'(x)\wp'(\alpha) + \frac{1}{2}\,\wp'(\alpha)^2.\end{aligned}$$

Hence,

$$\begin{aligned}\wp(x+\alpha)-\wp(\alpha) &= -\frac{1}{2}\,\frac{\wp'(x)}{(\wp(x)-\wp(\alpha))^2}\,\wp'(\alpha) \\ &\quad + \frac{3\wp(\alpha)^2 - g_2/4}{\wp(x)-\wp(\alpha)} + \frac{1}{2}\left(\frac{\wp'(\alpha)}{\wp(x)-\wp(\alpha)}\right)^2.\end{aligned} \tag{54}$$

This gives:

$$\wp'(\alpha) = \frac{1}{4}\,c_0c_3, \qquad 3\wp(\alpha)^2 - \frac{1}{4}\,g_2 = \frac{1}{8}\,c_1c_3.$$

Formula (51) follows from equation (54) by dividing by $c_3/4$. The Lemma is proved.

COROLLARY 13. *The general solution of equation* (40) *has the form*

$$u_*(x) = c_1\left(\frac{\cosh 2\sqrt{c_2}\,x - 1}{(2\sqrt{c_2})^2}\right) + c_0\,\frac{\sinh 2\sqrt{c_2}\,x}{2\sqrt{c_2}} \tag{55}$$

as $c_3 \to 0$.

PROOF. Let

$$u_*(x) = \lim_{c_3\to 0} u(x), \quad \psi_*(x) = \lim_{c_3\to 0}\psi(x), \quad \wp_*(x) = \lim_{c_3\to 0}\wp(x).$$

By Lemma 12, the function $\wp_*(x)$ satisfies the equation

$$\begin{aligned}(\wp_*'(x))^2 &= 4\wp_*(x)^3 - 3\Big(\frac{2c_2}{3}\Big)^2\wp_*(x) + \Big(\frac{2c_0}{3}\Big)^3\\ &= 4\Big(\wp_*(x) - \frac{c_2}{3}\Big)^2\Big(\wp_*(x) + \frac{2}{3}c_2\Big).\end{aligned}$$

Therefore

$$(\psi_*'(x))^2 = \frac{1}{4}\left(\frac{-\wp_*'(x)}{(\wp_*(x) - c_2/3)^2}\right)^2 = \frac{\wp_*(x) + 2c_2/3}{(\wp_*(x) - c_2/3)^2} = 2\psi_*(x) + 4c_2\psi_*(x)^2. \tag{56}$$

Differentiating (56) with respect to x, one obtains

$$\psi_*''(x) = 4c_2\psi_*(x) + 1, \qquad \psi_*(0) = 0,\ \psi_*'(0) = 0.$$

Therefore

$$\psi_*(x) = \frac{\cosh 2\sqrt{c_2}\,x - 1}{(2\sqrt{c_2})^2}.$$

In view of (51), it follows that

$$u_*(x) = c_1\psi_*(x) + c_0\psi_*'(x).$$

Corollary 13 is proved.

Note that according to Lemma 11, if the functions (φ, τ, A) satisfy equation (34), then the function $\tau(x)$ is determined uniquely by the equation

$$\frac{\tau'(x)}{\tau(x)} = \frac{1}{2}\,\frac{u'(x) + c_0}{u(x)},$$

subject to the initial conditions $\tau(0) = 0$, $\tau'(0) = 1$, and the function $A(x)$ is determined by equation (43):

$$A(x) = u(x)/\tau(x).$$

Hence we may regard the functions $\varphi(x)$ as solutions of the equation (34).

THEOREM 14. *The general solution of equation* (34)

$$\varphi(x+y) = \varphi(x) + \varphi(y) + \tau(x)\tau(y)A(x+y)$$

is given by the function

$$\varphi(x) = \frac{4}{c_3}\,(\zeta(\alpha) - \zeta(x+\alpha) - \wp(\alpha)x), \qquad \varphi(0) = \varphi'(0) = 0, \tag{57}$$

where $\zeta(x)$ *and* $\wp(x)$ *are the Weierstrass* ζ*-function and* $\wp$*-function with parameters* g_2 *and* g_3 *(see Lemma* 12*).*

PROOF. According to Lemmas (11) and (12), it is sufficient to prove that any function $\varphi(x)$ given by formula (57) is a solution of equation (14). It is convenient to consider two different cases.

Case 1. $c_3 = 0$.

$$\varphi_*(x) = \lim_{c_3 \to 0} \varphi(x).$$

In this case $\varphi_*(x) = \int_0^\infty u_*(x)\,dx$ and hence, using Corollary 13, we obtain

$$\varphi_*(x) = c_1 \frac{\sinh 2\sqrt{c_2}\,x - 2\sqrt{c_2}\,x}{(2\sqrt{c_2})^3} + c_0 \frac{\cosh 2\sqrt{c_2}\,x - 1}{(2\sqrt{c_2})^2}. \tag{58}$$

Using the elementary identity

$$e^{x+y} - 1 = (e^x - 1) + (e^y - 1) + (e^{x/2} - e^{-x/2})(e^{y/2} - e^{-y/2})\,e^{(x+y)/2}, \tag{59}$$

we obtain

$$\begin{aligned}\sinh 2\sqrt{c_2}(x+y) &= \sinh 2\sqrt{c_2}\,x + \sinh 2\sqrt{c_2}\,y \\ &\quad + 4\sinh\sqrt{c_2}\,x \sinh\sqrt{c_2}\,y \sinh\sqrt{c_2}(x+y), \\ \cosh 2\sqrt{c_2}(x+y) &= \cosh 2\sqrt{c_2}\,x + \cosh 2\sqrt{c_2}\,y \\ &\quad + 4\sinh\sqrt{c_2}\,x \sinh\sqrt{c_2}\,y \cosh\sqrt{c_2}(x+y).\end{aligned}$$

Hence

$$\varphi_*(x+y) = \varphi_*(x) + \varphi_*(y) + \tau_*(x)\tau_*(y)A_*(x+y),$$

where

$$\tau_*(x) = \frac{\sinh\sqrt{c_2}\,x}{\sqrt{c_2}}, \qquad A_*(x) = \frac{c_1}{2}\frac{\sinh\sqrt{c_2}\,x}{\sqrt{c_2}} + c_0\cosh\sqrt{c_2}\,x. \tag{60}$$

Case 2. $c_3 \neq 0$. Then without any restriction we may take $c_3 = 2$. According to the Frobenius-Stickelberger formula [**FS**], the functions $f(x)$, $g(y)$, $h(z)$ constitute a solution of equation (1):

$$f(x) = \zeta(\alpha_1 - \alpha/2 - x) - \wp(\alpha)x - \zeta(\alpha_1 - \alpha/2), \tag{61}$$
$$g(y) = \zeta(-\alpha_1 - \alpha/2 - y) - \wp(\alpha)y + \zeta(\alpha_1 + \alpha/2), \tag{62}$$
$$h(z) = \zeta(\alpha - z) - \wp(\alpha)z - \zeta(\alpha). \tag{63}$$

Using the reduction of (1) to equation (34) described above, we obtain

$$\varphi(x) = -2h(-x) = 2(\zeta(\alpha) - \zeta(x+\alpha) - \wp(\alpha)x)$$

which gives the solution of equation (34). The theorem is proved.

COROLLARY 15. *The general normalized solution of equation* (32) *is given by the formulas*

$$\begin{aligned}\varphi(x) &= \frac{4}{c_3}(\zeta(\alpha) - \zeta(x+\alpha) - \wp(\alpha)x), \\ \xi(x) &= \frac{2u(x)}{c_0 - 2b_3u(x) + u'(x)},\end{aligned} \tag{64}$$

where

$$u(x) = \varphi'(x) = \frac{4}{c_3}(\wp(x+\alpha) - \wp(\alpha))$$

and b_3 is free parameter,

$$\eta(x) = \varphi(x) - \varphi'(x)\xi(x), \qquad \gamma(x) = -\varphi'(x)\xi(x)^2.$$

The proof follows from Theorem 14, formula (61) and Lemma 10. Let us recall that in the proof of Lemma 10 we gave an explicit construction of the solution to equation (12) using the solution of equation (14).

Thus it is already proved that if $(f(x), g(y), h(z))$ is the nondegenerate solution of equation (1) satisfying the additional conditions

$$f(x) \neq g(x), \qquad f(0) = g(0) = h(0) = h'(0) = 0, \tag{65}$$

then it is necessary to have

$$h(x) = \frac{2}{c_3}\,(\zeta(\alpha - x) - \wp(\alpha)x - \zeta(\alpha)), \tag{66}$$

where c_3, α and the parameters g_2, g_3 of the $\wp$-Weierstrass function satisfy the condition of Lemma 12. Moreover, if $c_3 \neq 0$, then for the functions

$$f(x) = \frac{2}{c_3}\Big(\zeta\Big(\alpha_1 - \frac{\alpha}{2} - x\Big) - \wp(\alpha)x - \zeta\Big(\alpha_1 - \frac{\alpha}{2}\Big)\Big), \tag{67}$$

$$g(x) = \frac{2}{c_3}\Big(\zeta\Big(-\alpha_1 - \frac{\alpha}{2} - x\Big) - \wp(\alpha)x + \zeta\Big(\alpha_1 + \frac{\alpha}{2}\Big)\Big), \tag{68}$$

where α_1 is free parameter, the function $h(x)$ of the form (66) gives the solution of equation (1). Hence, there are two unsolved problems.

1. Are the functions $f(x)$ and $g(x)$ for $c_3 \neq 0$ the only functions that give the solution of equation (1) for a fixed function $h(x)$?

2. How can we find sufficient conditions for $c_3 = 0$ on the parameters of the function $h_*(x) = \lim_{c_3 \to 0} h(x)$ such that there exist functions $f(x)$ and $g(x)$ for which $(f(x), g(x), h_*(x))$ is the solution of equation (1) and how can we find all such functions $(f(x), g(x))$?

Let us note that in the case $c_3 = 0$, the main problem is that we cannot pass to the limit as $c_3 \to 0$ in formulas (67), (68) (in contrast to (66)).

To solve these two problems we shall first consider the reduction of equation (1) to equation (12) and shall use the general analytic solution of equation (12) (see Lemma 9 and Corollary 15).

Let us begin with the case $c_3 \neq 0$.

LEMMA 16. *Let the functions $(f_1(x), g_1(x), h_1(x))$ satisfy equation (1) and the initial conditions under consideration. If $h_1(x) = h(x)$ is the function from equation (66), then*

$$f_1(x) = s_1 f(x) + s_2 g(x), \tag{69}$$

$$g_1(x) = t_1 f(x) + t_2 g(x), \tag{70}$$

where $f(x)$ and $g(x)$ are given by equations (67) and (68), and $s_1 + s_2 = 1$, $t_1 + t_2 = 1$.

PROOF. For the functions given by equations (47) and (48), we have

$$\xi(x) = f(x) - g(x) = \frac{2}{c_3}\Big[\zeta\Big(\alpha_1 - \frac{\alpha}{2} - x\Big) + \zeta\Big(\alpha_1 + \frac{\alpha}{2} + x\Big) - \zeta\Big(\alpha_1 - \frac{\alpha}{2}\Big) - \zeta\Big(\alpha_1 + \frac{\alpha}{2}\Big)\Big]. \tag{71}$$

Then

$$\xi'(x) = \frac{2}{c_3}\Big[\wp\Big(\alpha_1 - \frac{\alpha}{2} - x\Big) - \wp\Big(\alpha_1 + \frac{\alpha}{2} + x\Big)\Big],$$
$$\xi''(x) = \frac{2}{c_3}\Big[-\wp'\Big(\alpha_1 - \frac{\alpha}{2} - x\Big) - \wp'\Big(\alpha_1 + \frac{\alpha}{2} + x\Big)\Big].$$

We see that if the parameters α and α_1 are sufficiently close to the point $x = 0$, then $\xi'(0) \neq 0$, and the value of $\xi''(0)$ gives the value of free parameter b_3 required to construct the general normalized solution of equation (32). Therefore, in this case the general solution of the equation has the form

$$\varphi(x) = -2h(-x), \quad \eta(x) + b_1\xi(x), \quad b_2\xi(x),$$

where $h(x)$ is the function (66), $\xi(x) = f(x) - g(x)$ and $\eta(x) = f(x) + g(x)$ for the functions (47) and (48).

Now if we introduce

$$f_1(x) + g_1(x) = \eta(x) + b_1\xi(x), \qquad f_1(x) - g_1(x) = b_2\xi(x),$$

we have

$$f_1(x) = \frac{1}{2}\eta(x) + \frac{b_1 + b_2}{2}\xi(x) = s_1 f(x) + s_1 g(x),$$
$$g_1(x) = \frac{1}{2}\eta(x) + \frac{b_1 - b_2}{2}\xi(x) = t_1 f(x) + t_2 g(x),$$

where

$$s_1 = \frac{1}{2} + \frac{b_1 + b_2}{2}, \quad s_2 = \frac{1}{2} - \frac{b_1 + b_2}{2}, \quad t_1 = \frac{1}{2} + \frac{b_1 - b_2}{2}, \quad t_2 = \frac{1}{2} - \frac{b_1 - b_2}{2}.$$

The Lemma is proved.

Now it remains to find the values of parameters s_1 and t_1 for which the set of functions $(f_1(x), g(x), h(x))$ from Lemma 16 gives the solution of equation (1).

Let us introduce the notation

$$\det(f, g, h) = \det\begin{pmatrix} f''(x) & g''(y) & h''(z) \\ f'(x) & g'(y) & h'(z) \\ 1 & 1 & 1 \end{pmatrix}$$

and use the following formula (see [**WW**], p. 458)

$$\frac{1}{2}\det(\wp(x), \wp(y), \wp(z)) = \frac{\sigma(x+y+z)\sigma(x-y)\sigma(y-z)\sigma(z-x)}{\sigma^3(x)\sigma^3(y)\sigma^3(z)}.$$

If the conditions of Lemma 16 are satisfied, we have

$$\det(s_1 f(x) + s_2 g(x), t_1 f(y) + t_2 g(y), h(z)) = s_1 t_1 \det(f(x), f(y), h(z)) + s_2 t_2 \det(g(x), g(y), h(z)). \tag{72}$$

On the other hand,

$$\begin{aligned}(73)\qquad &\frac{c_3^3}{8}\det\left(f(x),f(y),h(z)\right)\\ &\quad=\frac{c_3^3}{8}\det\Big(\wp\Big(\alpha_1-\frac{\alpha}{2}-x\Big),\wp\Big(\alpha_1-\frac{\alpha}{2}-y\Big),\wp(\alpha-z)\Big)\\ &\quad=\frac{c_3^3}{4}\,\frac{\sigma(2\alpha_1)\,\sigma(y-x)\,\sigma(z-y+\alpha_1-3\alpha/2)\,\sigma(x-z+3\alpha/2-\alpha_1)}{\sigma^3(\alpha_1-\alpha/2-x)\,\sigma^3(\alpha_1-\alpha/2-y)\,\sigma^3(\alpha-z)},\end{aligned}$$

$$\begin{aligned}(74)\qquad &\frac{c_3^3}{8}\det\left(g(x),g(y),h(z)\right)\\ &\quad=\frac{c_3}{8}\det\Big(\wp\Big(-\alpha_1-\frac{\alpha}{2}-x\Big),\wp\Big(-\alpha_1-\frac{\alpha}{2}-y\Big),\wp(\alpha_1-z)\Big)\\ &\quad=\frac{c_3^3}{4}\,\frac{\sigma(2\alpha_1)\,\sigma(y-x)\,\sigma(y-z+\alpha_1+3\alpha/2)\,\sigma(x-z+\alpha_1+3\alpha/2)}{\sigma^3(\alpha_1+\alpha/2+x)\,\sigma^3(\alpha_1+\alpha/2+y)\,\sigma^3(\alpha-z)}.\end{aligned}$$

The comparison of expressions (73) and (74) shows that if $\alpha_1=\omega_k$ is the one of the three halfperiods of the Weierstrass-function $\wp(x)$, then $\det(\cdot)$ given by formula (72) is equal to zero identically for all values of s_1 and t_1. If, however, $\alpha_1\neq\omega_k$, $k=1,2,3$, then this determinant is equal to zero if and only if $s_1t_1=s_2t_2=0$. So we have proved our main result.

Now let us consider the case $c_2\to 0$.

The general normalized solution in this case is given by the function (58).

Let us denote

$$\varphi_{**}(x)=\lim_{c_2\to 0}\varphi_*(x),\quad \tau_{**}(x)=\lim_{c_2\to 0}\tau_*(x),\quad A_{**}(x)=\lim_{c_2\to 0}A_*(x).$$

From (58) we obtain

$$(75)\qquad \varphi_{**}(x)=c_1\,\frac{x^3}{3!}+c_0\,\frac{x^2}{2},$$

and according to formulas (40), we have

$$\tau_{**}(x)=x,\qquad A_{**}(x)=c_1\frac{x}{2}+c_0.$$

Further,

$$(76)\qquad \xi_{**}(x)=\frac{x}{1-b_3x},$$

$$(77)\qquad \eta_{**}(x)=c_1\,\frac{x^3}{3!}+c_0\,\frac{x^2}{2}-\left(c_1\frac{x^2}{2}+c_0x\right)\frac{x}{1-b_3x},$$

$$(78)\qquad \gamma_{**}(x)=-\left(c_1\,\frac{x^2}{2}+c_0x\right)\frac{x^2}{(1-b_3x)^2}.$$

Hence the general solution of equation (32) is given in this case by the functions

$$\varphi_{**}(x),\quad \eta_{**}+b_1\xi_{**},\quad b_2\xi_{**},\quad b_2(\gamma_{**}+b_1\xi_{**}^2).$$

COROLLARY 17. *The general solution of equation* (14) *in the class of entire functions has the form*

$$f(x) = \alpha_1 e^{\lambda x} + \beta x + \gamma_1, \qquad F(x) = (\alpha_1 e^{\lambda x} + \gamma/\sqrt{3})^2 + 2\alpha_2\alpha_3 e^{-\lambda x},$$
$$g(x) = \alpha_2 e^{\lambda x} + \beta x + \gamma_2, \qquad G(x) = (\alpha_2 e^{\lambda x} + \gamma/\sqrt{3})^2 + 2\alpha_1\alpha_3 e^{-\lambda x},$$
$$h(x) = \alpha_3 e^{\lambda x} + \beta x + \gamma_3, \qquad H(x) = (\alpha_3 e^{\lambda x} + \gamma/\sqrt{3})^2 + 2\alpha_1\alpha_2 e^{-\lambda x},$$

where $\gamma = \gamma_1 + \gamma_2 + \gamma_3$.

For the case in which $\lambda \to 0$ *and for the corresponding* α_k, $k + 1, 2, 3$, β, *we obtain the solution*

$$f(x) = \alpha x^2 + \beta_1 x + \gamma_1, \qquad F(x) = 2\alpha^2(x - a_1)^4 + 2\alpha\tilde{\gamma}(x - a_1)^2 + \tilde{\gamma}^2/3,$$
$$g(x) = \alpha x^2 + \beta_2 x + \gamma_2, \qquad G(x) = 2\alpha^2(x - a_2)^4 + 2\alpha\tilde{\gamma}(x - a_2)^2 + \tilde{\gamma}^2/3,$$
$$h(x) = \alpha x^2 + \beta_3 x + \gamma_3, \qquad H(x) = 2\alpha^2(x - a_3)^4 + 2\alpha\tilde{\gamma}(x - a_3)^2 + \tilde{\gamma}^2/3,$$

where

$$a_1 = \frac{1}{6\alpha}(\beta_2 + \beta_3 - 2\beta_1), \qquad a_2 = \frac{1}{6\alpha}(\beta_1 + \beta_3 - 2\beta_2),$$
$$a_3 = \frac{1}{6\alpha}(\beta_1 + \beta_2 - 2\beta_3), \qquad \tilde{\gamma} = \gamma_1 + \gamma_2 + \gamma_3 - \frac{1}{4\alpha}(\beta_1^2 + \beta_2^2 + \beta_3^2).$$

Appendix

It is interesting to note that the general solution of the functional equation (1) has found applications in another physical context. In [**BB**] a Lax representation for the system of equations

$$\ddot{q}_j = \sum_{k \neq j} (a + b\dot{q}_j)(a + b\dot{q}_k) V_{jk}(q_j - q_k), \qquad j = 1, \dots, n, \tag{79}$$

was constructed; this system describes the motion of n particles on the line. Particular cases of this system are integrable relativistic ($b \neq 0$) and nonrelativistic ($b = 0$) Calogero–Moser systems, as well as Toda systems. Within the framework of this paper, let us consider in more detail, following [**BB**], the case of three particles.

For the system of equations

$$\ddot{q}_j = \sum_{k \neq j} V_{jk}(q_j - q_k), \qquad j = 1, 2, 3,$$

let us search for a Lax representation $\dot{L} = [L, M]$ in the form

$$L(q) = \dot{q}_d + A, \qquad M(q) = (B\tau)_d + C,$$

where L, M, A, B, C are (3×3)-matrices and $A = (A_{jk}(q_j - q_k))$, $A_{jj} \equiv 0$, $C = (C_{jk}(q_j - q_k))$, $C_{jj} \equiv 0$, $B = (B_{jk}(q_j - q_k))$, $\tau = (1, 1, 1)$, and $\dot{q}_d$, $(B\tau)_d$ are diagonal matrices whose diagonals contain the coordinates of the vectors $\dot{q} = (\dot{q}_1, \dot{q}_2, \dot{q}_3)$ and $B\tau$, respectively.

The Lax representation leads to the equation

$$\ddot{q}_d + [\dot{q}_d, A'] = [A, (B\tau)_d] + [A, C] + [\dot{q}_d, C], \tag{80}$$

where $A' = (A'_{jk}(q_j - q_k))$. Therefore

$$\sum_{k\neq j} V_{jk}(q_j - q_k) = [A, C]_{jj}, \tag{81}$$

$$([\dot{q}_d, C - A'] + [A, (B\tau)_d + C])_{jk} = 0. \tag{82}$$

From (82) we obtain $C = A'$ and hence, by (81),

$$V_{jk}(q_j - q_k) = A_{jk}A'_{kj} - A_{jk}A'_{kj} = -V_{kj}(q_k - q_j). \tag{83}$$

Further, (82) yields

$$\sum_{l=1}^{3} A_{jk}(B_{jl} - B_{kl}) + A'_{jl}A_{lk} - A_{jl}A'_{lk} = 0. \tag{84}$$

Now let us set

$$\Phi_{jk} = (A'_{jl}A_{lk} - A_{jl}A'_{lk})/A_{jk}. \tag{85}$$

A direct verification shows that (84) implies the condition

$$\Phi_{jk} + \Phi_{km} + \Phi_{mj} = 0. \tag{86}$$

Let us introduce the functions

$$b_1(x) = -A_{23}(x)A_{32}(-x), \quad b_2(y) = -A_{31}(y)A_{13}(-y),$$
$$b_3(z) = -A_{12}(z)A_{21}(-z).$$

Taking $x = q_2 - q_3$, $y = q_3 - q_1$, $z = q_1 - q_2$ and using the condition

$$\Phi_{21} + \Phi_{13} + \Phi_{32} = 0,$$

we immediately obtain

$$\begin{aligned} &b_2(y)A_{23}A'_{32} - b_1(x)A_{31}A'_{13} + b_3(z)A_{31}A'_{13} \\ &\quad - b_2(y)A_{12}A'_{21} + b_1(x)A_{12}A'_{21} - b_3(z)A_{23}A'_{32} = 0. \end{aligned} \tag{87}$$

Similarly, from the condition

$$\Phi_{12} + \Phi_{23} + \Phi_{31} = 0$$

we get

$$\begin{aligned} &b_1(x)A_{13}A'_{31} - b_2(y)A_{32}A'_{23} + b_3(z)A_{32}A'_{23} \\ &\quad - b_1(x)A_{21}A'_{12} + b_2(y)A_{21}A'_{12} - b_3(z)A_{13}A'_{31} = 0. \end{aligned} \tag{88}$$

Taking into account the relation

$$\begin{aligned} b'_1(x) = A_{23}A'_{32} - A_{32}A'_{23}, \quad b'_2(y) = A_{31}A'_{13} - A_{13}A'_{31}, \\ b'_3(z) = A_{12}A'_{21} - A_{21}A'_{12}, \end{aligned} \tag{89}$$

and adding equations (87) and (88), we obtain

$$b_2(y)b'_1(x) - b_1(x)b'_2(y) + b_3(z)b'_2(y) - b_2(y)b'_3(z) + b_1(x)b'_3(z) - b_3(z)b'_1(x) = 0. \tag{90}$$

Equation (90) may be rewritten in the form

$$\det\begin{pmatrix} b_1'(x) & b_2'(y) & b_3'(z) \\ b_1(x) & b_2(y) & b_3(z) \\ 1 & 1 & \end{pmatrix} \equiv 0. \tag{91}$$

Recall that in our notation $x+y+z=0$ and therefore, by Lemma 1, the functions

$$f(x)=\int b_1(x)\,dx, \quad g(y)=\int b_2(y)\,dy, \quad h(z)=\int b_3(z)\,dz \tag{92}$$

satisfy equation (1) ($\equiv$(14)) for the corresponding functions $F(x)$, $G(y)$, and $H(z)$.

Thus we obtain the following result.

THEOREM. *The system of equations*

$$\ddot{q}_j=\sum_{k\neq j} V_{jk}(q_j-q_k), \qquad j=1,2,3,$$

has a Lax representation $\dot{L}=[L,M]$ *of the form indicated above if and only if*

$$V_{jk}(q_j-q_k)=-a\wp'(q_j-q_k+\lambda_j-\lambda_k). \tag{93}$$

PROOF. Suppose the system of equations does have the indicated Lax representation. Then, by (83), (91), and (92),

$$V_{23}(x)=f''(x), \quad V_{31}(y)=g''(y), \quad V_{12}(z)=h''(z),$$

and therefore by the main theorem

$$V_{jk}(q_j-q_k)=-\alpha\wp'(q_j-q_k-q_l), \qquad l\notin(j,k).$$

In view of condition (21), i.e., $a_1+a_2+a_3=0$, we see that the a_l's may be presented in the form $\lambda_j-\lambda_k$, which means that we have (92).

The proof of the converse statement is in fact contained in [**Cal2**].

References

[A] N. H. Abel, *Méthode genérale pour trouver des fonctions d'une seule quantité variable, lorsqu'une propriété de ces fonctions est exprimée par une équation entre deux variables*, Magazin for Naturvidenskaberne, Aargang I, Bind 1, Christiania, Oeuvres completes **1** (1823), Christiania, 1881, 1–10.

[BB] H. W. Braden and V. M. Buchstaber, *Integrable systems with pairwise interactions and functional equations*, Preprint, hep-th/9411240 (1994).

[BC] M. Bruschi and F. Calogero, *General analytic solution of certain functional equations of addition type*, SIAM J. Math. Anal. **21** (1990), 1019–1030.

[BFV] V. Buchstaber, G. Felder, and A. Veselov, *Elliptic Dunkl operators, root systems and functional equations*, Duke Math. J. **76** (1994), no. 3, 885–911.

[BK] V. Buchstaber and I. Krichever, *Vector addition theorems and Baker–Akhiezer functions*, Teoret. Mat. Fiz. **94**, no. 2, 200–212; English transl. in Theoret. and Math. Phys. (1993).

[BP] V. Buchstaber and A. Perelomov, *On the functional equation related to the quantum three-body problem*, Preprint MPI/93–17 (1993).

[Cau] A. L. Cauchy, *Cours d'Analyse de l'Ecole Polyt, 1. Analyse algebrique,* **103**, Oeuvres completes (2) **3** (1821), 98–105.

[Cal1] F. Calogero, *One-dimensional many-body problems with pairwise interactions whose exact ground-state wave function is of product type*, Lett. Nuovo Cimento **13** (1975), 507–511.

[Cal2] ———, *Exactly solvable one-dimensional many-body problems*, Lett. Nuovo Cimento (2) **13** (1975), 411–416.

[FS] G. Frobenius and L. Stickelberger, *Über die Addition und Multiplication der elliptischen Functionen*, J. Reine Angew. Math. **88** (1880), 146–184.

[OP] M. A. Olshanetsky and A. M. Perelomov, *Quantum integrable systems related to Lie algebras*, Phys. Rep. **94** (1983), 313–404.

[P] A. M. Perelomov, *Integrable systems of classical mechanics and Lie algebras*, Birkhäuser, 1990.

[S] B. Sutherland, *Exact ground-state wave function for a one-dimensional plasma*, Phys. Rev. Lett. **34** (1975), 1083–1085.

[WW] E. Whittaker and G. Watson, *Course of modern analysis*, Cambridge Univ. Press, 1927.

Translated by THE AUTHORS

Amer. Math. Soc. Transl.
(2) Vol. **175**, 1996

The Differential Calculus on Quantum Linear Groups

L. D. Faddeev and P. N. Pyatov

ABSTRACT. The non-commutative differential calculus on the quantum groups $SL_q(N)$ is constructed. The quantum external algebra proposed contains the same number of generators as in the classical case. The exterior derivative defined in a constructive way obeys a modified version of the Leibniz rules.

§1. Introduction

Recent interest in constructing differential calculi on quantum groups stems from Woronowicz's pioneering work [**33**]. In it he formulated the general algebraic framework for dealing with the problem. In subsequent investigations the emphasis was on two main directions. First, experience in dealing with such algebras was accumulated while considering the simplest low dimensional examples (see, e.g., [**32, 23, 27**]). It was soon recognized that the true quantum group differential calculus should be bicovariant, and that this condition is very restrictive. Indeed, only the use of this condition allows one to obtain the unique external algebra construction for the $SL_q(2)$ Cartan 1-forms [**13**]. Next, a very close connection was established with the theory of quadratic quantum algebras (quantum spaces) [**19, 10, 31**]. It was then realized that the condition of unique ordering of higher order monomials (the so-called diamond condition) is very important [**21, 28**], and that in fact it must only be checked for cubic monomials [**20**].

Another direction of investigation was the search for an adequate technique for dealing with quantum differential algebras. Here the close connections between the quantum differential calculi and the R-matrix formulation for quantum groups and algebras [**10**] were soon established [**16, 11, 34**] (for further considerations see [**6**]). It turns out that the R-matrix technique is highly appropriate in treating the arising problems.

The next stage of investigations was to combine both lines of research to obtain concrete differential algebra constructions for known series of quantum groups. Here substantial progress was achieved for the $GL_q(N)$ case. Namely, in the series of

---

1991 *Mathematics Subject Classification.* Primary 81R50.

The first author was supported by the Russian Academy of Sciences and the Academy of Finland. The second author was supported in part by RFBR (grant No. 93-02-3827), ISF (grant RFF-000) and INTAS (grant No. 93-127).

papers [**18, 29, 26, 28, 30**] a pair of nice-looking differential algebras on $GL_q(N)$ was constructed. But the situation with the q-deformed series of simple Lie groups appears to be much more complex. The natural way of obtaining the $SL_q(N)$, $SO_q(N)$, and $SP_q(N)$ differential calculi by performing reduction from the $GL_q(N)$ calculi failed in the quantum case, because one cannot consistently reduce the number of the generating elements in the $GL_q(N)$ differential algebras constructed (see the discussion in [**8, 35**]). In principle one can treat these nonreduced (or partially reduced) differential calculi as a quantizations of nonstandard classical calculi on the special groups (see [**22**]), but the problem of finding the deformations of the ordinary calculi still remained open. It is rather natural in this situation to revise once again the basic postulates involved in the construction scheme. The only postulate that seems too restrictive is the classical Leibniz rule for the exterior derivative [**11**]

$$d(f\cdot g) = df\cdot g + (-1)^{|f|} f\cdot dg\,.$$

Indeed, let us recall that the basic vector fields after quantization correspond to finite shifts rather than to infinitesimal differentiations. The natural Leibniz rule for them is multiplicative rather than being additive. Correspondingly, the Leibniz rule for the differential must take into account this shift property of vector fields.

In this paper we propose a construction of the differential algebra with the appropriately modified Leibniz rule. We consider the case closest to $GL_q(N)$—the $SL_q(N)$ differential algebra. Here only one Cartan 1-form and one basic vector field must be reduced. The reduction scheme for vector fields was already developed in [**28**]. We propose the reduction scheme for Cartan 1-forms. Here we do not discuss the involution leading to the unitary reduction of our system. As was shown in [**2**], this can be done for q on the circle ($|q| = 1$) for the algebra of vector fields and functions on the quantum group. We believe that the involution found in [**2**] can be extended to the differential forms as well.

The paper is organized as follows. In §2 we fix the notation of the R-matrix technique and formulate the basic postulates of our construction. We believe that it was the consistent use of the R-matrix technique that allowed us to carry through the construction. This not only simplified the calculations, but played an important heuristic role. In §3 we present the external algebra on $SL_q(N)$. This algebra is also supplied with the action of the basic vector fields (or Lie derivatives). We refer to this extended algebra as the differential algebra on $SL_q(N)$. Section 4 is devoted to construction of the exterior derivative operator d. Note that the proposed scheme can be equally applied to $GL_q(N)$. In this way one can recover a wide variety of differential algebras on $GL_q(N)$. It seems to us that such a nonuniqueness is due to the fact that $GL_q(N)$ is not semisimple.

§2. The basic principles and notation

The starting point for our consideration is the Hopf algebras $\mathrm{Fun}\,(GL_q(N))$ and $\mathrm{Fun}\,(SL_q(N))$ [**10**]. We present here some facts and definitions related to these algebras.

We choose the corresponding R-matrix [**15**] $R \in \mathrm{Mat}_N(\mathbb{C})^{\otimes 2}$ in the form

$$R = q\sum_i e_{ii}\otimes e_{ii} + \sum_{i\neq j} e_{ji}\otimes e_{ij} + \lambda\sum_{j<i} e_{jj}\otimes e_{ii}, \tag{2.1}$$

where $i, j = 1, \dots, N$ and $\lambda = q - 1/q$. In what follows we shall also use the shorthand notation R for the matrix $R \otimes I \in \mathrm{Mat}_N(\mathbb{C})^{\otimes 3}$, where $I \in \mathrm{Mat}_N(\mathbb{C})$ is the unit matrix. One can easily distinguish, in the context of each formula, whether R belongs to $\mathrm{Mat}_N(\mathbb{C})^{\otimes 2}$ or to $\mathrm{Mat}_N(\mathbb{C})^{\otimes 3}$. The R-matrix (2.1) satisfies the Yang–Baxter equation and the Hecke condition, respectively,

$$RR'R = R'RR', \tag{2.2}$$

$$R^2 = \mathbf{I} + \lambda R. \tag{2.3}$$

Here $R' = I \otimes R$, and $\mathbf{I} = I \otimes I$. It is worthwhile to establish the connection with other frequently used R-matrix conventions:

$$\text{our } R \text{ equals } \widehat{R}_{12} = P_{12}R_{12} = R^+_{12}P_{12},$$
$$\text{our } R^{-1} \text{ equals } R^-_{12}P_{12}.$$

Here $P \in \mathrm{Mat}_N(\mathbb{C})^{\otimes 2}$ is the permutation matrix and the notation $\widehat{R}_{12}$, R_{12}, $R^{\pm}_{12}$ is presented in [**10**].

The unital associative algebra $\mathrm{Fun}\,(GL_q(N))$ is generated by N^2 elements $T = (t_{ij})_{i,j=1}^N$. Multiplication and comultiplication in it are defined, respectively, by

$$RTT' = TT'R, \tag{2.4}$$

$$\Delta(t_{ij}) = t_{ik} \otimes t_{kj}, \tag{2.5}$$

where T means $T \otimes I$ in (2.4) and $T' = I \otimes T$.

The q-deformed Levi–Civita tensor $\varepsilon_q^{i_1 \dots i_N}$ ($= \varepsilon_q^{1 \dots N}$ in shorter notation) satisfies the following characteristic relations:

$$\varepsilon_q^{1\dots N} R_i = -q^{-1} \varepsilon_q^{1 \dots N}, \qquad 1 \leqslant i \leqslant N, \tag{2.6}$$
$$\varepsilon_q^{i_1 \dots i_N}|_{i_1 = 1, \dots, i_k = k, \dots, i_N = N} = 1.$$

Here $R_i = I^{\otimes(i-1)} \otimes R \otimes I^{\otimes(N-i-1)}$ (note: $R_1 = R$, $R_2 = R'$). The quantum determinant of T, $\det_q T$, defined via the relation

$$\varepsilon_q^{1\dots N} T_1 T_2 \cdots T_N = T_1 T_2 \cdots T_N \varepsilon_q^{1 \dots N} = \varepsilon_q^{1\dots N} \cdot \det\nolimits_q T, \tag{2.7}$$

where $T_k = I^{\otimes(k-1)} \otimes T \otimes I^{\otimes(N-k)}$, is a central element of the algebra $\mathrm{Fun}\,(GL_q(N))$. This can be checked by means of the following formula

$$\Psi^{N+1} \varepsilon_q^{1 \dots N} R_N^{\pm 1} \cdots R_1^{\pm 1} = q^{\pm 1} \Psi^1 \varepsilon_q^{2 \dots N+1}, \tag{2.8}$$

where $\Psi = (\psi^i)_{i=1}^N \in \mathbb{C}^N$ is an arbitrary vector. The Hopf algebra $\mathrm{Fun}\,(SL_q(N))$ is then obtained by adding one more relation

$$\det\nolimits_q T = 1 \tag{2.9}$$

to (2.4). Finally, the antipodal mapping $S(\,\cdot\,)$ on $\mathrm{Fun}\,(GL_q(N))$ and $\mathrm{Fun}\,(SL_q(N))$ (for its explicit form see [**10**]) satisfies the relations

$$S(T)T = TS(T) = I; \tag{2.10}$$

therefore, in what follows we prefer the notation T^{-1} to $S(T)$.

Now let us turn to the differential algebra of extensions of $\mathrm{Fun}\,(GL_q(N))$ and $\mathrm{Fun}\,(SL_q(N))$. First, we must fix the basic principles of our construction:

A. *The bicovariance condition.* Following [**33**], we require that a differential algebra should possess the bicomodule structure with respect to the underlying quantum group. By this condition we guarantee that the left and right translations in a quantum group do not affect the structure of its differential calculus. From this viewpoint it looks most natural to use, say, right-invariant and left-adjoint vector fields $L = (l_{ij})_{i,j=1}^{N}$ and Cartan 1-forms $\Omega = (\omega_{ij})_{i,j=1}^{N}$ in addition to the T's as the generating elements for differential algebra[1]. The left and right $\mathrm{Fun}\,(GL_q(N))$-coactions in this case read:

$$\delta_L(x_{ij}) = t_{ik}t_{lj}^{-1} \otimes x_{kl}, \qquad \delta_R(x_{ij}) = x_{ij} \otimes 1, \tag{2.11}$$

where by $X = (x_{ij})_{i,j=1}^{N}$ we understand either L or Ω.

In the case of the $SL_q(N)$-differential algebra, the number of independent Cartan 1-forms should be reduced by 1. This can only be achieved in a bicovariant manner by use of the q-deformed trace [**10, 24**] (see also [**34, 28, 12**]). Here we define this operation and present several useful formulas

$$\mathrm{Tr}_q(X) = \mathrm{Tr}\,(\mathcal{D}X), \qquad \mathcal{D} = \mathrm{diag}\,\{q^{-N+1}, q^{-N+3}, \ldots, q^{N-1}\}. \tag{2.12}$$

The Tr_q-operation possesses the invariance property

$$\mathrm{Tr}_q\, \delta_L(X) = 1 \otimes \mathrm{Tr}_q\, X, \tag{2.13}$$

and also satisfies the relations

$$\begin{gathered} \mathrm{Tr}_{q(2)}(RXR^{-1}) = \mathrm{Tr}_{q(2)}(R^{-1}XR) = I \cdot \mathrm{Tr}_q\, X, \\ \mathrm{Tr}_{q(1,2)}(Rf(X,R)\,R^{-1}) = \mathrm{Tr}_{q(1,2)}\, f(X,R), \\ \mathrm{Tr}_{q(2)}\, R^{\pm 1} = q^{\pm N} I, \qquad \mathrm{Tr}_q\, I = [N]_q\,. \end{gathered} \tag{2.14}$$

Here the index in parentheses denotes the number of the matrix space in which the operation Tr_q acts, and $[N]_q = (q^N - q^{-N})/\lambda$.

B. *The ordering condition.* We suppose that multiplication in the differential algebra is defined by relations quadratic in T, Ω and L and that these relations allow us to order lexicographically any quadratic monomial of the generators. Moreover, they must yield a *unique* ordering for any higher order monomial of T, Ω and L. The latter is the so-called *diamond* (*or confluence*) *condition* (see, e.g., [**7**]). It guarantees us that the Poincaré series of the classical differential algebra does not change under quantization. The direct verification of this condition consists in the use of the Diamond Lemma [**7**]. Such calculations appear to be very cumbersome already in the $N = 2$ case (see the discussion in subsection 3.8 of [**21**]) and it seems hard to generalize them to an arbitrary N. The alternative approach that we shall advocate here consists in noticing that the quadratic relations for T, Ω, L express in fact the action of some representation of the braid group on the differential algebra. The diamond condition is then the consequence of the braid group defining relations, so that it should follow from the general properties (2.2), (2.3) of the R-matrix. Examples of such formal R-matrix manipulations are presented in [**14**] and in §3 of the present paper.

C. The last but not least condition is that the differential algebra is to be supplied with a differential complex structure. In other words, we must define the

[1] For left-invariant and right-adjoint generators all the constructions proceed similarly.

$\mathbb{C}$-linear differential mapping d on it. Taking into account the discussion above, we choose the following set of its characteristic properties:

- d is of degree 1 with respect to the natural $\mathbb{Z}$-grading on the algebra of differential forms;
- d satisfies the *nilpotence condition*: $d^2 = 0$.

Now let us proceed to the construction of such a differential algebra.

§3. The differential algebra

We summarize the main result of this section in

THEOREM 1. *For general values of the deformation parameter* q ($[2]_q \neq 0$, $[N]_q \neq 0$, $[N]_q \neq -\lambda q^N$, $[N \pm 1]_q \neq \pm q^{N \mp 4}$) *the* $GL_q(N)$*-differential algebra defined as*

$$RTT' = TT'R, \tag{3.1}$$

$$R\Omega R\Omega + \Omega R\Omega R^{-1} = \kappa_q(\Omega^2 + R\Omega^2 R), \tag{3.2}$$

$$R\Omega R^{-1}T = T\Omega', \tag{3.3}$$

$$RLRL = LRLR, \tag{3.4}$$

$$RLRT = q^{2/N}TL', \tag{3.5}$$

$$R^{-1}\Omega RL = LR\Omega R^{-1}, \tag{3.6}$$

where

$$\kappa_q = \frac{\lambda q^N}{[N]_q + \lambda q^N}, \tag{3.7}$$

admits a consistent reduction to $SL_q(N)$. *This reduction is achieved by adding three more relations*

$$\det_q T = 1, \qquad \mathrm{Tr}_q\, \Omega = 0, \qquad \mathrm{Det}\, L = 1, \tag{3.8}$$

to (3.1)–(3.6). *Here*

$$\begin{aligned} \varepsilon_q^{1\ldots N}\, \mathrm{Det}\, L &= q^{1-N}(R_1R_2\cdots R_{N-1}L_1)^N \varepsilon_q^{1\ldots N} \\ &= q^{1-N}(L_1R_1R_2\cdots R_{N-1})^N \varepsilon_q^{1\ldots N}. \end{aligned} \tag{3.9}$$

PROOF. It is not difficult to check the bicovariance condition for (3.1)–(3.8) by using the commutation properties of T's (2.4) and the definitions for left and right transitions on the quantum group (2.5), (2.11). Here we only mention the transformation properties of $\mathrm{Det}\, L$:

$$\delta_L(\mathrm{Det}\, L) = 1 \otimes \mathrm{Det}\, L, \qquad \delta_R(\mathrm{Det}\, L) = \mathrm{Det}\, L \otimes 1.$$

The validity of the ordering condition for the quadratic monomials of T, Ω, L can be verified by rewriting relations (3.1)–(3.6) in matrix components. Instead, we can convince ourselves of its validity by noticing that relations (3.1), (3.2), (3.4) contain the correct number of the commutation relations for the T's, Ω's, and L's because of their symmetry properties

$$P_q^{\pm}(RTT' - TT'R)P_q^{\pm} \equiv P_q^{\pm}(RLRL - LRLR)P_q^{\pm} \equiv 0,$$

$$P_q^{\pm}(R\Omega R\Omega + \Omega R\Omega R^{-1} - \kappa_q(\Omega^2 + R\Omega^2 R))P_q^{\mp} \equiv 0.$$

Here $P_q^{\pm} = (\pm R + q^{\mp 1})/[2]_q$ are the quantum symmetrizer and antisymmetrizer, respectively (see [**15, 10**]).

Now, let us concentrate on checking the diamond condition for monomials cubic in T, Ω, and L. First, we choose a suitable complete set of such monomials:

$$
\begin{aligned}
&(R'R\Omega)^3,\ T(R'\Omega)^2,\ RTT'\Omega'',\ R'R^{-1}\Omega R'^{-1}R\Omega R'RL,\ TT'T'',\\
&(R'RL)^3,\ T(R'L)^2,\ RTT'L'',\ R'^{-1}R^{-1}\Omega(R'RL)^2,\ T\Omega'R'LR'.
\end{aligned} \tag{3.10}
$$

Here $T'' = T_3 = I^{\otimes 2}\otimes T$ and the same is true for Ω'' and L''. The combinations (3.10) are constructed so that one can apply the "commutation rules" (3.1)–(3.6) to any adjacent pair of generators entering into them. We interpret this operation as the (q-)permutation of a pair of generators. Applying the q-permutations three times to the monomials (3.10), we arrange their entries in the inverse order. Obviously, this reordering can be performed in two different ways, depending on whether we first permute the left pair of generators or the right one. The diamond condition states that in both cases the result will be the same. We demonstrate how the calculations proceed in the most complex case of the $(R'R\Omega)^3$-reordering. This example was already considered in [**14**] and here we present a simpler derivation.

The calculations proceed as follows:

$$
\begin{aligned}
(R'R\Omega)^3 &= RR'\underline{R\Omega R\Omega}R'R\Omega\\
&\quad \downarrow \ 1\leftrightarrow 2 \text{ perm.}\\
&- R\Omega RR'\underline{R\Omega R\Omega}R'^{-1} + \kappa_q RR'(\Omega^2 + R\Omega^2 R)\,R'R\Omega\\
&\quad \downarrow \ 2\leftrightarrow 3 \text{ perm.}\\
&\underline{R\Omega R\Omega}R'R\Omega R^{-1}R'^{-1} - \kappa_q R\Omega RR'(\Omega^2 + R\Omega^2 R)\,R'^{-1}\\
&\quad \downarrow \ 1\leftrightarrow 2 \text{ perm.}\\
&- \Omega RR'\Omega RR'^{-1}\Omega R^{-1}R'^{-1} + \kappa_q(\Omega^2 + R\Omega^2 R)\,R'R\Omega R^{-1}R'^{-1},
\end{aligned} \tag{3.11}
$$

and in another way

$$
\begin{aligned}
(R'R\Omega)^3 &= R'R\Omega RR'\underline{R\Omega R\Omega}\\
&\quad \downarrow \ 2\leftrightarrow 3 \text{ perm.}\\
&- R'\underline{R\Omega R\Omega}R'R\Omega R^{-1} + \kappa_q R'R\Omega RR'(\Omega^2 + R\Omega^2 R)\\
&\quad \downarrow \ 1\leftrightarrow 2 \text{ perm.}\\
&\Omega RR'\underline{R\Omega R\Omega}R'^{-1}R^{-1} - \kappa_q R'(\Omega^2 + R\Omega^2 R)\,R'R\Omega R^{-1}\\
&\quad \downarrow \ 2\leftrightarrow 3 \text{ perm.}\\
&- \Omega RR'\Omega RR'^{-1}\Omega R^{-1}R'^{-1} + \kappa_q \Omega RR'(\Omega^2 + R\Omega^2 R)\,R'^{-1}R^{-1}.
\end{aligned} \tag{3.12}
$$

Here we use (2.2) and (3.2) through all the calculations. It remains to compare the κ_q-terms arising under transformations (3.11) and (3.12). Here we need one more formula (see [**14**]), namely

$$
R\Omega^2 R\Omega - \Omega R\Omega^2 R = 0\,. \tag{3.13}
$$

It is derived as follows: denoting the left-hand side of (3.13) by U and using (3.2) twice, we obtain

$$
U + \kappa_q RUR = 0\,. \tag{3.14}
$$

Now, dividing U into the sum of q-symmetric and q-antisymmetric parts $U_\pm$,

$$U_\pm = U \pm RUR^{\pm 1}, \quad P_q^\pm U_+ P_q^\mp = P_q^\pm U_- P_q^\pm = 0,$$
$$U = \frac{1+R^{-2}}{[2]_q^2}\,(U_+ + U_-),$$

we transform (3.14) into the following pair of relations

$$(I + \kappa_q R^2)U_+ = 0, \qquad (1-\kappa_q)U_- = 0\,.$$

Then, under restrictions $(1+\kappa_q R^2) \not\sim P_q^\pm$, $\kappa_q \neq 1$, or, equivalently, $[N\pm 1]_q \neq \pm q^{N\mp 4}$, $[N]_q \neq 0$, we get the desired relation (3.13).

Now one can compare the κ_q-terms in (3.11) and (3.12), moving all the Ω^2 entries to the left. The result is the same in both cases and, thus, the diamond condition on $(R'R\Omega)^3$ is satisfied. The same calculations, although simpler, can be carried out for all other monomials of (3.10), and we leave them as an exercise.

It remains to check the consistency of the $SL_q(N)$-reduction. The centrality of $\det_q T$ is easily proved by using relation (2.8). Next, the application of $\mathrm{Tr}_{q(2)}$ to (3.2) and the subsequent use of the Hecke relation (2.3) give

$$[\mathrm{Tr}_q \Omega, \Omega]_+ + \lambda q^N \Omega^2 = \kappa_q([N]_q + \lambda q^N)\Omega^2 + \kappa_q \mathrm{Tr}_q \Omega^2,$$

from which we conclude that $\mathrm{Tr}_q \Omega$ anticommutes with Ω under the conditions that

- the parameter κ_q is chosen as in (3.7);
- the quadratic scalar combination $\mathrm{Tr}_q \Omega^2$ identically vanishes.

The latter statement is a direct consequence of (3.2). It is derived as follows. Applying $\mathrm{Tr}_{q(1,2)}(\dots)$ and $\mathrm{Tr}_{q(1,2)}(\dots R^{-1})$ operations to (3.2) and using (2.14), (2.3) we obtain a system of linear relations on the quadratic scalars $(\mathrm{Tr}_q \Omega)^2$ and $\mathrm{Tr}_q \Omega^2$:

$$2\,(\mathrm{Tr}_q \Omega)^2 - [N]_q \kappa_q \mathrm{Tr}_q \Omega^2 = 0,$$
$$-\lambda(\mathrm{Tr}_q \Omega)^2 + (2q^N - \kappa_q(q^N + q^{-N}))\,\mathrm{Tr}_q \Omega^2 = 0\,.$$

The determinant of this system, $q^N[2]_q^2[N]_q/([N]_q + \lambda q^N)$, does not vanish under the conditions of the theorem and, hence, we conclude that

$$(\mathrm{Tr}_q \Omega)^2 = \mathrm{Tr}_q \Omega^2 = 0\,.$$

Then, applying the $\mathrm{Tr}_{q(2)}$ operation to (3.3), (3.6), we see that $\mathrm{Tr}_q \Omega$ is the (graded) central element in the algebra (3.1)–(3.7).

Finally, to construct the central element from the L's, we use the following trick suggested in [**2, 28, 9**] (see also [**35**]). Consider the matrix $Z = LT$. It behaves like T under left and right transitions in $GL_q(N)$. Moreover, it possesses similar algebraic properties:

$$RZZ' = ZZ'R, \qquad R^{-1}\Omega RZ = Z\Omega', \qquad RLRZ = q^{2/N}ZL'.$$

Hence, $\mathrm{Det}\, L = \det_q Z \cdot (\det_q T)^{-1}$ is central in the algebra (3.1)–(3.6). Now, let us show that $\mathrm{Det}\, L$ indeed depends only on L:

$$\begin{aligned}\varepsilon_q^{1\dots N} \cdot \mathrm{Det}\, L &= (L_1T_1)(L_2T_2)\cdots(L_NT_N)\varepsilon_q^{1\dots N} \cdot (\det_q T)^{-1} \\ &= q^{N-1}L_1(R_1L_1R_1)\cdots(R_N\cdots R_1L_1R_1\cdots R_N)\varepsilon_q^{1\dots N}.\end{aligned}$$

The expression (3.9) for $\operatorname{Det} L$ is then extracted by using (3.4), (2.2) and performing induction in N. □

COMMENT. Among the relations (3.1)–(3.6), only (3.3) is a completely new relation. Formula (3.2) was proposed in the $N = 2$ case in [**13**] and for general N in [**14**] as commutation relations for Cartan 1-forms on $SL_q(N)$. Formulas (3.4), (3.5) appeared in [**1**] as the algebra of functions on the cotangent bundle of $GL_q(N)$. The algebra of vector fields (3.4)–(3.6) was suggested in [**28, 35**] for the differential calculus on $GL_q(N)$ and $SL_q(N)$. The definition of quantum determinant $\operatorname{Det} L$ can also be found in these works and in [**9**]. Note also the recent paper [**4**], where the external algebra (3.1)–(3.3) was given in components in the $N = 2$ case. The really new point in our approach is that all these formulas are consistently combined into a single algebra.

REMARK 1. Besides the algebra (3.1)–(3.8), there exist three more differential algebras on $SL_q(N)$. They can be obtained from (3.1)–(3.6) by substitutions of two types:

$$\text{(3.15)} \qquad \mathbf{S1}: \quad R \leftrightarrow R^{-1},\ \kappa_q \leftrightarrow \kappa_{1/q} \quad \text{in (3.2);}$$

$$\text{(3.16)} \qquad \mathbf{S2}: \quad R \leftrightarrow R^{-1},\ q \leftrightarrow q^{-1} \quad \text{in (3.3)–(3.6), (3.9).}$$

For $N = 2$, the substitution (3.15) is trivialized. Indeed, the relations (3.2) and $\mathbf{S1}$·(3.2) in the case $N = 2$ differ by a term proportional to $P_q^-(\Omega^2 + R\Omega^2 R) \sim P_q^-\Omega^2 P_q^- \sim \operatorname{Tr}_q \Omega^2 \cdot P_q^-$, and, since the scalar relation $\operatorname{Tr}_q \Omega^2$ is contained both in (3.2) and $\mathbf{S1}$·(3.2), it follows that relations (3.2) and $\mathbf{S1}$·(3.2) for $N = 2$ are identical. This result agrees with the statement of [**4**] that there exist only two different external algebra structures on $SL_q(2)$. We should stress here that this mechanism does not work for $N > 2$, where we have 4 noncoinciding differential algebras.

REMARK 2. The very limited number of q-deformations for the differential calculus on $SL(N)$ seems to be a consequence of the simplicity property of $SL(N)$. In contrast, one can derive a lot of quantized versions in the $GL(N)$ case. For instance, if we omit the condition of the existence of the $SL_q(N)$-reduction, then there is no need of fixing the parameters κ_q and $q^{2/N}$ in (3.2), (3.5). Another possibility is to use the following commutation rules for T and Ω, which differ from (3.3),

$$\text{(3.17)} \qquad R\Omega RT = T\Omega'.$$

Algebras of that type were considered in [**21, 18, 29, 26, 22, 28, 30, 35, 14**].

REMARK 3. A few words on the interpretation of the basic vector fields L are in order. It is very natural to suppose that the algebra of classical vector fields V behaves under quantization like $U_q g$ and, hence, is not quadratic. On the other hand, simple quadratic relations are achieved for different types of generators, namely L^+, L^- [**10**] and L [**25, 3, 2, 27**]. These generators constitute finite shifts on the quantum group and can be viewed as a kind of "exponentiated" form of infinitesimal vector fields $L = I + \lambda V + O(\lambda^2)$. That is why the $SL_q(N)$ reduction for L is not performed by the Tr_q-like condition, but by its exponentiated Det-like form. It is also natural from this point of view that the quantities $Z = LT$ obtained from the T's by finite L-shifts behave algebraically like T's.

§4. The exterior derivative

We shall define the differential mapping d on the external algebra (3.1)–(3.3) in a constructive way.

1. We define the action of d on the generators T and Ω by setting

$$dT = \Omega T, \qquad d\Omega = \Omega^2. \tag{4.1}$$

2. For Cartan 1-forms, we postulate that the ordinary Leibniz rule is satisfied

$$d \cdot \Omega = \Omega^2 - \Omega \cdot d. \tag{4.2}$$

Using (3.13), it is straightforward to check that this prescription agrees with the commutation relations for the Ω's (3.2). Besides, due to (4.2), the action of the exterior derivative on T and on any function F of Ω is nilpotent: $d^2T = d^2F(\Omega) = 0$. Leaving aside the mathematical reasonings, we would like to stress that it is rather natural to retain the classical Leibniz picture for infinitesimal objects like Ω.

Using rules 1. and 2. above, we can calculate the action of the exterior derivative on any monomial in T and Ω of first order in T. Namely, we must first move all the Ω's to the left by using the commutation relations (3.3), and then apply (4.1) and (4.2) to get $d(F(\Omega)T) = dF(\Omega)T + F(-\Omega)\Omega T$. In this way we automatically obtain the consistency of the differential mapping with the algebraic relations (3.3) and the nilpotence of d on any monomial of that type.

The next step is to construct the differential mapping for the general quadratic monomial of T: TT'. We stress here that since under quantization we obtain finite shifts L acting on T rather than differentiation, it is reasonable to expect modified Leibniz rules for T. The action of d should take into account the algebraic relations (3.1):

$$R\, d(T, T') = d(TT')\, R.$$

Note also that the expression for $d(TT')$ must be of first order in Ω. The general ansatz satisfying both these conditions reads

$$d(TT') = f(R)(\Omega + R\Omega R)TT'. \tag{4.3}$$

Here $f(R)$ is a function of R and the combination $\Omega + R\Omega R$ commutes with the R-matrix due to the Hecke conditions (2.3). The exact form of the function $f(R)$ is dictated by the nilpotence condition:

$$0 = d^2(TT') = f\{(\Omega^2 + R\Omega^2 R) - f(\Omega + R\Omega R)^2\}TT'.$$

Using (3.2), it is straightforward to obtain

$$(\Omega + R\Omega R)^2 = (I + \kappa_q R^2)(\Omega^2 + R\Omega^2 R),$$

and, hence, d is nilpotent on TT' if we put[2]

$$f(R) = (I + \kappa_q R^2)^{-1}. \tag{4.4}$$

Using (3.3), (4.1)–(4.3), we can now see how d acts on any monomial of T and that Ω is quadratic in T, and again the nilpotence of d is guaranteed by (4.2).

Thus, we have given a detailed consideration of the first few steps in the construction of the differential mapping d. Generalizing this procedure to monomials of any order in T, we get

[2]Note that under the restrictions of Theorem 1 the matrix $(I + \kappa_q R^2)$ is invertible.

THEOREM 2. *For the external algebra (3.1)–(3.3) presented in Theorem 1 there exists a differential mapping d acting from the left, and defined by (4.1), (4.2), and*

$$d(T_1T_2\cdots T_k) = \{I + \kappa_q(S_k(I) - I)\}^{-1} S_k(\Omega) T_1T_2\cdots T_k, \tag{4.5}$$

where

$$S_k(X) = X + \sum_{i=1}^{k-1} R_i \cdots R_2R_1XR_1R_2\cdots R_i. \tag{4.6}$$

In particular,

$$d(\det_q T) = \frac{\operatorname{Tr}_q \Omega \det_q T}{q^{N-1}(1-\kappa_q) + [N]_q\kappa_q}, \tag{4.7}$$

$$d(\operatorname{Tr}_q \Omega) = \operatorname{Tr}_q \Omega^2 = 0, \tag{4.8}$$

which guarantees the compatibility of d with the reduction conditions (3.8). This differential mapping commutates with the action of the basic vector fields L:

$$[d, L] = 0. \tag{4.9}$$

PROOF. As in the case $k = 2$, we start with the following general ansatz:

$$d(T_1T_2\cdots T_k) = f_k S_k(\Omega) T_1T_2\cdots T_k. \tag{4.10}$$

Here f_k is a function of $R_1, \dots, R_k$ to be specified below, and

$$R_i f_k = f_k R_i, \quad R_i S_k(\Omega) = S_k(\Omega) R_i, \qquad i = 1, \dots, k-1. \tag{4.11}$$

The first of the relations (4.11) is the restriction on the possible form of f_k, while the last is a direct consequence of the Yang–Baxter equation (2.2) and the Hecke condition (2.3). By virtue of (4.11), we have

$$R_i\, d(T_1\cdots T_k) = d(T_1\cdots T_k)\, R_i, \qquad i = 1, \dots, k-1,$$

and thus ansatz (4.10) is compatible with the relations (3.1) of the external algebra. The nilpotence condition $d^2(T_1\cdots T_k) = 0$ leads to the relation

$$S_k(\Omega^2) - f_k(S_k(\Omega))^2 = 0.$$

It remains to compute the quantity $(S_k(\Omega))^2$. This calculation, based on the essential use of (3.2), (2.2), and (2.3), is rather lengthy. Here we only present the result

$$(S_k(\Omega))^2 = \{I + \kappa_q(S_k(I) - I)\} S_k(\Omega^2).$$

Hence, the function f_k must be chosen as in (4.5). Note that with this choice f_k satisfies conditions (4.11).

In order to obtain formula (4.7), one must use properties (2.6) of the q-deformed Levi–Civita tensor, and also the relation

$$\varepsilon_q^{1\dots N} S_N(X) = q^{1-N} \operatorname{Tr}_q X \varepsilon_q^{1\dots N}.$$

The verification of the compatibility of condition (4.9) with the algebra (3.4)–(3.6) is straightforward. □

REMARK 1. Using (4.1), (4.2), (4.5), and (3.2), (3.3), one can derive the explicit form of the modified Leibniz rules. These rules appear in modified form for T and Ω-polynomials for which the exterior derivative acting from the left must cross T under evaluation. For quadratic polynomials we have

$$d(TT') = (I + \kappa_q R^2)^{-1}\{R^2\, dTT' + T\, dT'\},$$
$$d(T\Omega') = (1 - \kappa_q)T\, d\Omega' + dT\Omega' + \{(1-\kappa_q)R^2 - I\}\Omega^2 T.$$

Here the term $\Omega^2 T$ may be treated either as $d\Omega T$ or as $\Omega\, dT$. Note that the operator R^2, being the generating element of the braid group B_2, plays a special role in these formulas. This observation is given further support if we evaluate the action of d on the monomials of any order in T:

$$d(T_1 \cdots T_k) = \left\{I + \kappa_q \sum_{i=1}^{k-1} B_{k,i}\right\}^{-1} \sum_{i=1}^{k} B_{k,i} T_1 \cdots dT_i \cdots T_k, \tag{4.12}$$
$$B_{k,i} = (R_i R_{i+1} \cdots R_{k-1})(R_{k-1} R_{k-2} \cdots R_i), \qquad i = 1, \dots, k-1,$$
$$B_{k,k} = I^{\otimes k}.$$

Here $\{B_{k,i}\}_{i=1}^{k}$ is the set of generating elements of the braid group B_k.

REMARK 2. Note that in constructing the differential mapping d, the self-commutation relations for T (3.1) and Ω (3.2) are essential. The explicit form of the cross-commutation relations for T and Ω (3.3) is not relevant. We should only be aware of the fact that these relations allow us to move all the Ω's to the left in any monomial of T and Ω. Thus, the algorithm described can be applied equally well to the external algebras considered in **[21, 18, 29, 26, 22, 28, 30, 35, 14]** and to those satisfying cross-multiplication relations of the type (3.17). In this way one can search for all the external algebraic structures on $GL_q(N)$ compatible with the ordinary Leibniz prescriptions. It turns out that only two external algebras obtained in the references above satisfy this condition. The first of these algebras is defined by relations (3.1), (3.17) and (3.2) in which one must put $\kappa_q = 0$. The second algebra is obtained from the first if one makes the substitution $R \leftrightarrow R^{-1}$ in all the formulas. This result agrees with the quasiclassical considerations of [**5**].

Acknowledgments. We would like to thank A. P. Isaev for fruitful collaboration and stimulating discussions.

References

1. A. Yu. Alekseev and L. D. Faddeev, $(T^*G)_t$: *A toy model for conformal field theory*, Comm. Math. Phys. **141** (1991), 413–422.
2. ______, *Involution and dynamics in the system q-deformed top*, Zap. Nauchn. Sem. Leningrad. Otdel. Mat. Inst. Steklov. (LOMI) **200** (1992), 3–16; English transl. in J. Soviet Math. (to appear).
3. A. Yu. Alekseev, L. D. Faddeev, and M. A. Semenov-Tyan-Shansky, *Hidden quantum groups inside Kac-Moody algebra*, Comm. Math. Phys. **149** (1992), 335–345.
4. I. Ya. Aref′eva, G. E. Arutyunov, and P. B. Medvedev, *Poisson-Lie structures on the external algebra of SL(2) and their quantization*, Journ. Math. Phys. **35** (1994), 6658–6671.
5. G. E. Arutyunov and P. B. Medvedev, *Quantization of external algebra on a Poisson-Lie group*, Preprint SMI-11-93 and hep-th/9311096.
6. P. Aschieri and L. Castellani, *An introduction to non-commutative differential geometry on quantum groups*, Internat. J. Modern. Phys. A **8** (1993), 1667–1706; L. Castellani, M. A. R-Monteiro, *A note on quantum structure constants*, Phys. Lett. B **314** (1993), 25–30.

7. G. M. Bergman, *The diamond lemma for ring theory*, Adv. in Math. **29** (1978), 178–218.
8. U. Carow-Watamura, M. Schlieker, S. Watamura, and W. Weich, *Bicovariant differential calculus on quantum groups $SU_q(N)$ and $SO_q(N)$*, Comm. Math. Phys. **142** (1991), 605–641.
9. B. Drabant, B. Jurčo, M. Schlieker, W. Weich, and B. Zumino, *The Hopf algebra of vector fields on complex quantum groups*, Lett. Math. Phys. **26** (1992), 91–96.
10. L. D. Faddeev, N. Yu. Reshetikhin, and L. A. Takhtadzhyan, *Quantization of Lie groups and Lie algebras*, Algebra i Analiz **1** (1989), 178–206; English transl., Leningrad Math. J. **1** (1990), 193–226.
11. L. D. Faddeev, *Lectures on Int. Workshop "Interplay between Mathematics and Physics"*, Vienna 1992 (unpublished).
12. A. P. Isaev and Z. Popowicz, *q-trace for quantum groups and q-deformed Yang-Mills theory*, Phys. Lett. B **281** (1992), 271–278; A. P. Isaev and R. P. Malik, *Deformed traces and covariant quantum algebras for quantum groups $GL_{qp}(2)$ and $GL_{qp}(1|1)$*, Phys. Lett. B **280** (1992), 219–226.
13. A. P. Isaev and P. N. Pyatov, *$GL_q(N)$-covariant quantum algebras and covariant differential calculus*, Phys. Lett. A **179** (1993), 81–90.
14. ———, *Covariant differential complexes on quantum linear groups*, Journ. Phys. A: Math. Gen. **28** (1995), 2227–2246.
15. M. Jimbo, *A q-analogue of $U(gl(N+1))$, Hecke algebra, and the Yang-Baxter equation*, Lett. Math. Phys. **11** (1986), 247–252.
16. B. Jurčo, *Differential calculus on quantized simple Lie groups*, Lett. Math. Phys. **22** (1991), 177–186.
17. P. P. Kulish and R. Sasaki, *Covariance properties of reflection equation algebras*, Progr. Theoret. Phys. **89** (1993), 741–761.
18. G. Maltsiniotis, *Groupes quantuques et structures différentielles*, C. R. Acad. Sci. Paris Sér. I Math. **331** (1990), 831–834; *Calcul diffe'rentiel sur le groupe line'arie quantique*, Preprint ENS (1990); *Le langage des espaces et des groupes quantiques*, Comm. Math. Phys. **151** (1993), 275–302.
19. Yu. I. Manin, *Quantum groups and noncommutative geometry*, Université de Montreal, Montreal, 1988.
20. ———, *Multiparametric quantum deformation of the general linear supergroup*, Comm. Math. Phys. **123** (1989), 163–175.
21. ———, *Notes on quantum groups and quantum de Rahm complexes*, Teoret. Mat. Fiz. **92** (1992), 425–450; English transl. in Theoret. and Math. Phys. **92** (1992).
22. F. Müller-Hoissen, *Differential calculi on the quantum group $GL_{p,q}(2)$*, J. Phys. A: Math. Gen. **25** (1992), 1703–1734; F. Müller-Hoissen and C. Reuten, *Bicovariant differential calculi on $GL_{p,q}(2)$ and quantum subgroups*, J. Phys. A: Math. Gen. **26** (1993), 2955–2976.
23. P. Podleś and S. L. Woronowicz, *Quantum deformation of Lorentz group*, Comm. Math. Phys. **130** (1990), 381–431.
24. N. Yu. Reshetikhin, *Quasitriangular Hopf algebras and invariants of tangles*, Algebra i Analiz **1** (1989), no. 2, 169–188; English transl. in Leningrad Math. J. **1** (1990).
25. N. Yu. Reshetikhin and M. A. Semenov-Tyan-Shansky, *Central extensions of quantum current groups*, Lett. Math. Phys. **19** (1990), 133–142.
26. A. Schirrmacher, *Remarks on the use of R-matrices*, Groups and Related Topics. Proceedings, Wroclaw 1991 (R. Gielerak, J. Lukierski, and Z. Popowicz, eds.), Kluwer Academic Publishers, 1992, pp. 55–65.
27. W. M. Schmidke, S. P. Vokos, and B. Zumino, *Differential geometry of the quantum supergroup $GL_q(1|1)$*, Z. Phys. C **48** (1990), 249–255.
28. P. Schupp, P. Watts, and B. Zumino, *Differential geometry on linear quantum groups*, Lett. Math. Phys. **25** (1992), 139–147; *Bicovariant quantum algebras and quantum Lie algebras*, Comm. Math. Phys. **157** (1993), 305–329.
29. A. Sudbery, *Canonical differential calculus on quantum general linear groups and supergroups*, Phys. Lett. B **284** (1992), 61–65; *Erratum*, Phys. Lett. B **291** (1992), 519; *The algebra of differential forms on a full matric bialgebra*, Math. Proc. Cambridge Philos. Soc. **114** (1993), 111–130.
30. B. Tzygan, *Notes on differential forms on quantum groups*, Penn. Univ. Preprint (1992).

31. J. Wess and B. Zumino, *Covariant differential calculus on the quantum hyperplane*, Nuclear Phys. B Proc. Suppl. **18** (1990), 302–312.
32. S. L. Woronowicz, *Twisted $SU(2)$ group. An example of noncommutative differential calculus*, Publ. Res. Inst. Math. Sci. Kyoto Univ. **23** (1987), 117–181.
33. ———, *Differential calculus on compact matrix pseudogroups (quantum groups)*, Comm. Math. Phys. **122** (1989), 125–170.
34. B. Zumino, *Introduction to the differential geometry of quantum groups*, Proc. of Xth IAMP Conf. Leipzig 1991, Springer-Verlag, Berlin–Heidelberg, 1992.
35. ———, *Differential calculus on quantum spaces and quantum groups*, Proc. XIX-th ICGTMP Salamanca 1992, CIEMAT/RSEF Madrid, 1993.

St. Petersburg Branch of the Steklov Mathematical Institute, Fontanka 27, St. Petersburg 191011, Russia

Bogolyubov Theoretical Laboratory, Joint Institute for Nuclear Research, 141980 Dubna, Moscow Region, Russia

Translated by THE AUTHORS

Amer. Math. Soc. Transl.
(2) Vol. **175**, 1996

The Penrose Transform on Flag Domains in $\mathcal{F}(\mathbb{C}P^2)$

Simon Gindikin

I would like to explain the choice of the subject of this paper. Alik Berezin had a strong interest in geometry and analysis on symmetric manifolds, and we discussed this subject many times, especially when he was working on his project related to quantization on classical domains. In the corresponding papers, he was the first to consider extensions of holomorphic discrete series of unitary representations realized in Hardy spaces on different components of the boundaries. Alik was very excited by his discovery, and we had extremely interesting talks about the nature of these results. At his request, I proved his conjecture about invariant positive distributions in classical domains and generalized this construction to arbitrary homogeneous bounded domains in $\mathbb{C}^n$ (not only symmetric). I think that experts in representation theory did not give enough credit to Berezin for these outstanding results (which later were remarkably developed in the work of Rossi–Vergne, Wallach and others). In the last years of Alik's life, we talked a lot about the Penrose transform, about twistors, and about the possibility of developing a quantization on the $\overline{\partial}$-cohomology in higher dimensions for nonholomorphic discrete series. Here I shall explain the Penrose transform for such representations on the example of the simplest case of nonholomorphic discrete series for $\mathrm{SU}(2,1)$. I believe that it would have been natural to talk about these things with Alik.

§1. Geometrical background (complex picture)

Let $\mathbb{C}P^2_z$, $\mathbb{C}P^2_Z$ be dual projective planes with homogeneous coordinates $z = (z_0, z_1, z_2)$, $Z = (Z_0, Z_1, Z_2)$ and the duality

$$\langle Z, z\rangle = Z_0 z_0 + Z_1 z_1 + Z_2 z_2 .$$

We shall identify $Z \in \mathbb{C}P^2_Z$ with the line $\langle Z, \cdot\rangle = 0$ in $\mathbb{C}P^2_z$. We shall also denote by $l(u,v)$ the line going through $u, v \in \mathbb{C}P^2_z$. Let $F = \mathcal{F}(\mathbb{C}P^2_z)$ be the flag manifold for $\mathbb{C}P^2_z$, i.e., the manifold of pairs

$$(z, Z),\ z \in \mathbb{C}P^2_z,\ Z \in \mathbb{C}P^2_Z,\ z \in Z \quad (\langle Z, z\rangle = 0).$$

1991 *Mathematics Subject Classification.* Primary 32F10, 32M15, 22E46.
Research partially supported by NSF Grant DMS 92 02049.

Then $F = \mathrm{SL}(3;\mathbb{C})/B \cong \mathrm{SU}(3)/T$, where B is the Borel subgroup and T is a maximal torus in $\mathrm{SU}(3)$, $\dim F = 3$.

Let M be the manifold of pairs (w, W), $w \in \mathbb{C}P^2_z$, $W \in \mathbb{C}P^2_Z$, $w \notin W$ (i.e., $\langle w, W\rangle \neq 0$). Then

$$M = \mathrm{SL}(3;\mathbb{C})/\,\mathrm{GL}(2;\mathbb{C}), \qquad \dim M = 4$$

(M is a pseudo-Riemannian symmetric Stein manifold). Let us define the incidence relation between F and M:

$$(z, Z) \in (w, W) \iff (z, Z) \in F,\ (w, W) \in M,\ z \in W,\ w \in Z.$$

The manifold X of incident pairs $(z, Z \mid w, W)$ is the homogeneous manifold $\mathrm{SL}(3;\mathbb{C})/\widetilde{B}$, where $\widetilde{B}$ is the Borel subgroup in $\mathrm{GL}(2,\mathbb{C})$, $\dim X = 3$, and we have the dual fibering

$$\text{(1)} \qquad \begin{array}{ccc} & X & \\ {}^{\pi}\swarrow & & \searrow^{\rho} \\ F & & M \end{array}$$

corresponding to the incidence relation. The fibers of π are isomorphic to $\mathbb{C}^2$ and the fibers of ρ to $\mathbb{C}P^1$. Let $E_{(w,W)} = \pi(\rho^{-1}(w, W))$, $S_{(z,Z)} = \rho(\pi^{-1}(z, Z))$. So $E_{(w,W)}$ is the set of flags incident to $(w, W) \in M$. The manifolds $E_{(w,W)}$ are rational curves on F. It turns out that the normal bundles of these curves are isomorphic to $\mathcal{O}(1) \oplus \mathcal{O}(1)$, so we can interpret F as a (curved) twistor space in the sense of Penrose.

So on the three-dimensional (complex) manifold F we have a four-dimensional family of rational curves $E_{(w,W)}$, which are parametrized by points of the manifold M. This interpretation of M induces a conformal structure on M: the conformal distance between (w, W) and (u, U) is 0 iff the curves $E_{(w,W)}$, $E_{(u,U)}$ have a nonempty intersection (when the line $l(w, u)$ in $\mathbb{C}P^2_z$ passes through the intersection point $W \cap U$). An easy computation shows that this condition is quadratic and we indeed have a conformal metric.

Let us consider a coordinate chart $w = (1, x, y)$, $W = (1, X, Y)$ on M. Then we have the conformal metric

$$\text{(2)} \qquad g = dx\, dX + dy\, dY + (y\, dx - x\, dy)(Y dX - X\, dY).$$

Let $V_{(w,W)} \subset T_{(w,W)}M$ be the isotropic cone in the tangent space. The two-dimensional manifolds $S_{(z,Z)}$ (consisting of curves $E_{(w,W)}$ containing the flag (z, Z)) are isotropic submanifolds on M (α-submanifolds in Penrose's terminology). For any point (w, W) there is a one-parameter family of $S_{(z,Z)}$ ($(z, Z) \in E_{(w,W)}$) passing through it. Let $\alpha_{(z,Z)}$ be the corresponding tangent two-planes in $T_{(w,W)}M$ (α-planes). Their union is $V_{(w,W)}$, they do not intersect outside the vertex, and they are parametrized by $\mathbb{C}P^1$. The existence of α-submanifolds (integrability of α-distributions) implies the selfduality of the conformal metric g (the self-dual part W_+ of the Weyl curvature is zero).

On M there are also two two-parameter families of isotropic two-dimensional submanifolds:

$$R_{w^0} = \{(w, W),\, w = w^0\}, \qquad T_{W^0} = \{(w, W),\, W = W^0\}.$$

For each (w^0, W^0) we have one submanifold from each family (R_{w^0} and T_{W^0}) passing through (w^0, W^0). The corresponding tangent subspaces β_{w^0}, β_{W^0} lie on the cone $V_{(w^0,W^0)}$ and are representatives of a second one-parameter family of two-planes on V (β-planes). There are no isotropic submanifolds, different from R_w, T_W which are tangent β-planes. This means that the anti-self-dual part of the Weyl curvature W_- does not vanish and the metric g is not conformally flat.

§2. Spinor bundles on M

Let Y be the manifold of generic pairs of flags (u, U), (v, V). This means that

$$\begin{gathered} \langle u, V\rangle \neq 0, \quad \langle v, U\rangle \neq 0, \quad \dim Y = 6, \\ Y = \mathrm{SL}(3; \mathbb{C})/D, \quad D \text{ is the diagonal subgroup.} \end{gathered} \tag{3}$$

Y is isomorphic to the manifold of generic triples of points of $\mathbb{C}P^2$. Let us put the normalization condition

$$\langle u, V\rangle + \langle v, U\rangle = 0. \tag{4}$$

We have the natural double fibering

$$\begin{array}{ccccc} & & Y & & \\ & \overset{\sigma}{\swarrow} & & \overset{\tau}{\searrow} & \\ F & & & & M \end{array} \tag{5}$$

$$\sigma(u, U \mid v, V) = (u, U), \qquad \tau(u, U \mid v, V) = (U \cap V, l(u, v)).$$

In other words, Y gives a characterization of $E_{(w,W)}$ by pairs of flags (u, U), (v, V) on it. Condition (4) allows us to describe $E_{(w,W)}$ as

$$E_{(w,W)} = \{(z, Z) \in F,\ z = \tau_0 u + \tau_1 w,\ Z = \tau_0 U + \tau_1 V,\ (\tau_0, \tau_1) \in \mathbb{C}P^1_\tau\}. \tag{6}$$

Now we define spinor bundles $\Sigma(\alpha, \beta)$ on M, where α, β are nonnegative integers. Let us denote by $T_q(g)$ the representations of $\mathrm{SL}(2; \mathbb{C})$ in the space of monomials $\tau_0^k \tau_1^l$, $k + l = q$, induced by the transformation

$$\tau = \begin{pmatrix} \tau_0 \\ \tau_1 \end{pmatrix} \mapsto \tau g, \qquad g \in \mathrm{SL}(2; \mathbb{C}).$$

The sections F of the $(q + 1)$-dimensional bundle $\Sigma(\alpha, \beta)$ ($F \in \Gamma(\Sigma(\alpha, \beta))$) are vector-functions $F^{k,l}$, $k + l = q$, on Y satisfying the conditions

$$\begin{gathered} F^{k,l}(\lambda u, \lambda v, \mu U, \mu V) = \lambda^{-\alpha-1} \mu^{-\beta-1} F^{k,l}(u, v, U, V), \\ F^{k,l}(\rho u, \sigma v, \rho U, \sigma V) = \rho^{-k-1} \sigma^{-l-1} F^{k,l}(u, v, U, V), \\ F(g(u, v), g(U, V)) = T_q(g) F(u, v, U, V). \end{gathered} \tag{7}$$

Let us define differential operators of the first order on $\Gamma(\Sigma(\alpha,\beta))$:

$$\begin{aligned} \partial_u &= w\cdot\frac{\partial}{\partial u} = w_0\frac{\partial}{\partial u_0} + w_1\frac{\partial}{\partial u_1} + w_2\frac{\partial}{\partial u_2}, & \partial_v &= w\cdot\frac{\partial}{\partial v}, \\ D_U &= W\cdot\frac{\partial}{\partial U}, & D_V &= W\cdot\frac{\partial}{\partial V}. \end{aligned} \tag{8}$$

The operators ∂ are the derivatives in the directions of the β-plane β_w (respectively D are the derivatives in the directions of β_W), ∂_u corresponds to the direction $\alpha_{(u,U)}\cap\beta_w$ (we infinitesimally rotate the curve $E_{(w,W)}$ around (u,U) for a fixed w).

§3. Geometric background (real forms)

We shall connect a real geometry with $\mathrm{SU}(2,1)$, the real form of the complex group $\mathrm{SL}(3;\mathbb{C})$. On $\mathbb{C}P^2$ it has two open orbits: the ball B_+ and the complement B_-:

$$B_+ = \{|z_0|^2 - |z_1|^2 - |z_2|^2 > 0\}, \qquad B_- = \{|z_0|^2 - |z_1|^2 - |z_2|^2 < 0\}. \tag{9}$$

The domain B_+ is the Hermitian symmetric manifold $\mathrm{SU}(2,1)/U(2)$, the domain B_- is the pseudo-Hermitian symmetric manifold $\mathrm{SU}(2,1)/U(1,1)$. On the flag manifold F, we have three open orbits: the flags (z,Z) with $z\in B_+$, or with the line $Z\subset B_-$, or with $z\in B_-$ and $Z\cap B_+\neq\varnothing$. We are interested in the last domain $F_+\cong\mathrm{SU}(2,1)/T$. Correspondingly, on M we consider the domain

$$M_+ = \{(w,W),\ w\in B_+,\ W\subset B_-\}.$$

The connection between the two domains F_+, M_+ is the following: the curves $E_{(w,W)}$ lie in F_+ if and only if $(w,W)\in M_+$. The group $\mathrm{SU}(2,1)$ acts on M_+, but not transitively. Of course, $\mathrm{SU}(2,1)\times\mathrm{SU}(2,1)$ acts on M_+, but this action cannot be extended to M and we shall not consider it.

On M_+ we have the $\mathrm{SU}(2,1)$-orbit

$$\widetilde{B}_+ = \{(w,W),\ w=(w_0,w_1,w_2)\in B_+,\ W=\sigma(w)=(\overline{w}_0,-\overline{w}_1,-\overline{w}_2)\}. \tag{10}$$

$\widetilde{B}_+$ is not a complex submanifold in M and the distributions β_w, β_W define a complex structure on B_+. A very important point is that the restriction of the conformal metric g to $\widetilde{B}_+$ has a Riemann signature. The geometrical nature of this fact is that the curves $E_{(w,W)}$, $(w,W)\in\widetilde{B}_+$, produce a fibering of F_+: they cover the entire set F_+ and do not intersect or coincide. As a result, we have a (noncomplex) fibering

$$F_+\to B_+ \qquad (\mathrm{SU}(2,1)/T\to\mathrm{SU}(2,1)/U(2),\ U(2)>T). \tag{11}$$

We have the double fibering

$$\begin{array}{ccccc} & & Y_+ & & \\ & \overset{\sigma}{\swarrow} & & \overset{\tau}{\searrow} & \\ F_+ & & & & M_+ \end{array} \tag{5'}$$

which is the restriction of the double fibering (5) to F_+.

§4. The Penrose transform

The domain F_+ is 1-pseudoconcave. We consider the one-dimensional $\overline{\partial}$-cohomology $H^{(0,1)}(F_+, \mathcal{O}(-\alpha-1, -\beta-1))$ with coefficients in the line bundles $\mathcal{O}(-\alpha-1,-\beta-1)$. To represent this cohomology, let us consider $\overline{\partial}$-closed $(0,1)$-forms $\omega(z, Z, \overline{z}, \overline{Z}, d\overline{z}, d\overline{Z})$ on F_+ that have homogeneity degree $(-\alpha-1,-\beta-1)$ in (z, Z) and $(0,0)$ in $(\overline{z},\overline{Z})$. Let us define the Penrose transform as

$$\begin{gathered}\mathcal{P}\omega^{k,l}(u,U,v,V) = \int_{\mathbb{C}P^1_\tau} \omega|_{z=\tau_0 u+\tau_1 v, \tau_0 U + \tau_1 V}\, \tau_0^k \tau_1^l \wedge (-\tau_0\, d\tau_1 + \tau_1\, d\tau_0), \\ k+l=\alpha+\beta=q, \quad (u,v,U,V)\in Y_+,\end{gathered} \tag{12}$$

where for each (u, U, v, V) the integrand is a $(1,1)$-form on the projective line $E_{(w,W)} \cong \mathbb{C}P^1_\tau$ and the result of the integration is a section of the spinor bundle $\Sigma(\alpha,\beta)$. All properties of the Penrose transform $\mathcal{P}\omega$ can be checked by direct computation (as, for example, the fact that $\mathcal{P}\omega$ is a spinor field) or in the same way as for the usual Penrose transform in the conformally flat case (as in [**2**]). Now we list some properties of $\mathcal{P}$.

1. It is evident that $\mathcal{P}\omega \equiv 0$ if ω is $\overline{\partial}$-exact. As a result, we can regard the Penrose transform as an operator on $H^{(0,1)}(F_+)$.

2. It is possible to prove that if $\mathcal{P}\omega \equiv 0$, then ω is $\overline{\partial}$-exact (cf. [**2**]). This means that $\mathcal{P}$ is injective on $H^{(0,1)}(F_+)$.

3. The field $\mathcal{P}\omega$ satisfies the system of differential equations

$$\partial_u(\mathcal{P}\omega)^{k-1,l} = \partial_v(\mathcal{P}\omega)^{k,l-1}, \qquad D_U(\mathcal{P}\omega)^{k-1,l} = D_V(\mathcal{P}\omega)^{k-1,l} \tag{13}$$

if $q>0$. If $\alpha=\beta=q=0$, then

$$\square(\mathcal{P}\omega) = 0, \qquad \square = \partial_u D_V - \partial_v D_U \tag{14}$$

In the general case, all the components $\mathcal{P}\omega$ satisfy the equation

$$\square(\mathcal{P}\omega)^{k,l} = 0, \tag{14'}$$

which is a direct consequence of (13).

Systems (13) are analogs of massless equations. The connection between (13) and (14) is similar to the connection between the Laplace and Cauchy–Riemann equations.

The verification of (13) is direct. The action of the operator ∂_u on (12) is equivalent to the action of the operator $\tau_0\partial_z = \tau_0 w \cdot (\partial/\partial z)$ on the integrand. The action of ∂_v is equivalent to the action of $\tau_1\partial_z$ on the integrand.

4. The restriction of $\square$ to $\widetilde{B}_+$ is an elliptic operator and the restriction of (13) is an elliptic system. This corresponds to the fact that $g|_{\widetilde{B}_+}$ is a positive metric.

§5. The inverse Penrose transform

The crucial point of this construction is a version of the Gelfand–Graev–Shapiro operator $\varkappa$. Let us consider the operator $\varkappa$ from sections φ on Y to one-forms on

Y with differentials only along the fibers of σ:

$$\begin{aligned}\varphi(u,U,v,V) &\mapsto \varkappa\varphi(u,U,v,V;dv,dV)\\ &= \frac{1}{\langle w,W\rangle}\{\partial_u\varphi\cdot\langle W,dv\rangle + D_U\varphi\cdot\langle w,dV\rangle - (q+1)\varphi\langle v,dV\rangle\},\\ &\qquad (w,W)=\tau(u,U,v,V).\end{aligned}\tag{15}$$

So $\varkappa\varphi$ is a (holomorphic) one-form on flags (v,V) that depends on the flags (u,U) as parameters. If the section φ is a solution of the equation

$$\Box\varphi = 0\tag{16}$$

and

$$\varphi(u+\lambda v, U+\lambda V, v, V) = \varphi(u,U,v,V),\tag{17}$$

then the form $\varkappa\varphi$ is closed for each (u,U); moreover, the fact that $\varkappa\varphi$ is closed under condition (17) is equivalent to equation (16). The proof is quite direct. Under condition (17), we have

$$d_{(v,V)}\varphi = \frac{1}{\langle w,W\rangle}\left\{\partial_v\varphi\cdot\langle W,dv\rangle + D_V\varphi\langle w,dV\rangle + \left(u\frac{\partial}{\partial v}+U\frac{\partial}{\partial V}\right)\langle V,dv\rangle\right\}\tag{18}$$

(it follows from (17) that $d_{(v,V)}\varphi$ is a combination of $\langle V,dv\rangle$, $\langle W,dv\rangle$, and $\langle w,dV\rangle$) and we need only to express the coefficients in terms of

$$\frac{\partial}{\partial v}\varphi,\quad \frac{\partial}{\partial V}\varphi.$$

Let us remark that

$$\langle V,dv\rangle = -\langle v,dV\rangle,\quad \langle u,V\rangle = -\langle v,U\rangle = \langle w,W\rangle.$$

It follows from the last property, (7), and (14′) that $\varphi = (\mathcal{P}\omega)^{q,0}$ satisfies (16) and (17) on Y_+.

Now let us consider a section γ of the fibering σ in (5′) (of course holomorphic sections do not exist, but real sections do exist, so the fibers are contractible) and let

$$\omega_\gamma = \{\varkappa(\mathcal{P}\omega)^{q,0}|_\gamma\}^{(0,1)}.\tag{19}$$

Informally, we can regard the restriction to $\gamma((v,V)=\gamma(u,U))$ as a one-form on F_+ and take the $(0,1)$-part of this form. The result is a $\bar\partial$-closed $(0,1)$-form on F_+.

MAIN THEOREM. *For a normalized constant $c\neq 0$, the form $\omega - c\omega_\gamma$ is $\bar\partial$-exact.*

So this is the way to reconstruct the cohomology class of w; the forms ω_γ for different sections γ differ by $\bar\partial$-exact forms. The proof can be carried out as in [**2**]. We must prove that

$$\mathcal{P}\omega_\gamma = \mathcal{P}\omega \qquad \text{for each } (u,U,v,V)\in Y_+,\tag{20}$$

where c is a constant. We can select the section γ in the most convenient way and do it only in the neighborhood of the curve $E_{(w,W)}$ along which we integrate. We can take $\gamma(u,U)$ for $(u,U) \in E_{(w,W)}$ so that

$$\tau((u,U),\gamma(u,U)) = (w,W).$$

Then $\mathcal{P}\omega_\gamma(u,U,v,V)$ for any $\tau(u,U,v,V) = (w,W)$ differs from $\mathcal{P}\omega$ only by a factor independent of ω.

It is clear that here we only use the fact that $\mathcal{P}\omega$ is a solution of (13) and (7). So the Penrose transform determines an isomorphism between $H^{(0,1)}(F_+)$ and the space of sections of spinor bundles that satisfy (13). Of course it is an intertwining operator. It is known that in $H^{(0,1)}(F_+)$ one can realize representations of nonholomorphic discrete series of $\mathrm{SU}(2,1)$. So we have obtained the realizations of such representations in solutions of (13) on M_+.

We can restrict these solutions to $\widetilde{B}_+ \subset M_+$. As we mentioned already, the system is elliptic there and all the solutions on $\widetilde{B}_+$ admit holomorphic extensions to (complex) neighborhoods of $\widetilde{B}_+$. In fact, they all extend to M_+. The operator $\varkappa$ gives the explicit extension: we need to take $\gamma(u,U)$ on F_+ so that $\tau(u,U,\gamma(u,U)) \in \widetilde{B}_+$ $(E_{(w,W)}$, $(w,W) \in \widetilde{B}_+$, yields a fibering of F_+. Then we can reconstruct $\varkappa\varphi|_{F_+}$ (and hence $\mathcal{P}\varkappa\varphi_\gamma)$ only in terms of φ. But $\mathcal{P}(\varkappa\varphi_\gamma)$ is the extension of φ to M_+. Restrictions of system (13) to $\widetilde{B}_+$ are equivalent to Schmid systems [**3**]. For other properties of the Penrose transform on M_+, see [**1**].

§6. Final remarks

The problem that we considered in this paper illustrates the following general situation. We have a flag manifold

$$F = G_{\mathbb{C}}/P \cong K/C,$$

where $G_{\mathbb{C}}$ is a complex semisimple Lie group, P is a parabolic subgroup, K is the compact form of $G_{\mathbb{C}}$, C is the centralizator of a torus in K. On the other hand, let

$$M = G_{\mathbb{C}}/L$$

be a homogeneous Stein manifold, $L \supset C$ be a reductive subgroup in $G_{\mathbb{C}}$ and suppose we have the double fibering

$$\begin{array}{ccc} & X & \\ {}^{\pi}\swarrow & & \searrow^{\rho} \\ F & & M \end{array}$$

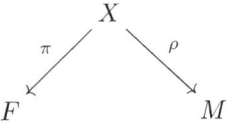

where $X = G_{\mathbb{C}}/\widetilde{P}$, $\widetilde{P} = P \cap L$, that gives a realization of M as the manifold of compact submanifolds E_w, $w \in M$, in F that are isomorphic to $L/\widetilde{P} \cong U/C$, U being a maximal compact subgroup in L, $U \supset C$. Then on M we have a canonical generalized conformal structure. Namely, for each $w^0 \in M$ we consider the conical subvariety M_{w^0} of $w \in M$ such that E_w is not in general position with respect to E_{w^0} and we let $V_w \subset T_w M$ be the algebraic cone tangent to M_w. So we have an invariant field of (linearly equivalent) algebraic cones $V_w \subset T_w$ that define the generalized conformal structure [**4**]. In our example, the cones V_w were quadratic and the generalized conformal structure was the ordinary conformal structure. In the general case, the (nonquadratic) cones V_w possess analogs of α-planes and β-planes and we can define generalizations of conformal invariant operators ∂, D, and $\Box$.

Let G be a real form of $G_{\mathbb{C}}$ having an open orbit on F

$$F_+ \cong G/C,$$

which is the union of submanifolds E_w, $w \in M_+$ (E_w is concave). We have the fibering

$$F_+ = G/C \to B = G/U,$$

where U is a maximal compact subgroup in G and G/U is a Riemannian symmetric manifold. The fibers have the structure of U/C and the corresponding E_w, $w \in \widetilde{B}$, give a homogeneous embedding $\widetilde{B}$ of B in M_+. The restrictions of the conformal operators ∂, D, $\Box$ on $\widetilde{B}$ are elliptic Schmid operators. As we saw, $\widetilde{B}$ can be noncomplex when B is Hermitian.

Finally, we consider the corresponding homogeneous vector bundles on F_+ and M_+. Then the generalized Penrose transform is the integration of $H^{(0,q)}(F_+)$, $q = \dim E_w$, with coefficients in the appropriate vector bundles on F_+ along submanifolds E_w. The result satisfies the corresponding generalized conformally invariant system of complex Schmid operators. All these computations can be done explicitly.

In this way, on the symmetric space $B = G/U$ we define a generalized conformal structure of elliptic type (the isotropy cones are imaginary).

References

1. S. Gindikin, *The holomorphic Cauchy–Szegö kernel for nonholomorphic representations of* SU(2,1), Representation Theory and Harmonic Analysis, Contemp. Math., vol. 191, Amer. Math. Soc., Providence, RI, 1994, pp. 75–82.
2. S. G. Gindikin and G. M. Henkin, *The Penrose transform and complex integral geometry*, Itogi Nauki i Tekhniki. Sovremennye Problemy Matematiki, vol. 17, VINITI, Moscow, 1981, pp. 57–111; English transl. in J. Soviet Math. **21** (1983), no. 4.
3. W. Schmid, *On the realization of the discrete series of a semisimple Lie group*, Rice Univ. Studies **56** (1970), no. 2, 99–108.
4. S. Gindikin, *Generalized conformal structures*, Twistors in Mathematics and Physics, Cambridge Univ. Press, New York, 1990, pp. 36–52.

Department of Mathematics, Rutgers University, New Brunswick, New Jersey 08903, U.S.A

E-mail address: gindikin@math.rutgers.edu

Translated by THE AUTHOR

Amer. Math. Soc. Transl.
(2) Vol. **175**, 1996

Defining Relations Associated with Principal $\mathfrak{sl}(2)$-subalgebras

P. Grozman and D. Leites

ABSTRACT. The notion of defining relations is, clearly, well defined for a nilpotent Lie algebra. Therefore, a conventional way to present a simple Lie algebra $\mathfrak{g}$ is by splitting it into the direct sum of a commutative Cartan subalgebra and two-maximal nilpotent subalgebras $\mathfrak{g}_\pm$ (positive and negative). The relations obtained among the $2\operatorname{rk}\mathfrak{g}$ generators of $\mathfrak{g}_\pm$ are neat; they are called *Serre relations*. The generators of $\mathfrak{g}_\pm$ generate $\mathfrak{g}$ as well.

It is possible to define the notion of relation for generators of different type. For instance, with the principal embeddings of $\mathfrak{sl}(2)$ into $\mathfrak{g}$ one can associate only *two* elements that generate $\mathfrak{g}$. We explicitly describe the presentations associated with the principal embeddings of $\mathfrak{sl}(2)$ of simple Lie algebras, finite-dimensional ones and certain infinite-dimensional ones; namely, the Lie algebra of matrices of complex size realized as a subalgebra of the Lie algebra of either Hamiltonian vector fields in two indeterminates or differential operators in one indeterminate.

The obtained relations are rather simple, especially for nonexceptional algebras. Our results might be of interest in applications to integrable systems (like the Leznov–Saveliev equation, i.e., the two-dimensional Toda lattice or the vector-valued Liouville equation) based on the principal $\mathfrak{sl}(2)$-subalgebras. But first of all they indicate how to q-quantize the Lie algebras of matrices of complex size.

Introduction

This paper continues the description of presentations of simple Lie superalgebras. It is the direct continuation of [**LSe**], and [**LP**], where the case of the simplest (for computations) base is considered and where non-Serre relations are first described, though in a different setting.

In what follows we describe some "natural" generators and relations for *classical Lie algebras* over $\mathbb{C}$. The answer is important in questions when one must identify an algebra given its generators and relations. (Examples are Eastbrook–Vahlquist

1991 *Mathematics Subject Classification*. Primary 17B01.
Key words and phrases. defining relations, Lie algebras, principal embeddings.
We are thankful to G. Post for timely information on [**PH**] and related papers. Financial support of the Swedish Institute and NFR is gratefully acknowledged.

prolongations, Drinfeld's quantum algebras, symmetries of differential equations, integrable systems, etc.).

If $\mathfrak{g}$ is nilpotent, the problem of its presentation has a natural and unambiguous solution: representatives of $\mathfrak{g}/[\mathfrak{g},\mathfrak{g}]$ are generators of $\mathfrak{g}$ and $H_2(\mathfrak{g})$ describes the relations.

On the other hand, if $\mathfrak{g}$ is simple, then $\mathfrak{g}=[\mathfrak{g},\mathfrak{g}]$ and there is no "most natural" way to select the generators of $\mathfrak{g}$. The choice of generators is not unique.

Still, among algebras with the property $\mathfrak{g}=[\mathfrak{g},\mathfrak{g}]$, simple algebras are distinguished by the fact that their structure is very well known. By trial and error, people discovered that for finite-dimensional simple Lie algebras, there are certain "first among equal" sets of generators:

1) Chevalley generators corresponding to positive and negative simple roots;

2) a pair of elements that generate any finite-dimensional simple Lie algebra associated with the principal $\mathfrak{sl}(2)$ subalgebra.

The relations associated with Chevalley generators are well known, see e.g., [**OV**]. These relations are called *Serre relations.*

The possibility of generating any simple finite-dimensional Lie algebra by two elements was first claimed by N. Jacobson (an exercise in [**J**]) and first (as far as we know) proved in [**BO**]. We do not know what generators Jacobson had in mind; in [**BO**] for these two generators linear combinations (with generic coefficients) of positive and negative root vectors, respectively, are taken; nothing like the "natural" choice of what we suggest calling *Jacobson's generators* was ever proposed.

To generate a simple algebra by only two elements is tempting, but nobody has done this explicitly yet. To check whether the relations between these elements are nice-looking is impossible without a computer (cf. an implicit description in [**F**]). As far as we could test, the relations for any other pair of generators chosen in a way distinct from ours are more complicated.

One of our aims was to decipher [**F**]. Certain statements from [**F**] are clarified (also with the help of a computer) in [**PH**]; we use some of these clarifications in §3.

In what follows we explicitly list the relations among the Jacobson generators; actually, in order to write out the relations, we introduce a third generator. Throughout the paper $\mathfrak{g}$ is a simple Lie algebra.

§1. The case of a finite-dimensional $\mathfrak{g}$

1.1. Principal embeddings. There exists only one (up to equivalence) embedding $\rho\colon \mathfrak{sl}(2)\to\mathfrak{g}$ such that $\mathfrak{g}$, regarded as an $\mathfrak{sl}(2)$-module, splits into $\operatorname{rk}\mathfrak{g}$ irreducible modules. (The reader may consider this statement as an exercise or consult [**D**], [**LS**] or [**OV**].) This embedding is called *principal* and, sometimes, *minimal* because for other embeddings (there are many) the number of irreducible $\mathfrak{sl}(2)$-modules is greater than $\operatorname{rk}\mathfrak{g}$. Example: for $\mathfrak{g}=\mathfrak{sl}(n)$, $\mathfrak{sp}(2n)$, or $\mathfrak{o}(2n+1)$ the principal embedding is the one corresponding to the irreducible representation of $\mathfrak{sl}(2)$ of dimension n, $2n$, $2n+1$, respectively.

For completeness, let us recall what the irreducible finite-dimensional $\mathfrak{sl}(2)$-modules look like. Select the following basis in $\mathfrak{sl}(2)$:

$$X_- = \begin{pmatrix} 0 & 0 \\ -1 & 0 \end{pmatrix}, \quad H = \begin{pmatrix} 1 & 0 \\ 0 & -1 \end{pmatrix}, \quad X_+ = \begin{pmatrix} 0 & 1 \\ 0 & 0 \end{pmatrix}.$$

Then any finite-dimensional irreducible $\mathfrak{sl}(2)$-module is of the form L_n for $n \in \mathbb{N}$. Such a representation is illustrated by a graph whose nodes are eigenvectors l_i of H with the weight indicated; the edges depict the action of $X_\pm$ (the action of X_+ is directed to the right, that of X_- to the left: $X_+ l_k = k(n-k)\, l_{k+2}$ and $X_- l_k = l_{k-2}$; in the module L_n below $X_+(l_n) = X_-(l_{-n}) = 0$):

$$\overset{-n}{\circ} - \overset{-n+2}{\circ} - \cdots - \overset{n-2}{\circ} - \overset{n}{\circ}.$$

Regarded as the $\mathfrak{sl}(2)$-module corresponding to the principal embedding, a simple finite-dimensional Lie algebra is as follows (cf. [**OV**, Table 4]).

TABLE 0. $\mathfrak{g}$ as the $\mathfrak{sl}(2)$-module

$\mathfrak{g}$	the $\mathfrak{sl}(2)$-spectrum of $\mathfrak{g} = L_2 \oplus L_{k_1} \oplus L_{k_2} \dots$
$\mathfrak{sl}(n)$	$L_2 \oplus L_4 \oplus L_6 \cdots \oplus L_{2n-2}$
$\mathfrak{o}(2n+1)$, $\mathfrak{sp}(2n)$	$L_2 \oplus L_6 \oplus L_{10} \cdots \oplus L_{4n-2}$
$\mathfrak{o}(2n)$	$L_2 \oplus L_6 \oplus L_{10} \cdots \oplus L_{4n-2} \oplus L_{2n-2}$
$\mathfrak{g}_2$	$L_2 \oplus L_{10}$
$\mathfrak{f}_4$	$L_2 \oplus L_{10} \oplus L_{14} \oplus L_{22}$
$\mathfrak{e}_6$	$L_2 \oplus L_8 \oplus L_{10} \oplus L_{14} \oplus L_{16} \oplus L_{22}$
$\mathfrak{e}_7$	$L_2 \oplus L_{10} \oplus L_{14} \oplus L_{18} \oplus L_{22} \oplus L_{26} \oplus L_{34}$
$\mathfrak{e}_8$	$L_2 \oplus L_{14} \oplus L_{22} \oplus L_{26} \oplus L_{34} \oplus L_{38} \oplus L_{46} \oplus L_{58}$

One can show that $\mathfrak{g}$ can be generated by two elements: $x := X_+ \in L_2 = \mathfrak{sl}(2)$ and a lowest weight vector $z := l_{-r}$ from the module L_r. For L_r we take either L_{k_1} if $\mathfrak{g} \neq \mathfrak{o}(2n)$ or the last module L_{2n-2} in the table above if $\mathfrak{g} = \mathfrak{o}(2n)$. (Clearly, z is defined up to proportionality; we shall assume that a basis of L_r is fixed and denote $z = t \cdot l_{-r}$ for $t \in \mathbb{C}$.)

The exceptional choice for $\mathfrak{o}(2n)$ is occasioned by the fact that by choosing $z \in L_{k_1}$ instead, we generate $\mathfrak{o}(2n-1)$.

We call the above x and z and also $y := X_- \in L_2$ the *Jacobson generators*. The presence of y considerably simplifies the form of the relations, though slightly increases their number. (One can, clearly, take the element l_n symmetric to z but it so happens that this only complicates the relations.)

1.2. Relations among the Jacobson generators. First, observe that if an ideal of a free Lie algebra is homogeneous (with respect to the degrees of the generators of the algebra), then the number and the degrees of the defining relations (i.e., the generators of the ideal) are uniquely defined, provided the relations are homogeneous. This is obvious.

A simple Lie algebra $\mathfrak{g}$, however, is the quotient of a free Lie algebra $\mathfrak{F}$ modulo a nonhomogeneous ideal $\mathfrak{I}$, which is actually an ideal without homogeneous generators. Therefore, we can speak about the number and the degrees of relations only conditionally. Our conditions: any element $x \in \mathfrak{I}$ should be expressible in terms of the generators $g_1, \dots$ of $\mathfrak{I}$ by a formula of the form

$$(*) \qquad x = \sum c_i g_i, \quad \text{where } c_i \in \mathfrak{F} \text{ and } \deg c_i + \deg g_i \leqslant \deg x \text{ for all } i.$$

(The degree is calculated with respect to that of the generators of $\mathfrak{F}$.)

Under condition $(*)$, the number and the degrees of the relations are uniquely defined. Now we can explain the necessity of the generator y: without it, the weight relations would have been of very high degree.

We divide the relations among Jacobson's generators into the following types:

0. relations in $L_2 = \mathfrak{sl}(2)$;
1. relations coming from $L_2 \otimes L_{k_1}$;
2. relations coming from $L_{k_1} \wedge L_{k_1}$;
3. relations coming from $L_{k_1} \wedge L_{k_1} \wedge L_{k_1}$;
∞. relations bounding the dimension (for small $\operatorname{rk}\mathfrak{g}$, relations of type ∞ can be of the above types.)

For $\mathfrak{e}_7$ there is also a stray relation of type 4.

There is an easy explanation of the types: the relation of type 0 are the well-known relations in the algebra $\mathfrak{sl}(2)$, those of type **1** express that the space L_{k_1} is the following (k_1+1)-dimensional $\mathfrak{sl}(2)$-module:

$$(0) \qquad \mathbf{0.1.}\ \ [[x,y],x] = 2x, \qquad \mathbf{0.2.}\ \ [[x,y],y] = -2y,$$

$$(1) \quad \mathbf{1.1.}\ \ [y,z] = 0, \qquad \mathbf{1.2.}\ \ [[x,y],z] = -k_1 z, \qquad \mathbf{1.3.}\ \ (\operatorname{ad} x)^{k_1+1} z = 0.$$

1.3. THEOREM. *For the Lie algebras indicated above, all the relations among the Jacobson generators are the above relations of types* **0, 1** *and the relations from Table* 1; *the parameter t depends on the choice of z and can be arbitrary.*

In what follows E_{ij} are the matrix units; $X_i^{\pm}$ stand for the standard Chevalley generators of $\mathfrak{g}$. To simplify notation we denote: $z_i = (\operatorname{ad} x)^i z$.

TABLE 1. Relations among Jacobson's generators

$\mathfrak{sl}(n)$ for $n \geqslant 3$. Generators:

$$x = \sum_{1 \leqslant i \leqslant n-1} i(n-i) E_{i,i+1}, \quad y = \sum_{1 \leqslant i \leqslant n-1} E_{i+1,i}, \quad z = t \sum_{1 \leqslant i \leqslant n-2} E_{i+2,i}.$$

Relations:

2.1. $3[z_1, z_2] - 2[z, z_3] = 24t^2(n^2-4)y$,
3.1. $[z,[z,z_1]] = 0$,
3.2. $4[z_3,[z,z_1]] - 3[z_2,[z,z_2]] = 576t^2(n^2-9)z$.
∞. $(\operatorname{ad} z_1)^{n-2} z = 0$.

For $n = 3, 4$ the degree of the last relation is lower than the degree of some other relations, which yields a simplification:

$n = 4$:

2.1. $3[z_1, z_2] - 2[z, z_3] = 288t^2 y$,
3.1. $[z,[z,z_1]] = 0$,
3.2. $[z_3,[z,z_1]] = -576t^2 z$,
∞. $(\operatorname{ad} z_1)^2 z = 0$.

$n = 3$:

∞. $[z_1, z] = 0$,

2.1. $[z_1, z_2] = 24t^2 y$.

$\mathfrak{o}(2n+1)$ for $n \geqslant 3$. Generators:

$$x = n(n+1)(E_{n+1,2n+1} - E_{n,n+1}) + \sum_{1\leqslant i\leqslant n-1} i(2n+1-i)(E_{i,i+1} - E_{n+i+2,n+i+1}),$$

$$y = (E_{2n+1,n+1} - E_{n+1,n}) + \sum_{1\leqslant i\leqslant n-1} (E_{i+1,i} - E_{n+i+1,n+i+2}),$$

$$z = t\Big((E_{2n-1,n+1} - E_{n+1,n-2}) - (E_{2n+1,n-1} - E_{2n,n}) + \sum_{1\leqslant i\leqslant n-3} (E_{i+3,i} - E_{n+i+1,n+i+4})\Big).$$

Relations:

2.1. $2[z_1, z_2] - [z, z_3] = 144t(2n^2 + 2n - 9)z$,

2.2. $9[z_2, z_3] - 5[z_1, z_4] = 432t(2n^2 + 2n - 9)z_2 + 1728t^2(n-1)(n+2)(2n-1)(2n+3)y$,

3.1. $[z, [z, z_1]] = 0$,

3.2. $7[z_3, [z, z_1]] - 6[z_2, [z, z_2]] = 2880t(n-3)(n+4)[z, z_1]$,

∞. $(\operatorname{ad} z_1)^{n-1} z = 0$.

$\mathfrak{sp}(2n)$ for $n \geqslant 3$. Generators:

$$x = n^2 E_{n,2n} + \sum_{1\leqslant i\leqslant n-1} i(2n-i)(E_{i,i+1} - E_{n+i+1,n+i}),$$

$$y = E_{2n,n} + \sum_{1\leqslant i\leqslant n-1} (E_{i+1,i} - E_{n+i,n+i+1}),$$

$$z = t\Big((E_{2n,n-2} + E_{2n-2,n}) - E_{2n-1,n-1} + \sum_{1\leqslant i\leqslant n-3} (E_{i+3,i} - E_{n+i,n+i+3})\Big).$$

Relations:

2.1. $2[z_1, z_2] - [z, z_3] = 72t(4n^2 - 19)z$,

2.2. $9[z_2, z_3] - 5[z_1, z_4] = 216t(4n^2 - 19)z_2 + 1728t^2(n^2 - 1)(4n^2 - 9)y$,

3.1. $[z, [z, z_1]] = 0$,

3.2. $7[z_3, [z, z_1]] - 6[z_2, [z, z_2]] = 720t(4n^2 - 49)[z, z_1]$,

∞. $(\operatorname{ad} z_1)^{n-1} z = 0$.

$\mathfrak{g}_2$. Generators:

$$x = 6X_1^+ + 10X_2^+, \quad y = X_1^- + X_2^-, \quad z = \frac{t}{129600}[[X_1^-, X_2^-], [X_1^-, [X_1^-, X_2^-]]].$$

Relations:

2.1. $[z, z_1] = 0$,

2.2. $[z_1, z_2] = 0$,

2.3. $[z_2, z_3] = -6tz$,

2.4. $[z_3, z_4] = -8tz_2$,

2.5. $[z_4, z_5] = -8tz_4 + 6t^2y$.

$\underline{\mathfrak{f}_4}$. Generators:

$$\begin{aligned} x &= 16X_1^+ + 30X_2^+ + 42X_3^+ + 22X_4^+, \\ y &= X_1^- + X_2^- + X_3^- + X_4^-, \\ z &= \frac{t}{907200}\big(2\,[[X_1^-, X_2^-], [X_3^-, [X_1^-, X_2^-]]] \\ &\quad + 2\,[[X_1^-, X_2^-], [X_4^-, [X_2^-, X_3^-]]] - [[X_3^-, X_4^-], [X_2^-, [X_2^-, X_3^-]]]\big). \end{aligned}$$

Relations:

2.1. $[z, z_1] = 0$,
2.2. $4[z_2, z_3] - 9[z_1, z_4] = 42tz$,
2.3. $5[z_3, z_4] - 6[z_2, z_5] = 28tz_2$,
2.4. $13[z_4, z_5] - 14[z_3, z_6] = 56tz_4 + 306t^2y$.

$\underline{\mathfrak{e}_6}$. Generators:

$$\begin{aligned} x &= 16X_1^+ + 30X_2^+ + 42X_3^+ + 30X_4^+ + 16X_5^+ + 22X_6^+, \\ y &= X_1^- + X_2^- + X_3^- + X_4^- + X_5^- + X_6^-, \\ z &= \frac{t}{8!}\big([[X_1^-, X_2^-], [X_3^-, X_4^-]] - [[X_1^-, X_2^-], [X_3^-, X_6^-]] \\ &\quad + [[X_2^-, X_3^-], [X_4^-, X_5^-]] + [[X_3^-, X_6^-], [X_4^-, X_5^-]]\big). \end{aligned}$$

Relations:

2.1. $50[z_2, z_3] + 14[z, z_5] - 35[z_1, z_4] = 0$,
2.2. $20[z_3, z_4] - 15[z_2, z_5] + 7[z_1, z_6] = 14t^2y$,
3.1. $[z_1, [z, z_1]] = 0$,
3.2. $[z_2, [z, z_1]] = 0$,
3.3. $4[z_3, [z, z_1]] + 7[z_1, [z_1, z_2]] = 0$,
3.4. $5[z_3, [z, z_2]] + [z_4, [z, z_1]]$,
3.5. $8[z_4, [z, z_2]] + 5[z_3, [z_1, z_2]] = 0$,
3.6. $3[z_4, [z_1, z_2]] + 4[z_4, [z, z_3]] = 0$,
3.7. $51[z_5, [z_1, z_2]] + 4[z_5, [z, z_3]] = -384t^2z$.

$\underline{\mathfrak{e}_7}$. Generators:

$$\begin{aligned} x &= 27X_1^+ + 52X_2^+ + 75X_3^+ + 96X_4^+ + 66X_5^+ + 34X_6^+ + 49X_7^+, \\ y &= X_1^- + X_2^- + X_3^- + X_4^- + X_5^- + X_6^- + X_7^-, \\ z &= \frac{7}{10!}\big([[X_2^-, X_3^-], [X_7^-, [X_4^-, X_5^-]]] + [[X_4^-, X_7^-], [X_5^-, [X_3^-, X_4^-]]] \\ &\quad + [[X_5^-, X_6^-], [X_7^-, [X_3^-, X_4^-]]] + 2\,[[X_4^-, X_5^-], [X_3^-, [X_1^-, X_2^-]]] \\ &\quad + 2\,[[X_5^-, X_6^-], [X_4^-, [X_2^-, X_5^-]]] - 3\,[[X_4^-, X_7^-], [X_3^-, [X_1^-, X_2^-]]]\big). \end{aligned}$$

Relations:

2.1. $3[z, z_5] - 9[z_1, z_4] + 14[z_2, z_3] = -2868z$,
2.2. $18[z_1, z_6] - 50[z_2, z_5] + 75[z_3, z_4] = -9560z_2$,
2.3. $14[z_2, z_7] - 35[z_3, z_6] + 50[z_4, z_5] = -4780z_4 + 49335y$;
3.1. $[z, [z, z_1]] = 0$,
3.2. $9[z_1, [z, z_2]] - 4[z_2, [z, z_1]] = 0$,
3.3. $330[z_2, [z, z_2]] - 425[z_3, [z, z_1]] - 1458[z_1, [z_1, z_2]] = 0$,

3.4. $665[z_3,[z,z_2]]-640[z_4,[z,z_1]]-1134[z_2,[z_1,z_2]]=0$,
3.5. $5485[z_3,[z,z_3]]-3910[z_4,[z,z_2]]-3182[z_3,[z_1,z_2]]=2527815[z,z_1]$,
3.6. $825[z_4,[z,z_3]]-598[z_5,[z,z_2]]-876[z_4,[z_1,z_2]]=338422[z,z_2]$,
3.7. $1525[z_5,[z,z_3]]-7524[z_5,[z_1,z_2]]+2415[z_4,[z_1,z_3]]=1106875[z,z_3]+2734746[z_1,z_2]$,
3.8. $25250[z_6,[z,z_4]]-94920[z_6,[z_1,z_3]]+44252[z_5,[z_1,z_4]]=-1305480[z,z_5]+41398712[z_1,z_4]-1117925005z$,
4.1. $12[[z,z_2],[z_1,z_2]]-5[[z,z_2],[z,z_3]]=0$,
∞. $[[z,z_2],[z_1,[z,z_1]]]=0$.

$\underline{\mathfrak{e}_8}$. Generators:

$$x=58X_1^++114X_2^++168X_3^++220X_4^++270X_5^++182X_6^++92X_7^++136X_8^+,$$
$$y=X_1^-+X_2^-+X_3^-+X_4^-+X_5^-+X_6^-+X_7^-+X_8^-,$$
$$\begin{aligned}z=\frac{1}{13!}\big(&[[X_7^-,[X_5^-,X_6^-]],[[X_3^-,X_4^-],[X_5^-,X_8^-]]]\\&+[[X_8^-,[X_4^-,X_5^-]],[[X_3^-,X_4^-],[X_5^-,X_6^-]]]\\&+[[X_8^-,[X_5^-,X_6^-]],[[X_1^-,X_2^-],[X_3^-,X_4^-]]]\\&+[[X_8^-,[X_5^-,X_6^-]],[[X_2^-,X_3^-],[X_4^-,X_5^-]]]\\&+[[X_8^-,[X_5^-,X_6^-]],[[X_4^-,X_5^-],[X_6^-,X_7^-]]]\\&+2[[X_4^-,[X_2^-,X_3^-]],[[X_5^-,X_8^-],[X_6^-,X_7^-]]]\\&-3[[X_7^-,[X_5^-,X_7^-]],[[X_1^-,X_2^-],[X_3^-,X_4^-]]]\big).\end{aligned}$$

Relations:
2.1. $91[z,z_5]-325[z_1,z_4]+550[z_2,z_3]=0$,
2.2. $13[z_1,z_6]-45[z_2,z_5]+75[z_3,z_4]=-268814z$,
2.3. $33[z_2,z_7]-11[z_3,z_6]+180[z_4,z_5]=-682374z_2$,
2.4. $11[z_3,z_8]-35[z_4,z_7]+56[z_5,z_6]=-186102z_4$,
2.5. $3[z_4,z_9]-9[z_5,z_8]+14[z_6,z_7]=-41356z_6+2686866y$;
3.1. $[z,[z,z_1]]=0$,
3.2. $13[z_1,[z,z_2]]-6[z_2,[z,z_1]]=0$,
3.3. $542[z_2,[z,z_2]]-639[z_3,[z,z_1]]-2236[z_1,[z_1,z_2]]=0$,
3.4. $1067[z_3,[z,z_2]]-950[z_4,[z,z_1]]-1892[z_2,[z_1,z_2]]=0$,
3.5. $7255[z_3,[z,z_3]]-4995[z_4,[z,z_2]]-4527[z_3,[z_1,z_2]]=0$,
3.6. $105460[z_4,[z,z_3]]-69597[z_5,[z,z_2]]-119430[z_4,[z_1,z_2]]=0$,
3.7. $844277[z_5,[z,z_3]]+1556775[z_4,[z_1,z_3]]-4442058[z_5,[z_1,z_2]]=-17362538193[z,z_1]$,
3.8. $334453[z_6,[z,z_4]]+746586[z_5,[z_1,z_4]]-1414050[z_6,[z_1,z_3]]=1120518212[z,z_3]+3082429152[z_1,z_2]$,
∞. $[[z,z_1],[z,z_2]]=0$.

§2. The Lie algebras of matrices of complex size as subalgebras of $\mathfrak{diff}(1)$

The Poincaré–Birkhoff–Witt theorem states that, as spaces,

$$U(\mathfrak{sl}(2))\cong\mathbb{C}[X_-,H,X_+].$$

We also know that to study representations of $\mathfrak{g}$ is the same as to study representations of $U(\mathfrak{g})$. Still, if we are interested in irreducible representations, we do not need all of $U(\mathfrak{g})$ and can do with a smaller algebra, which is easier to study.

This observation is used now and again; Feigin applied it in [**F**] saying, actually, that by setting

$$X_- = -\frac{\partial}{\partial u}, \qquad H = 2u\frac{\partial}{\partial u} - (\lambda - 1), \qquad X_+ = u^2\frac{\partial}{\partial u} - (\lambda - 1)u$$

we obtain a morphism of $\mathfrak{sl}(2)$-modules and, moreover, of associative algebras: $U(\mathfrak{sl}(2)) \to \mathbb{C}[u, \partial/\partial u]$. The kernel of this morphism is the ideal generated by $\Delta - (\lambda^2 - 1)$, where $\Delta = 2(X_+X_- + X_-X_+) + H^2$.

REMARK. In their proof of certain statements from [**F**] that we shall recall, the authors of [**PH**] make use of the well-known fact that the Casimir operator Δ acts on the irreducible $\mathfrak{sl}(2)$-module L_μ with highest weight μ (i.e., $H \cdot l_\mu = \mu \cdot l_\mu$ and $X_+ l_\mu = 0$) as the scalar operator of multiplication by $\mu^2 + 2\mu$. The passage from [**PH**]'s λ to [**F**]'s μ is done with the help of a shift by the weight ρ, which for $\mathfrak{sl}(2)$ can be identified with 1.

With the associative algebra $U(\mathfrak{sl}(2))$ let us associate the Lie algebra $U(\mathfrak{sl}(2))_L$ (we denote by the subscript $_L$ the functor that sends an associative algebra to the Lie algebra with the bracket determined by the commutator). It is easy to see that this algebra, regarded as a $\mathfrak{sl}(2)$-module, satisfies

$$U(\mathfrak{sl}(2))_L/(\Delta - (\lambda^2 - 1)) = L_0 \oplus L_2 \oplus L_4 \oplus \cdots \oplus L_{2n} \oplus \cdots . \tag{2.1}$$

It is not difficult to show (see [**PH**] for details) that for $\lambda = n \in \mathbb{N} \setminus \{0, 1\}$, the Lie algebra $U(\mathfrak{sl}(2))_L/(\Delta - (n^2 - 1))$ contains an ideal I_n and the quotient $(U(\mathfrak{sl}(2))_L/(\Delta - (n^2 - 1)))/I_n$ is the ordinary algebra $\mathfrak{gl}(n)$. In [**PH**] it is proved that for $\lambda \neq \mathbb{Z}$ the Lie algebra $U(\mathfrak{sl}(2))_L/(\Delta - (n^2 - 1))$ has only one ideal—the space of constants. This justifies Feigin's suggestive notation

$$\begin{aligned} \mathfrak{gl}(\lambda) &= \begin{cases} U(\mathfrak{sl}(2))_L/(\Delta - (\lambda^2 - 1)) & \text{for } \lambda \neq \mathbb{N} \setminus \{0, 1\}, \\ (U(\mathfrak{sl}(2))_L/(\Delta - (n^2 - 1)))/I_n & \text{otherwise,} \end{cases} \\ \mathfrak{sl}(\lambda) &= \mathfrak{gl}(\lambda)/\langle 1 \rangle . \end{aligned} \tag{2.2}$$

The definition directly implies that $\mathfrak{sl}(-\lambda) \cong \mathfrak{sl}(\lambda)$; therefore, speaking of the real values of λ, we can confine ourselves to nonnegative values.

Now, consider another realization of $\mathfrak{sl}(2)$, before the factorization (2.2). This realization was the starting point of the whole activity, so we present it here for completeness. Take the Lie algebra $\mathfrak{po}(2)_{\text{ev}}$ of even degree polynomials $\mathbb{C}[q, p]_{\text{ev}}$ with respect to the Poisson bracket. Set

$$X_- = \frac{1}{2}q^2, \qquad X_+ = \frac{1}{2}p^2,$$

and note that $\langle q, p\rangle$ is the identity $\mathfrak{sl}(2)$-module. Observe that, as $\mathfrak{sl}(2)$-modules, the Lie algebras $\mathfrak{po}(2)_{\text{ev}}$ and its deform (the result of the quantization, a subalgebra of the Lie algebra $\mathfrak{diff}(1)$ of differential operators on the line) also have the spectrum (2.1). So it was natural to look at the deforms for various values of the parameter of deformation.

By analogy, Feigin defined $\mathfrak{o}(\lambda)$ ($\mathfrak{sp}(\lambda)$) as subalgebras of $\mathfrak{gl}(\lambda)$ that are invariant (respectively skew-invariant) with respect to the involution

$$X \to \begin{cases} -X & \text{if } X \in L_{4k}, \\ X & \text{if } X \in L_{4k+2}, \end{cases} \tag{2.3}$$

and the analog of the map

$$X \to -X^T \quad \text{for } X \in \mathfrak{g}(n). \tag{2.4}$$

Since $\mathfrak{o}(\lambda)$ and $\mathfrak{sp}(\lambda)$ differ by the value of the parameter, it is natural to denote them uniformly, but so as not to confuse them with the Lie superalgebras of the series $\mathfrak{osp}$. Set

$$\mathfrak{o}/\mathfrak{sp}(\lambda) = \begin{cases} \mathfrak{o}(\lambda) & \text{if } \lambda \in 2\mathbb{N}+1, \\ \mathfrak{sp}(\lambda) & \text{if } \lambda \in 2\mathbb{N}. \end{cases}$$

Let us list the relations among the Jacobson generators of $\mathfrak{sl}(\lambda)$ and $\mathfrak{o}/\mathfrak{sp}(\lambda)$.

TABLE 2. Relations among Jacobson's generators of $\mathfrak{sl}(\lambda)$ and $\mathfrak{o}/\mathfrak{sp}(\lambda)$

$\mathfrak{sl}(\lambda)$. Generators:

$$x = u^2\frac{\partial}{\partial u} - (\lambda-1)u, \qquad y = -\frac{\partial}{\partial u}, \qquad z = t\frac{\partial^2}{\partial u^2}.$$

Relations:

2.1. $3[z_1, z_2] - 2[z, z_3] = 24t^2(\lambda^2-4)y$,
3.1. $[z, [z, z_1]] = 0$,
3.2. $4[z_3, [z, z_1]] - 3[z_2, [z, z_2]] = 576t^2(\lambda^2-9)z$.

$\mathfrak{o}/\mathfrak{sp}(\lambda)$. Generators:

$$x = u^2\frac{\partial}{\partial u} - (\lambda-1)u, \qquad y = -\frac{\partial}{\partial u}, \qquad z = t\frac{\partial^3}{\partial u^3}.$$

Relations:

2.1. $2[z_1, z_2] - [z, z_3] = 72t(\lambda^2-19)z$,
2.2. $9[z_2, z_3] - 5[z_1, z_4] = 216t(\lambda^2-19)z_2 - 432t^2(\lambda^2-4)(\lambda^2-9)y$,
3.1. $[z, [z, z_1]] = 0$,
3.2. $7[z_3, [z, z_1]] - 6[z_2, [z, z_2]] = 720t(\lambda^2-49)[z, z_1]$.

§3. The Lie algebras $\mathfrak{sl}(\infty)$ and $\mathfrak{o}/\mathfrak{sp}(\infty)$

The parameter λ above runs, actually, over $\mathbb{CP}^1$, not just $\mathbb{C}$. The quantization of the above relations is performed by passing to the limit as $\lambda \to \infty$ under the change:

$$t \mapsto \begin{cases} t/\lambda & \text{for } \mathfrak{sl}(\lambda), \\ t/\lambda^2 & \text{for } \mathfrak{o}/\mathfrak{sp}(\lambda). \end{cases}$$

TABLE 3. Defining relations for $\mathfrak{sl}(\infty)$ and $\mathfrak{o}/\mathfrak{sp}(\infty)$

$\mathfrak{sl}(\infty)$.

2.1. $3[z_1, z_2] - 2[z, z_3] = 24t^2y$,
3.1. $[z, [z, z_1]] = 0$,
3.2. $4[[z, z_1], z_3] + 3[z_2, [z, z_2]] = -576t^2z$.

$\mathfrak{o}/\mathfrak{sp}(\infty)$.

2.1. $2[z_1,z_2]-[z,z_3]=72tz,$

2.2. $9[z_2,z_3]-5[z_1,z_4]=216tz_2-432t^2y,$

3.1. $[z,[z,z_1]]=0,$

3.2. $7[[z,z_1],z_3]+6[z_2,[z,z_2]]=-720t[z,z_1].$

3.1. THEOREM. *For the Lie algebras $\mathfrak{sl}(\lambda)$ and $\mathfrak{o}/\mathfrak{sp}(\lambda)$, $\lambda\in\mathbb{CP}^1$, all the relations among the Jacobson generators are relations of types 0, 1 for $\mathfrak{sl}$ or $\mathfrak{o}(2n+1)$ (same as those for $\mathfrak{sp}$), respectively, and the relations from Tables 2 and 3.*

§4. Remarks

4.1. Concerning proofs. Our theorems only make Proposition 2 of [**F**] explicit. The proof is straightforward: the quotient of the free Lie algebra generated by x, y and z modulo our relations is the finite-dimensional one needed for integer values of λ and in the exceptional cases. For the exceptional cases and small values of the parameter, this is checked on a computer; for generic λ and n this is still to be proved.

Feigin claimed [**F**] that for $\mathfrak{sl}(\infty)$ the type 3 relations follow from $L_{k_1}\wedge L_{k_2}\subset L_{k_1}\wedge L_{k_1}\wedge L_{k_1}$. We verified that this is true for all the above-considered algebras except $\mathfrak{e}_6-\mathfrak{e}_8$: for them, one should consider the whole $L_{k_1}\wedge L_{k_1}\wedge L_{k_1}$.

4.2. PROPOSITION. *For a principal embedding $\mathfrak{sl}(2)\to\mathfrak{g}$, where $\mathfrak{g}=\mathfrak{o}(2n+1)$, $\mathfrak{sp}(2n)$ or $\mathfrak{o}/\mathfrak{sp}(\lambda)$, $\lambda\in\mathbb{CP}^1$, there exists an embedding of $\sigma\colon\mathfrak{g}\to\mathfrak{sl}(k)$ for an appropriate $k\in\mathbb{C}$ such that the through map is principal. There is no such σ for exceptional Lie algebras or $\mathfrak{o}(2n)$.*

PROOF. The proposition immediately follows from definitions.

4.3. How to present $\mathfrak{o}(2n)$. Select z as in 1.1. Clearly, z (relations of type 1) and the number of relations of type 3 depend on n; the relations are not as neat as those for the above algebras.

$\mathfrak{o}(2n)$. Generators:

$$\begin{aligned} x&=\frac{n(n-1)}{2}(E_{n-1,n}-g_{2n,2n-1}+E_{n-1,2n}-E_{n,2n-1})\\ &\quad+\sum_{1\leqslant i\leqslant n-2} i(2n-1-i)(E_{i,i+1}-E_{n+i+1,n+i}),\\ y&=(E_{2n,n-1}-E_{2n-1,n})+\sum_{1\leqslant i\leqslant n-1}(E_{i+1,i}-E_{n+i,n+i+1}),\\ z&=\frac{1}{(2n-2)!}((E_{n,1}-E_{n+1,2n})+(E_{n+1,n}-E_{2n,1})). \end{aligned}$$

We cannot write out the relations in full generality; for small values of n they are as follows:

$n=4$. Relations:

2.1. $3[z,z_5]-5[z_1,z_4]+6[z_2,z_3]=\frac{1}{2}y;$

3.1. $[z,[z,z_1]]=0,$

3.2. $[z_1,[z,z_1]]=0,$

3.3. $[z_2, [z, z_1]] = 0,$
3.4. $[z_1, [z_1, z_2]] = 0,$
3.5. $[z_3, [z, z_1]] = 0,$
3.6. $[z_3, [z, z_2]] = 0,$
3.7. $[z_4, [z, z_2]] = z,$
3.8. $[z_4, [z_1, z_2]] = z_1,$
3.9. $[z_5, [z_1, z_2]] = z_2.$

$\underline{n=5}$. Relations:

2.1. $-4[z, z_7] + 7[z_1, z_6] - 9[z_2, z_5] + 10[z_3, z_4] = \frac{1}{2}y,$

There are 17 relations of type 3.

$\underline{n=6}$. The relation of type 2 is still more involved and there are 27 relations of type 3.

References

[BO] Yu. A. Bakhturin and A. Yu. Olshansky, *On approximation and characteristic subalgebras of free Lie algebras*, Trudy Seminara Petrovsk. **2** (1976), 145–150. (Russian)

[D] E. B. Dynkin, *Semi-simple subalgebras of semi-simple Lie algebras*, Mat. Sb. **30** (1952), 111–244; English transl. in *Amer. Math. Soc. Transl. Ser. 2 vol 6*, Amer. Math. Soc., Providence, RI, 1957.

[F] B. L. Feigin, *The Lie algebras* $\mathfrak{gl}(\lambda)$ *and cohomologies of Lie algebra of differential operators*, Uspekhi Mat. Nauk **43** (1988), no. 2, 157–158; English transl., Russian Math. Surveys **43** (1988), no. 2, 169–170.

[J] N. Jacobson, *Lie algebras*, Interscience, New York, 1962.

[LP] D. Leites and E. Poletaeva, *Defining relations for classical Lie algebras of polynomial vector fields*, Talk at Int. Conf. Euler IMI, 1990 (to appear).

[LS] A. Leznov and M. Saveliev, *Group-theoretical methods for integration of dynamical systems*, Birkhäuser, Basel, 1991.

[LSe] D. Leites and V. Serganova, *Defining relations for classical Lie superalgebras*. I, in Topological and Geometrical Methods in Field Theory (J. Mickelsson and O. Peckonnen, eds.), Proc. Conference (1991, Turku, Finland), World Sci., Singapore, 1992, pp. 194–201.

[OV] A. L. Onishchik and É. B. Vinberg, *Seminar on algebraic groups and Lie groups*, Springer-Verlag, Berlin and Heidelberg, 1990.

[PH] G. Post and N. van den Hijligenberg, *Explicit results on* $\mathfrak{gl}(\lambda)$ *and* $\mathfrak{sl}(\lambda)$, Memorandum 1143, Faculty of Applied Math., Univ of Twente (1993).

ROSLAGSV. 101, KRÄFTRIKET HUS 6, S-106 91, STOCKHOLM, SWEDEN

Translated by THE AUTHORS

Amer. Math. Soc. Transl.
(2) Vol. **175**, 1996

A Generalization of the Berezin–Lieb Inequality

A. Laptev and Yu. Safarov

In the early seventies F. Berezin [**B**] and independently E. Lieb [**L**] (see also [**S**]) obtained a Jensen's type inequality for convex functions of selfadjoint operators. This inequality turned out to be very useful, and it has been applied to various spectral problems, see for example [**BSh**].

If φ is a convex function, B_P is a selfadjoint operator (not necessarily bounded) in a Hilbert space H, and moreover the operator B_P can be represented as $B_P = PBP$, where P is an orthogonal projection in H, then the Berezin inequality reads

$$\operatorname{Tr} P\varphi(B_P)P \leqslant \operatorname{Tr} P\varphi(B)P,$$

provided that the right-hand side is finite.

Applying this inequality to the spectral analysis of pseudodifferential operators, we were interested in two-sided estimates of the trace $\operatorname{Tr} P\psi(B_P)P$ when the function ψ is not necessarily a convex function. In Theorem 12 of this paper we prove a trace estimate for such functions. This estimate implies a more general version of the Berezin inequality (see Corollary 13). In particular, we prove the inequality

$$\operatorname{Tr}(P\varphi(B)P - P\varphi(B_P)P) \geqslant 0,$$

assuming only that the difference $P\varphi(B)P - P\varphi(B_P)P$ is from the trace class. We also obtain inequalities when P is a contraction operator.

§1. The operator P^*BP

Let H and H_0 be Hilbert spaces, B be a selfadjoint operator in H, and $P\colon H_0 \to H$ be a bounded operator such that $\|P\|_{H_0\to H} \leqslant 1$. The operator B is allowed to be unbounded, and then we denote by $\mathcal{D}(B)$ its domain. We shall consider the operator P^*BP acting in the space H_0. When B is bounded, this operator is well defined and selfadjoint. However, when B is unbounded, the natural definition of P^*BP might make no sense (for example, if $\mathcal{D}(B) \cap PH_0 = \{0\}$). In this case we need some additional assumptions.

Let $(\,\cdot\,,\cdot\,)$, $\|\cdot\|$ and $(\,\cdot\,,\cdot\,)_0$, $\|\cdot\|_0$ be the scalar products and norms in H and H_0 respectively. We denote by $E_B(\lambda)$ the spectral measure of the operator B and

1991 *Mathematics Subject Classification.* Primary 47A63.

consider the skew-linear form

$$Q[\xi,\eta] \overset{\text{def}}{=} Q_{B,P}[\xi,\eta] = \int \lambda(dE_B(\lambda)P\xi, P\eta), \qquad \xi,\eta \in H_0,$$

and the corresponding quadratic form

$$Q[\xi] = Q_{B,P}[\xi] \overset{\text{def}}{=} \int \lambda(dE_B(\lambda)P\xi, P\xi), \qquad \xi \in H_0. \tag{1}$$

If B is bounded, then $Q[\xi,\eta] = (BP\xi, P\eta)$ and the form $Q[\xi]$ is defined on the whole space H_0. In the general situation, the domain of Q is the linear set

$$\mathcal{D}(Q) \overset{\text{def}}{=} \left\{\xi \in H_0 : \int |\lambda|(dE_B(\lambda)P\xi, P\xi) < \infty\right\}. \tag{2}$$

Obviously, we have

$$\mathcal{D}(Q) = \{\xi \in H_0 : P\xi \in \mathcal{D}(|B|^{1/2})\}, \tag{3}$$

$$\int |\lambda|(dE_B(\lambda)P\xi, P\xi) = \| |B|^{1/2}P\xi\|^2. \tag{4}$$

Generally speaking, the set (2) may also be very poor. Besides, even if that were not true, Q might not generate a selfadjoint operator. Therefore we introduce the following two conditions, which are assumed to be fulfilled throughout the paper.

(C_1) The set $\mathcal{D}(Q)$ is dense in H_0.

(C_2) The form $Q[\,\cdot\,]$ is semibounded and closable in H_0.

Let $\overline{Q}[\,\cdot\,]$ be the closure of the form $Q[\,\cdot\,]$. This closure is defined on some dense set $\mathcal{D}(\overline{Q}) \subset H_0$ containing $\mathcal{D}(Q)$, and it defines a Hilbert space structure on $\mathcal{D}(\overline{Q})$. We denote this Hilbert space by H_1, $H_1 \subset H_0$.

Let H' be a closed subspace of H_1 which is also dense in H_0, and $Q'[\,\cdot\,]$ be the restriction of the form $\overline{Q}[\,\cdot\,]$ to H'. Then $Q'[\,\cdot\,]$ is a closed quadratic form in H_0, and so it generates a selfadjoint operator B_P.

Obviously, if B is bounded, then $H' = H_1 = H_0$ and $B_P = P^*BP$. If B is an unbounded operator, then B_P is not defined uniquely. Each H' specifies a selfadjoint operator B_P, which can be regarded as a selfadjoint realization of P^*BP. All further results are valid for any such realization. Throughout the paper, we assume H' to be fixed and deal with the corresponding selfadjoint operator B_P.

The condition (C_2) is not effective. The following lemma gives an equivalent condition, which is more convenient to deal with.

LEMMA 1. *The condition* (C_2) *is fulfilled if and only if there exists a constant* C *such that*

$$\| |B|^{1/2}P\xi\|^2 \leqslant C(|Q[\xi]| + \|\xi\|_0^2), \qquad \forall \xi \in \mathcal{D}(Q). \tag{5}$$

PROOF. By Lemma 10.1.6 from [**BS**], the form $Q[\,\cdot\,]$ is closable if and only if for any sequence $\xi_k \in \mathcal{D}(Q)$, $k = 1, 2, \dots$, such that $\|\xi_k\|_0 \to 0$, $k \to \infty$, and

$$Q[\xi_k - \xi_j] \to 0, \qquad j,k \to \infty, \tag{6}$$

we have

$$Q[\xi_k, \eta] \to 0, \qquad \forall \eta \in \mathcal{D}(Q). \tag{7}$$

By (3) we can write

$$Q[\xi_k, \eta] = ((I + |B|)^{1/2} P\xi_k, B(I + |B|)^{-1/2} P\eta).$$

Therefore the form $Q[\cdot]$ is closable if and only if the sequence $(I + |B|)^{1/2} P\xi_k$ weakly tends to zero in H.

Condition (6) implies that the $Q[\xi_k]$'s are uniformly bounded. Hence, from (5) it follows that the numbers $\|(I + |B|)^{1/2} P\xi_k\|$ are also uniformly bounded. For any $u \in \mathcal{D}(|B|^{1/2})$, we have

$$((I + |B|)^{1/2} P\xi_k, u) = (P\xi_k, (I + |B|)^{1/2} u) \to 0.$$

Thus, the sequence $(I + |B|)^{1/2} P\xi_k$ is bounded and weakly tends to zero on the set $\mathcal{D}(|B|^{1/2})$, which is dense in H. This implies that this sequence weakly tends to zero. So (5) yields (C_2).

If the estimate (5) does not hold, then there exists a sequence $\xi_k \in \mathcal{D}(Q)$ such that $\|\xi_k\|_0 \to 0$, $Q[\xi_k] \to 0$, $k \to \infty$, but $\|(I + |B|)^{1/2} P\xi_k\| \to \infty$. For these ξ_k, the sequence $(I + |B|)^{1/2} P\xi_k$ does not weakly converge, and therefore the form $Q[\cdot]$ cannot be closed. The proof is complete.

§2. Function spaces

In what follows we always assume all functions to be measurable. Moreover, we shall only deal with functions from the class $BV^1(\mathbb{R})$, which is defined as follows.

DEFINITION 2. A complex function $\psi \in C(\mathbb{R})$ is from the class $BV^1(\mathbb{R})$ if its second derivative ψ'' coincides with a complex measure ρ_ψ on $\mathbb{R}$ in the sense of distribution theory.

Obviously, the complex measure ρ_ψ is defined uniquely by the function ψ. For example, the class $BV^1(\mathbb{R})$ contains all linear functions for which $\psi'' = \rho_\psi = 0$. Conversely, for each complex measure ρ there exists a function $\psi \in BV^1(\mathbb{R})$ such that $\rho = \rho_\psi$. This function is defined uniquely modulo a linear function. We denote by ψ^* the class of functions that differ from the function ψ by a linear function. Then we have a one-to-one correspondence between complex measures and classes ψ^*, $\psi \in BV^1(\mathbb{R})$.

REMARK 3. The first derivatives of functions from $BV^1(\mathbb{R})$ are functions of locally bounded variation, which explains the notation BV^1. In particular, for $\psi \in BV^1(\mathbb{R})$, the first derivative ψ' is a locally bounded function continuous almost everywhere and having the limits $\psi'(s - 0)$, $\psi'(s + 0)$ for every $s \in \mathbb{R}$. Therefore $BV^1(\mathbb{R}) \subset W^1_{\infty,\mathrm{loc}}(\mathbb{R})$, where $W^1_{\infty,\mathrm{loc}}(\mathbb{R})$ is the Sobolev space.

A real function φ defined on $\mathbb{R}$ is said to be *convex* if

$$\varphi(\alpha s_1 + (1 - \alpha) s_2) \leqslant \alpha\varphi(s_1) + (1 - \alpha)\varphi(s_2)$$

for any $s_1, s_2 \in \mathbb{R}$ and $\alpha \in [0, 1]$. This immediately implies that for convex functions

$$\varphi(\alpha s) \leqslant (1 - \alpha)\varphi(0) + \alpha\varphi(s), \tag{8}$$

$$\varphi(s + t) + \varphi(s - t) - 2\varphi(s) \geqslant 0 \tag{9}$$

for all $s, t \in \mathbb{R}$ and $\alpha \in [0, 1]$.

The next lemma characterizes the class of convex functions (see [**Hö,** Vol. 1, Theorem 4.1.6]). We prove it here for the sake of completeness.

LEMMA 4. *A function φ is convex if and only if $\varphi \in BV^1(\mathbb{R})$ and ρ_φ is a positive measure.*

PROOF. Let φ be convex. Then in view of (9), for a real nonnegative test function $f \in \mathcal{D}(\mathbb{R})$ we have

$$\begin{aligned} 0 &\leqslant \int [\varphi(s+t) + \varphi(s-t) - 2\varphi(s)]\, f(s)\, ds \\ &= \int \varphi(s)[f(s+t) + f(s-t) - 2f(s)]\, ds . \end{aligned}$$

Dividing by t^2, as $t \to 0$ we obtain $\langle \varphi'', f \rangle \geqslant 0$. Since a positive distribution is a positive measure, this proves the "only if" part of the lemma.

Now assume that $\varphi \in BV^1(\mathbb{R})$ and that φ'' coincides with a positive measure. Let $s_1 < s_2$, $\alpha \in [0,1]$, and

$$f(s) \stackrel{\text{def}}{=} \begin{cases} 0, & \text{for } s \leqslant s_1 \text{ and } s \geqslant s_2, \\ \alpha(s - s_1), & \text{for } s_1 \leqslant s \leqslant \alpha s_1 + (1-\alpha)\, s_2, \\ (1-\alpha)(s_2 - s), & \text{for } \alpha s_1 + (1-\alpha) s_2 \leqslant s \leqslant s_2. \end{cases}$$

The function f is nonnegative and continuous, and

$$f''(s) = \alpha\delta(s - s_1) + (1-\alpha)\,\delta(s - s_2) - \delta(s - \alpha s_1 - (1-\alpha) s_2),$$

where $\delta(\,\cdot\,)$ is the delta function. Therefore,

$$\begin{aligned} \alpha\varphi(s_1) + (1-\alpha)\,\varphi(s_2) - \varphi(\alpha s_1 + (1-\alpha)\, s_2) &= \int \varphi(s)\, f''(s)\, ds \\ &= \langle \varphi'', f \rangle = \int f\, d\rho_\varphi \geqslant 0 . \end{aligned}$$

This completes the proof.

Obviously Lemma 4 can be reformulated in the following way: a function φ is convex if and only if $\varphi \in BV^1(\mathbb{R})$ and the first derivative of φ is a nondecreasing function. Now we introduce the following

DEFINITION 5. Let $\psi \in BV^1(\mathbb{R})$ and φ be a convex function. We say that the function ψ is *dominated* by φ if $d\rho_\psi = g\, d\rho_\varphi$ for some density $g \in L_\infty(\mathbb{R}, \rho_\varphi)$. In this case, we denote $|\psi|_\varphi \stackrel{\text{def}}{=} \|g\|_{L_\infty(\mathbb{R},\rho_\varphi)}$.

Obviously, if ψ is dominated by φ, then any representative from the class ψ^* is dominated by every function from φ^*.

LEMMA 6. *Let $\psi \in BV^1(\mathbb{R})$ be dominated by a nonnegative convex function φ. Then there exists a linear function l such that*

$$|\psi(s) - l(s)| \leqslant |\psi|_\varphi \varphi(s), \qquad \forall\, s \in \mathbb{R}. \tag{10}$$

PROOF. Assume first that there exists a point s_0 such that $\varphi'(s_0 - 0) \leqslant 0$ and $\varphi'(s_0 + 0) \geqslant 0$. Without loss of generality we assume $|\psi|_\varphi = 1$; otherwise, we replace φ by $|\psi|_\varphi \varphi$. Then $|\rho_\psi(I)| \leqslant \rho_\varphi(I)$ for any bounded interval I. Therefore,

$$|\psi'(s \pm 0) - \psi'(s_0 + 0)| \leqslant \varphi'(s \pm 0) - \varphi'(s_0 + 0), \qquad s_0 < s, \tag{11}$$

$$|\psi'(s \pm 0) - \psi'(s_0 - 0)| \leqslant \varphi'(s_0 - 0) - \varphi'(s \pm 0), \qquad s < s_0, \tag{12}$$

and for arbitrary $s_1 \leqslant s_2$

$$|\psi'(s_2) - \psi'(s_1)| \leqslant \varphi'(s_2) - \varphi'(s_1). \tag{13}$$

Let us show that there is a constant $C \in \mathbb{R}$ such that

$$|\psi'(s) - C| \leqslant |\varphi'(s)|, \qquad \forall\, s \in \mathbb{R}. \tag{14}$$

We introduce two intervals I_1 and I_2 such that

$$\begin{aligned} I_1 &= [-\psi'(s_0+0) - \varphi'(s_0+0), -\psi'(s_0+0) + \varphi'(s_0+0)], \\ I_2 &= [-\psi'(s_0-0) + \varphi'(s_0-0), -\psi'(s_0-0) - \varphi'(s_0-0)]. \end{aligned} \tag{15}$$

If in (13) we substitute $s_2 = s_0 + 0$ and $s_1 = s_0 - 0$, we obtain

$$\begin{aligned} -\psi'(s_0-0) + \varphi'(s_0-0) &\leqslant -\psi'(s_0+0) + \varphi'(s_0+0), \\ -\psi'(s_0+0) - \varphi'(s_0+0) &\leqslant -\psi'(s_0-0) - \varphi'(s_0-0). \end{aligned}$$

In particular, this implies that the intersection of I_1 and I_2 is not empty. From (11) we infer that (14) is satisfied for any $s_0 < s$ and $C \in I_1$. Correspondingly, (14) follows from (12) for any $s < s_0$ and $C \in I_2$. If now $C \in I_1 \cap I_2$, then inequality (14) holds for all $s < s_0$, $s_0 < s$ and therefore for $s = s_0 - 0$ and $s = s_0 + 0$.

Inequality (14) implies

$$\begin{aligned} |\psi(s) - C(s - s_0) - \psi(s_0)| &= \left| \int_{s_0}^{s} (\psi'(t) - C)\, dt \right| \\ &\leqslant \int_{s_0}^{s} \varphi'(t)\, dt = \varphi(s) - \varphi(s_0) \leqslant \varphi(s), \qquad s > s_0, \\ |\psi(s) - C(s - s_0) - \psi(s_0)| &= \left| \int_{s}^{s_0} (\psi'(t) - C)\, dt \right| \\ &\leqslant \int_{s}^{s_0} \varphi'(t)\, dt = \varphi(s) - \varphi(s_0) \leqslant \varphi(s), \qquad s < s_0, \end{aligned}$$

and we obtain (10) with $l(s) = C(s - s_0) + \psi(s_0)$.

If there is no such point s_0, then either $\varphi(s) \to 0$ as $s \to -\infty$ or $\varphi(s) \to 0$ as $s \to +\infty$. Suppose, for example, that we have the first case. Then φ' is positive, $\varphi'(s) \to 0$ as $s \to -\infty$ and

$$\varphi(s) = \int_{-\infty}^{s} \varphi'(t)\, dt.$$

From the inequality obtained similarly to (11), we have

$$|\psi'(s) - \psi'(s_1 + 0)| \leqslant \varphi'(s) - \varphi'(s_1 + 0), \qquad s_1 \leqslant s.$$

This implies that the following limit $C = \lim_{s_1 \to -\infty} \psi'(s_1 + 0)$ exists and

$$|\psi'(s) - C| \leqslant \varphi'(s).$$

Therefore, if $C_1 = \lim_{s \to -\infty} (\psi(s) - Cs)$, we have

$$|\psi(s) - Cs - C_1| = \left| \int_{-\infty}^{s} (\psi'(t) - C)\, dt \right| \leqslant \int_{-\infty}^{s} \varphi'(t)\, dt = \varphi(s),$$

as well as (10) with $l(s) = Cs + C_1$. The lemma is proved.

The next proposition characterizes the domination property not in terms of measures, but in terms of functions themselves.

PROPOSITION 7. *A function* $\psi \in BV^1(\mathbb{R})$ *is dominated by the convex function* φ *if and only if*

$$|\psi(s+t)+\psi(s-t)-2\psi(s)| \leqslant C(\varphi(s+t)+\varphi(s-t)-2\varphi(s)), \quad \forall s,t\in\mathbb{R}, \tag{16}$$

for some constant C. *The minimal constant* C *satisfying* (16) *coincides with* $|\psi|_\varphi$.

PROOF. Let us assume first that (16) is fulfilled for some constant $C \geqslant 0$. Let $\psi_1 = \operatorname{Re}\psi$, $\psi_2 = \operatorname{Im}\psi$. Then for any real nonnegative test function f we have

$$\begin{aligned} -C_k \int [\varphi(s+t)+\varphi(s-t)-2\varphi(s)]\, f(s)\, ds \\ \leqslant \int [\psi_k(s+t)+\psi_k(s-t)-2\psi_k(s)]\, f(s)\, ds \\ \leqslant C_k \int [\varphi(s+t)+\varphi(s-t)-2\varphi(s)]\, f(s)\, ds, \end{aligned}$$

where $k = 1,2$ and C_k are some constants such that $C = \sqrt{C_1^2+C_2^2}$. Dividing by t^2 as $t \to 0$, we obtain

$$-C_k \int f\, d\rho_\varphi \leqslant \int f\, d\rho_{\psi_k} \leqslant C_k \int f\, d\rho_\varphi, \qquad k=1,2. \tag{17}$$

This implies that the measure $\rho_\psi = \rho_{\psi_1} + i\rho_{\psi_2}$ is absolutely continuous with respect to ρ_φ. Therefore, by the Radon–Nikodym theorem, we have $d\rho_\psi = g\, d\rho_\varphi$ with some complex density $g \in L_{1,\mathrm{loc}}(\mathbb{R},\rho_\varphi)$.

Now from (17) it also follows that

$$\left|\int f\, d\rho_\psi\right| = \left|\int fg\, d\rho_\varphi\right| \leqslant C \int |f|\, d\rho_\varphi$$

for any (not necessarily nonnegative) test function f. Hence, the function g defines a linear continuous functional on the space $L_1(\mathbb{R},\rho_\varphi)$ whose norm is estimated by C, and then $g \in L_\infty(\mathbb{R},\rho_\varphi)$, $\|g\|_{L_\infty(\mathbb{R},\rho_\varphi)} \leqslant C$.

It remains to prove the necessity. Let $d\rho_\psi = g\, d\rho_\varphi$ with $g \in L_\infty(\mathbb{R},\rho_\varphi)$, and

$$C_1 = \|\operatorname{Re} g\|_{L_\infty(\mathbb{R},\rho_\varphi)}, \qquad C_2 = \|\operatorname{Im} g\|_{L_\infty(\mathbb{R},\rho_\varphi)}.$$

Then the functions

$$\psi_1^\pm \stackrel{\mathrm{def}}{=} C_1\varphi \pm \operatorname{Re}\psi, \qquad \psi_2^\pm \stackrel{\mathrm{def}}{=} C_2\varphi \pm \operatorname{Im}\psi \tag{18}$$

are convex because their second derivatives are positive measures, and so for each of them (9) holds. These estimates together mean exactly that for all $s,t \in \mathbb{R}$

$$\begin{aligned} |\operatorname{Re}\psi(s+t)+\operatorname{Re}\psi(s-t)-2\operatorname{Re}\psi(s)| \leqslant C_1(\varphi(s+t)+\varphi(s-t)-2\varphi(s)), \\ |\operatorname{Im}\psi(s+t)+\operatorname{Im}\psi(s-t)-2\operatorname{Im}\psi(s)| \leqslant C_2(\varphi(s+t)+\varphi(s-t)-2\varphi(s)), \end{aligned}$$

which implies

$$|\psi(s+t)+\psi(s-t)-2\psi(s)| \leqslant C_0(\varphi(s+t)+\varphi(s-t)-2\varphi(s)), \qquad \forall\, s,t\in\mathbb{R}$$

with $C_0 = \sqrt{C_1^2+C_2^2} = \|g\|_{L_\infty(\mathbb{R},\rho_\varphi)}$. The proof is complete.

EXAMPLE 8. For the convex function $\varphi(s) = s^2/2$, the measure ρ_φ coincides with the Lebesgue measure on $\mathbb{R}$. In this case $\psi \in BV^1(\mathbb{R})$ is dominated by φ if only if $\psi \in W^2_{\infty,\mathrm{loc}}(\mathbb{R})$ and $\psi'' \in L_\infty(\mathbb{R})$, and $|\psi|_\varphi = \|\psi''\|_{L_\infty(\mathbb{R})}$.

Further on we use the following well-known result.

THEOREM 9 (Jensen inequality). *Let ν be a positive measure on $\mathbb{R}$ such that $\nu(\mathbb{R}) = 1$ and $\int s\,d\nu < \infty$, and φ be a convex function from $L_1(\mathbb{R},\nu)$. Then*

$$\int \varphi(s)\,d\nu - \varphi\Big(\int s\,d\nu\Big) \geqslant 0.$$

COROLLARY 10. *Suppose that in Theorem 8 $\nu(\mathbb{R}) \overset{\mathrm{def}}{=} c_\nu \leqslant 1$. Then*

$$(1 - c_\nu)\varphi(0) + \int \varphi(s)\,d\nu - \varphi\Big(\int s\,d\nu\Big) \geqslant 0. \tag{19}$$

PROOF. If we apply (8) and the Jensen inequality, we obtain

$$\varphi\Big(\int s\,d\nu\Big) \leqslant (1 - c_\nu)\varphi(0) + \varphi\Big(\int s c_\nu^{-1}\,d\nu\Big) \leqslant (1 - c_\nu)\varphi(0) + \int \varphi(s)\,d\nu,$$

which proves the corollary.

COROLLARY 11. *Suppose ν is a positive measure on $\mathbb{R}$ such that $\nu(\mathbb{R}) \overset{\mathrm{def}}{=} c_\nu \leqslant 1$, $\int s\,d\nu < \infty$, and $\psi \in BV^1(\mathbb{R}) \cap L_1(\mathbb{R},\nu)$ is dominated by a convex function $\varphi \in L_1(\mathbb{R},\nu)$. Then*

$$\begin{aligned}&\Big|(1 - c_\nu)\psi(0) + \int \psi(s)\,d\nu - \psi\Big(\int s\,d\nu\Big)\Big| \\ &\qquad \leqslant |\psi|_\varphi\Big((1 - c_\nu)\varphi(0) + \int \varphi(s)\,d\nu - \varphi\Big(\int s\,d\nu\Big)\Big).\end{aligned} \tag{20}$$

PROOF. As in the proof of Proposition 7, we introduce the convex function (18), and apply inequality (19) to each of them. Then we obtain the inequalities

$$\begin{aligned}&\Big|(1 - c_\nu)\operatorname{Re}\psi(0) + \int \operatorname{Re}\psi(s)\,d\nu - \operatorname{Re}\psi\Big(\int s\,d\nu\Big)\Big| \\ &\qquad \leqslant C_1\Big((1 - c_\nu)\varphi(0) + \int \varphi(s)\,d\nu - \varphi\Big(\int s\,d\nu\Big)\Big), \\ &\Big|(1 - c_\nu)\operatorname{Im}\psi(0) + \int \operatorname{Im}\psi(s)\,d\nu - \operatorname{Im}\psi\Big(\int s\,d\nu\Big)\Big| \\ &\qquad \leqslant C_2\Big((1 - c_\nu)\varphi(0) + \int \varphi(s)\,d\nu - \varphi\Big(\int s\,d\nu\Big)\Big),\end{aligned}$$

which are equivalent to (20).

§3. Berezin–Lieb inequality

We study operators of the form

$$G(B,P;\psi) \stackrel{\text{def}}{=} \psi(0)\,(I-P^*P)+P^*\psi(B)P-\psi(B_P),$$

where $\psi \in BV^1(\mathbb{R})$. Note that under the conditions (C_1) and (C_2) the operator $G(B,P;\psi)$ is well defined and equal to zero for linear functions ψ. When B is unbounded, for some functions ψ the expression $P^*\psi(B)P$ or $G(B,P;\psi)$ may not make sense. Therefore we introduce an additional restriction.

(C_3) The set

$$\mathcal{D}_\psi = \{\xi \in H_0 : P\xi \in \mathcal{D}(\psi(B))\} \cap \mathcal{D}(\psi(B_P)) \cap \mathcal{D}(B_P)$$

is dense in H_0 and the operator $G(B,P;\psi)$ defined on $\mathcal{D}_\psi$ is bounded.

Under these conditions we extend the operator $G(B,P;\psi)$ to the entire Hilbert space H_0, and then $P^*\psi(B)P$ is a well-defined selfadjoint operator with domain $\mathcal{D}(\psi(B_P))$. Obviously, if the condition (C_3) is satisfied for the function ψ, then it is also satisfied for any $\psi_1 \in \psi^*$ and $\mathcal{D}_{\psi_1} = \mathcal{D}_\psi$, $G(B,P;\psi_1) = G(B,P;\psi)$. Besides, if for some convex function φ the set $\mathcal{D}_\varphi$ is dense, then in view of Lemma 6 for any ψ dominated by φ the set $\mathcal{D}_\psi$ is also dense.

We denote by $\sigma(B_P)$ the spectrum of the selfadjoint operator B_P and by $\sigma_c(B_P)$ its continuous part. Let $\operatorname{ch}\sigma_c(B_P)$ be the closed convex hull of $\sigma_c(B_P)$, and $\operatorname{Int}\operatorname{ch}\sigma_c(B_P)$ be its interior. (The last set coincides with the interior of the minimal interval containing $\sigma_c(B_P)$.)

THEOREM 12. *Let the conditions* (C_1)–(C_2) *be fulfilled. Let* $\psi \in BV^1(\mathbb{R})$ *be dominated by a convex function* φ *such that* $\rho_\varphi(\operatorname{Int}\operatorname{ch}\sigma_c(B_P)) = 0$. *Assume that condition* (C_3) *is fulfilled for both* φ *and* ψ *and that the operators* $G(B,P;\varphi)$, $G(B,P;\psi)$ *are from the trace class* $\mathfrak{S}_1$. *Then*

$$|\operatorname{Tr} G(B,P;\psi)| \leqslant |\psi|_\varphi \operatorname{Tr} G(B,P;\varphi)\,. \tag{21}$$

PROOF. Let $\varphi_0 \in \varphi^*$ be a nonnegative representative, and $\psi_0 \in \psi^*$ be a representative such that $|\psi_0| \leqslant |\psi|_\varphi \varphi_0$ (see Lemma 6). If $\operatorname{Int}\operatorname{ch}\sigma_c(B_P)$ is not empty, we assume in addition that $\varphi = 0$ on $\operatorname{ch}\sigma_c(B_P)$. Then ψ is also equal to zero on $\operatorname{ch}\sigma_c(B_P)$.

For every $\xi \in \mathcal{D}_{\varphi_0}$ we have

$$\begin{aligned}\int \varphi_0(\lambda)(dE_B(\lambda)P\xi, P\xi) &= (\varphi_0(B)P\xi, P\xi)\\ &= (G(B,P;\varphi_0)\xi,\xi)_0 + (\varphi_0(B_P)\xi,\xi)_0 .\end{aligned} \tag{22}$$

Since the function φ_0 is nonnegative and the operator $G(B,P;\varphi_0)$ is bounded, it follows that (22) can be extended on $\xi \in \mathcal{D}(\varphi_0(B_P))$. For the chosen representative ψ_0, we have $\mathcal{D}(\varphi_0(B_P)) \subset \mathcal{D}(\psi_0(B_P))$ while

$$\begin{aligned}\int \psi_0(\lambda)(dE_B(\lambda)P\xi, P\xi) &= (\psi_0(B)P\xi, P\xi)_0\\ &= (G(B,P;\psi_0)\xi,\xi)_0 + (\psi_0(B_P)\xi,\xi)_0\end{aligned} \tag{23}$$

is also valid for $\xi \in \mathcal{D}(\varphi_0(B_P))$.

Let Π_c be the spectral projection of the operator B_P corresponding to the closed interval $\operatorname{ch}\sigma_c(B_P)$. Let us choose an orthonormed basis $\{\xi_k\}$ in the subspace $(I-\Pi_c)H_0$ formed by eigenfunctions ξ_k of the operator B_P with eigenvalues λ_k lying outside $\operatorname{ch}\sigma_c(B_P)$. It is clear that the ξ_k's are contained in $\mathcal{D}(\varphi_0(B_P)) \subset \mathcal{D}(\psi_0(B_P))$. Let us also choose an orthonormed basis $\{\eta_j\}$ in the subspace $\Pi_c H_0$ with $\eta_j \in \mathcal{D}(\varphi_0(B_P))$. Then the elements $\{\xi_k, \eta_j\}$ form an orthonormed basis in the whole space H_0.

Let ν_k be the positive measures with $d\nu_k = (dE_B(\lambda)P\xi_k, P\xi_k)$. Then

$$(\varphi_0(B_P)\xi_k, \xi_k)_0 = \varphi_0((B_P\xi_k, \xi_k)_0) = \varphi_0(\lambda_k),$$
$$(\psi_0(B_P)\xi_k, \xi_k)_0 = \psi_0((B_P\xi_k, \xi_k)_0) = \psi_0(\lambda_k),$$

and by (22), (23)

$$(\varphi_0(B)P\xi_k, P\xi_k) = \int \varphi_0(\lambda)\, d\nu_k, \qquad (\psi_0(B)P\xi_k, P\xi_k) = \int \psi_0(\lambda)\, d\nu_k.$$

Therefore, applying (20), we obtain

$$|(G(B,P;\psi_0)\xi_k, \xi_k)_0| \leqslant |\psi|_\varphi((G(B,P;\varphi_0)\xi_k, \xi_k)_0). \tag{24}$$

Since $\varphi_0(B_P)\eta_j = 0$ and $\psi_0(B_P)\eta_j = 0$, we have

$$(G(B,P;\varphi_0)\eta_j, \eta_j)_0 = \varphi(0)((I-P^*P)\eta_j, \eta_j)_0 + (\varphi_0(B)P\eta_j, P\eta_j),$$
$$(G(B,P;\psi_0)\eta_j, \eta_j)_0 = \psi(0)((I-P^*P)\eta_j, \eta_j)_0 + (\psi_0(B)P\eta_j, P\eta_j).$$

Then in view of (22), (23) and the inequality $|\psi_0| \leqslant |\psi|_\varphi \varphi_0$, we obtain

$$|(G(B,P;\psi_0)\eta_j, \eta_j)_0| \leqslant |\psi|_\varphi(G(B,P;\varphi_0)\eta_j, \eta_j)_0.$$

Summing these inequalities and inequalities (24), we obtain (21). The proof is complete.

If $\psi = \varphi$, then Theorem 12 is a generalization of the inequality obtained in [**B**] and [**L**].

COROLLARY 13 (generalized Berezin–Lieb inequality). *Suppose that conditions* (C_1)–(C_2) *are fulfilled. Let* φ *be a convex function such that*

$$\rho_\varphi(\operatorname{Int}\operatorname{ch}\sigma_c(B_P)) = 0.$$

Assume that (C_3) *holds for the function* φ *and that* $G(B,P;\varphi) \in \mathfrak{S}_1$. *Then*

$$\operatorname{Tr} G(B,P;\varphi) \geqslant 0. \tag{25}$$

The conditions of Theorem 12 are rather complicated. But most of them are needed only in order to define the unbounded operators. In particular, if B is bounded, then conditions (C_1)–(C_3) are fulfilled automatically, and Theorem 12 can be reformulated in the following way.

COROLLARY 14. *Let the operator* B *be bounded. Assume that* $\psi \in BV^1(\mathbb{R})$ *is dominated by a convex function* φ *such that* $\rho_\varphi(\operatorname{Int}\operatorname{ch}\sigma_c(B_P)) = 0$, *and* $G(B,P;\varphi)$, $G(B,P;\psi)$ *are from* $\mathfrak{S}_1$. *Then the estimate* (21) *holds.*

Let us denote by $\sigma_{\mathrm{ess}}(B_P)$ the essential spectrum of B_P. We have $\sigma_c(B_P) \subset \sigma_{\mathrm{ess}}(B_P)$, and therefore $\operatorname{ch}\sigma_c(B_P) \subset \operatorname{ch}\sigma_{\mathrm{ess}}(B_P)$. The following proposition gives another set of sufficient conditions for Theorem 12.

PROPOSITION 15. *Let conditions* (C_1)–(C_2) *hold and condition* (C_3) *be valid for a nonnegative convex function* φ *such that the operator* $\varphi(0)(I-P^*P)+P^*\varphi(B)P$ *is from the trace class* $\mathfrak{S}_1$. *Then*

1. φ *is equal to zero on the set* $\operatorname{ch}\sigma_{\mathrm{ess}}(B_P)$;
2. $\varphi(B_P) \in \mathfrak{S}_1$, *and, consequently,* $G(B,P;\varphi) \in \mathfrak{S}_1$;
3. *for any function* $\psi \in BV^1(\mathbb{R})$ *dominated by* φ, *the condition* (C_3) *holds and* $G(B,P;\psi) \in \mathfrak{S}_1$.

PROOF. Let θ_k be the eigenfunctions of the operator $\varphi(0)(I-P^*P)+P^*\varphi(B)P$ corresponding to eigenvalues μ_k, $|\mu_1| \leqslant |\mu_2| \leqslant \cdots$. By (19) for any $\xi \in H_0$ we have

$$\begin{aligned}
&\varphi(0)((I-P^*P)\xi,\xi)_0 + (P^*\varphi(B)P\xi,\xi)_0 \\
&\quad = \varphi(0)\Big(1 - \int (dE_B(\lambda)P\xi, P\xi)\Big) + \int \varphi(\lambda)(dE_B(\lambda)P\xi, P\xi) \\
&\quad \geqslant \varphi\Big(\int \lambda (dE_B(\lambda)P\xi, P\xi)\Big) = \varphi((B_P\xi,\xi)_0).
\end{aligned} \tag{26}$$

Since the operator $\varphi(0)(I-P^*P)+P^*\varphi(B)P$ is compact, (26) implies that there exists a positive sequence $\varepsilon_j \to 0$ such that

$$|\varphi((B_P\xi,\xi)_0)| \leqslant \varepsilon_j$$

for any normed vector ξ which is orthogonal to all the θ_k with $k \leqslant j$. By the minimax principle (see for example [**RS,** Theorem XIII.1]), it follows now that $\varphi(s) \to 0$ as $s \to \pm\infty$ if B_P is unbounded from above or from below respectively, and that $\varphi = 0$ on $\sigma_{\mathrm{ess}}(B_P)$. Obviously the set of zeros of a convex function is necessarily convex, and therefore we have proved (1).

Let ξ_k be the orthonormed eigenfunctions of B_P with eigenvalues λ_k lying outside $\operatorname{ch}\sigma_{\mathrm{ess}}(B_P)$. By (26) we have

$$\varphi(0)((I-P^*P)\xi_k,\xi_k)_0 + (P^*\varphi(B)P\xi_k,\xi_k)_0 \geqslant \varphi((B_P\xi_k,\xi_k)_0) = \varphi(\lambda_k).$$

Since $\varphi(0)(I-P^*P)+P^*\varphi(B)P \in \mathfrak{S}_1$, the positive series $\sum \varphi(\lambda_k)$ converges, which means that $\varphi(B_P) \in \mathfrak{S}_1$.

To prove the third assertion of the lemma, we use Lemma 6 to choose a function $\psi_0 \in \psi^*$ such that $|\psi| \leqslant |\psi|_\varphi \varphi$. Then for any orthonormed basis $\{\zeta_k\}$ in H_0 we have

$$\begin{aligned}
&|\psi_0(0)((I-P^*P)\zeta_k,\zeta_k)_0 + (P^*\psi_0(B)P\zeta_k,\zeta_k)_0| \\
&\qquad\qquad \leqslant |\psi_0(0)((I-P^*P)\zeta_k,\zeta_k)_0| + |(P^*\psi_0(B)P\zeta_k,\zeta_k)_0| \\
&\qquad\qquad \leqslant |\psi|_\varphi(\varphi(0)((I-P^*P)\zeta_k,\zeta_k)_0 + (P^*\varphi(B)P\zeta_k,\zeta_k)_0), \\
&\quad |(\psi_0(B_P)\zeta_k,\zeta_k)_0| \leqslant |\psi|_\varphi(\varphi(B_P)\zeta_k,\zeta_k)_0.
\end{aligned}$$

These estimates imply (see [**RS,** Chapter VI, Problem 26]) that the two operators $\psi_0(0)(I-P^*P)+P^*\psi_0(B)P$ and $\psi_0(B_P)$ are from the trace class. Since the operator $G(B,P;\,\cdot\,)$ is independent of the choice of representative from the factor-class ψ^*, this completes the proof.

REMARK 16. In fact, while proving (3), we have obtained a more precise result. Namely, if $\psi \in BV^1(\mathbb{R})$ is dominated by φ, then for a representative $\psi_0 \in \psi^*$ such that $|\psi_0| \leqslant |\psi|_\varphi \varphi$ both operators $\psi_0(0)(I - P^*P) + P^*\psi_0(B)P$ and $\psi_0(B_P)$ are from the trace class.

Proposition 15 with $\varphi(s) = s^2/2$ immediately implies

COROLLARY 17. *Let BP be from the Hilbert–Schmidt class $\mathfrak{S}_2$. Then*
1. *either $\sigma_{\mathrm{ess}}(B_P) = \{0\}$ or $\sigma_{\mathrm{ess}}(B_P) = \varnothing$;*
2. *for any function $\psi \in W^2_{\infty,\mathrm{loc}}(\mathbb{R})$ such that $\psi'' \in L_\infty(\mathbb{R})$, condition* (C$_3$) *is fulfilled and $G(B, P; \psi) \in \mathfrak{S}_1$.*

From Theorem 12 and Corollary 17 we obtain

COROLLARY 18. *Let $H_0 = H$ and $P\colon H \to H$ be an orthogonal projection in H. If the operator BP is from the Hilbert–Schmidt class, then for any function ψ from the Sobolev class $W^2_{\infty,\mathrm{loc}}(\mathbb{R})$ such that $\psi'' \in L_\infty(\mathbb{R})$ we have*

$$|\operatorname{Tr}(P\psi(B)P - P\psi(PBP)P)| \leqslant \frac{1}{2}\,\|\psi''\|_{L_\infty(K)}\|PB(I-P)\|^2_{\mathfrak{S}_2}.$$

REMARK 19. When we deal with a fixed operator B, it is sufficient to define the functions φ and ψ only on the set

$$\bigcup_{0\leqslant t\leqslant 1} t\sigma(B) \subset \mathbb{R}.$$

Then obviously all the conditions involving φ and ψ need only hold on this set.

References

[B] F. Berezin, *Convex functions of operators*, Mat. Sb. **88** (1972), no. 2, 268–276; English transl. in Math. USSR-Sb. **17** (1972).

[BS] M. Sh. Birman and M. Z. Solomyak, *Spectral theory of selfadjoint operators in Hilbert space*, D. Reidel Publ. Comp., 1987.

[BSh] F. Berezin and M. Shubin, *The Schrödinger equation*, Kluwer Acad. Publishers, Dordrecht–Boston, 1991.

[Hö] L. Hörmander, *The analysis of Linear Partial Differential operators*, vol. I, Springer-Verlag, Berlin–Heidelberg–New York–Tokyo, 1983.

[L] E. H. Lieb, *The classical limit of quantum spin systems*, Comm. Math. Phys. **31** (1973), 327–340.

[RS] M. Reed and B. Simon, *Methods of modern mathematical physics. Vol.* 4, Academic Press, New York–San Francisco–London, 1978.

[S] B. Simon, *The classical limit of quantum partition functions*, Comm. Math. Phys. **71** (1980), 247–276.

DEPARTMENT OF MATHEMATICS, ROYAL INSTITUTE OF TECHNOLOGY, S-100 44 STOCKHOLM, SWEDEN

E-mail address: `laptev@math.kth.se`

DEPARTMENT OF MATHEMATICS, KING'S COLLEGE LONDON, STRAND, LONDON WC2R 2LS, UK

E-mail address: `ysafarov@mth.kcl.ac.uk`

Translated by THE AUTHORS

Amer. Math. Soc. Transl.
(2) Vol. **175**, 1996

Quantization on Para-Hermitian Symmetric Spaces

V. F. Molchanov

One of the main goals that F. A. Berezin had set up for himself was to define a general concept of quantization. To this end he introduced some general notions and approaches: on the one hand, and obtained deep analytic and algebraic results; on the other hand, he constructed the quantization for an important class of symplectic manifolds, namely, for Hermitian symmetric spaces (classical), see [**2**–**5**]. Those are Kählerian (hence complex) manifolds and Riemannian semisimple symmetric spaces.

Now we can see that his methods, notions, constructions, etc., have turned out to be very fruitful. Some of them have already acquired his name: Berezin quantization, Berezin transform, Berezin kernel, etc.

We would like to outline a program for a quantization in the spirit of Berezin for another class of symplectic homogeneous manifolds, namely, for para-Hermitian symmetric spaces of the first category (see (c) below), and to present some results in this direction (proofs and details will be given elsewhere). These spaces, as well as Hermitian symmetric spaces, belong to the very wide class of semisimple symmetric spaces. Let us consider, in general, symplectic manifolds that are symmetric spaces with a simple group of motions. They are divided into four classes [**22**]: (a) Hermitian symmetric spaces; (b) semi-Kählerian irreducible symmetric spaces; (c) para-Hermitian symmetric spaces of the first category; (d) para-Hermitian symmetric spaces of the second category. The spaces of the three latter classes are not Riemannian, and each has a Riemannian form belonging to the class (a) of Hermitian symmetric spaces.

There is an inspiring analogy between (a) and (c), which starts at the coordinate level: $z, \overline{z} \leftrightarrow \xi, \eta$, see §4, and continues on the level of formulas and so on. On the other hand, it is well known that the passage from the Riemannian case to the non-Riemannian one drastically increases the difficulties. So, in this theory there are still many interesting open problems.

A local classification of the spaces in (c)+(d) can be obtained from Berger's list [**6**] by taking those that are reducible (which means that the stabilizer of a point acts in a reducible way on the tangent space), see, for example, [**15**]. Then deleting the complexifications of spaces in (a) (they form the class (d)), we obtain (c). The list of spaces in (c) is given in §4.

1991 *Mathematics Subject Classification.* Primary 22E45, 53C35.

I am grateful to the editors for the possibility to contribute to this volume and to honor the memory of F. A. Berezin—a remarkable mathematician. It was an unforgettable time when I studied at Moscow University and was a student of Berezin. His personality exerted a great influence on me.

I thank very much G. van Dijk who invited me to Leiden University and helped me to prepare the final version of this paper.

§1. Berezin quantization

Recall the concept of quantization proposed by Berezin, see [**2–4**]. We shall not give it in its full generality, but restrict ourselves to a rather simplified version.

Let M be a symplectic manifold. Then $C^\infty(M)$ is a Lie algebra with respect to the Poisson bracket $\{A, B\}$, $A, B \in C^\infty(M)$.

Quantization in the sense of Berezin consists of the following two steps:

(I) To construct a family $\mathcal{A}_h$ of associative algebras contained in $C^\infty(M)$ and depending on a parameter $h > 0$ (called the Planck constant), with a multiplication denoted by $*$ (depending on h also). These algebras must satisfy conditions (a) through (d):

$$\begin{aligned} &\text{(a)} \quad \lim_{h\to 0} A_1 * A_2 = A_1 A_2; \\ &\text{(b)} \quad \lim_{h\to 0} \frac{i}{h}\,(A_1 * A_2 - A_2 * A_1) = \{A_1, A_2\}; \end{aligned} \tag{1.1}$$

(c) the function $A_0 \equiv 1$ is the unit element of each algebra $\mathcal{A}_h$;

(d) the complex conjugation $A \mapsto \overline{A}$ is an anti-involution of any $\mathcal{A}_h$;

where the multiplication on the right-hand side of (a) is the pointwise multiplication and conditions (a) and (b) together are called the *correspondence principle* (CP).

(II) To construct representations $A \mapsto \widehat{A}$ of the algebras $\mathcal{A}_h$ by operators in a Hilbert space.

Berezin mainly considered the case when M is Kählerian, hence has a complex structure. The functions in question are functions $A(z, \overline{z})$ analytic in z and $\overline{z}$ separately. In this case complex conjugation reduces to the permutation of z and $\overline{z}$: $\overline{f(z,\overline{z})} = f(\overline{z}, z)$.

Note that for our theory we shall slightly change some of the conditions ((b), (d), (II), see below).

§2. Quantization on Hermitian symmetric spaces

In this section we recall Berezin's results about the quantization on Hermitian symmetric spaces G/K, see [**2–5**].

(a) Example: the Lobachevskiĭ plane $D = \{z \in C \mid z\overline{z} < 1\} = G/K$, where $G = SU(1,1)$, $K = U(1)$. The group G acts by fractional linear transformations (here and further groups act from the right):

$$z \mapsto \frac{az + \overline{b}}{bz + \overline{a}}, \qquad g = \begin{pmatrix} a & b \\ \overline{b} & \overline{a} \end{pmatrix},$$

the Bergman kernel is $b(z, \overline{z}) = \pi^{-1}(1 - z\overline{z})^{-2}$, the G-invariant measure is $dv(z) = b(z, \overline{z})\, dx\, dy$, where $z = x + iy$, the Laplace–Beltrami operator is

$$\Delta = \frac{1}{2}\,(1 - z\overline{z})^2 \frac{\partial^2}{\partial z\, \partial \overline{z}},$$

the rank is $r = 1$, the genus is $\kappa = 2$.

Let us consider the Fock space $\mathcal{F}$ consisting of functions $f(z)$ holomorphic on D with $(f, f) < +\infty$, where

$$(f_1, f_2) = c(\mu) \int_D f_1(z) \overline{f_2(z)} (1 - z\overline{z})^{-\mu} \, dv(z) \tag{2.1}$$

and $c(\mu) = -\mu - 1$. Assume $\mu < -1$. Then the integral converges absolutely when f_1 and f_2 are constants. The factor $c(\mu)$ is determined by the condition $(1, 1) = 1$. The Fock space becomes a Hilbert space if we endow it with the inner product (2.1).

For $w \in D$ define the function $\Phi_{\overline{w}}$ by setting

$$\Phi_{\overline{w}}(z) = \Phi(z, \overline{w}) = (1 - z\overline{w})^{\mu};$$

$\Phi_{\overline{w}}$ belongs to $\mathcal{F}$. Berezin called $\Phi(z, \overline{w})$ a *supercomplete system*. It has the reproducing property

$$(f, \Phi_{\overline{w}}) = f(w).$$

Let $\widehat{A}$ be a bounded operator on $\mathcal{F}$. Associate to it the function

$$A(z, \overline{w}) = \frac{(\widehat{A}\Phi_{\overline{w}})(z)}{\Phi(z, \overline{w})}.$$

The function $A(z, \overline{z})$ is called the (*covariant*) *symbol* of the operator $\widehat{A}$. The former function $A(z, \overline{w})$ is recovered from $A(z, \overline{z})$ using analyticity.

The operator $\widehat{A}$ is completely determined by its symbol

$$(\widehat{A}f)(z) = c(\mu) \int_D A(z, \overline{w}) \, f(w) \, \frac{\Phi(z, \overline{w})}{\Phi(w, \overline{w})} \, dv(w).$$

The multiplication of operators yields a multiplication of symbols, $\widehat{A}_1 \widehat{A}_2 = \langle \widehat{A_1 * A_2} \rangle$, where

$$(A_1 * A_2)(z, \overline{z}) = \int_D A_1(z, \overline{w}) \, A_2(w, \overline{z}) \, \mathcal{B}(z, \overline{z}; w, \overline{w}) \, dv(w). \tag{2.2}$$

Let us call the kernel $\mathcal{B}(z, \overline{z}; w, \overline{w})$ the *Berezin kernel* and the operator $\mathcal{B}$ with this kernel the *Berezin transform* (it acts on functions on D). It turns out that $\mathcal{B}$ can be expressed in terms of Δ:

$$\mathcal{B} = \frac{\Gamma(-\mu + \sigma)\Gamma(-\mu - \sigma - 1)}{\Gamma(-\mu)\Gamma(-\mu - 1)}, \tag{2.3}$$

where the right-hand side should be regarded as a function of $\sigma(\sigma + 1) = \Delta$. Using the asymptotic expansion [**10,** 1.18(4)] for the gamma function, we obtain from (2.3):

$$\mathcal{B} \sim 1 - \frac{1}{\mu} \Delta \qquad (\mu \to -\infty). \tag{2.4}$$

For the algebras $\mathcal{A}_h$ ($h = -1/\mu$) let us take the algebras consisting of symbols of bounded operators on $\mathcal{F}$ with the multiplication (2.2). Formula (2.4) tells us that the condition CP holds.

The Hermitian form on $L^2(D)$ defined by the kernel $\mathcal{B}(z, \overline{z}; w, \overline{w})$ is G-invariant, bounded, and positive definite ($\mu < -1$), hence can be regarded as an inner product.

Denote by U_μ the unitary representation of the group G acting by translations on functions on D with this inner product, and denote by U the quasiregular representation of G in $L^2(D)$. The representation U_μ first appeared in [**27**] as the "canonical representation".

Formula (2.3) is closely connected with the decomposition of U_μ into irreducible unitary representations. Indeed, the right-hand side of (2.3) is the ratio of Plancherel measures for U_μ and U.

(b) For the general case of a Hermitian symmetric space G/K similar results are true. Here we restrict ourselves to writing out a remarkable formula of Berezin expressing the Berezin transform $\mathcal{B}$ in terms of the Laplace operators $\Delta_1, \dots, \Delta_r$ on G/K, see [**5**], and also the recent preprint [**26**]. Let $G = NAK$ be the Iwasawa decomposition, $\mathfrak{a}$ the Lie algebra of A. Take a basis $X_1, \dots, X_r$ in $\mathfrak{a}$. Let $t_1, \dots, t_r$ be the coordinates in this basis. Denote $\rho(X_j) = \nu_j$, where ρ is half the sum of the positive restricted roots. Let $\Delta_j \in \mathbf{D}(G/K)$ be an operator such that its N-radial part is

$$\sum_{i=1}^{r} \left(\frac{\partial}{\partial t_i} - \rho_i \right)^{2j}.$$

In particular, Δ_1 is the Laplace–Beltrami operator. Then

$$\mathcal{B} = \prod_{j=1}^{r} \frac{\Gamma(-\mu + \nu_j - \frac{\kappa-1}{2})\Gamma(-\mu - \nu_j - \frac{\kappa-1}{2})}{\Gamma(-\mu + \rho_j - \frac{\kappa-1}{2})\Gamma(-\mu - \rho_j - \frac{\kappa-1}{2})},$$

where the right-hand side should be viewed as a function of $\Delta_j = \sum_{k=1}^{r} \nu_k^{2j}$, $j = 1, \dots, r$. This formula implies $\mathcal{B} \sim 1 - (\kappa/\mu)\Delta_1$ $(\mu \to -\infty)$ and that CP holds.

§3. Semisimple symmetric spaces

Let us recall some facts about semisimple symmetric spaces G/H. Here G is a connected semisimple Lie group with an involutive automorphism $\sigma \neq 1$. Denote by G^σ the subgroup of fixed points of σ. Then H is an open subgroup of G^σ. There exists a Cartan involution τ of G commuting with σ. Then $K = G^\tau$. For Lie groups $G, \dots$ we denote their Lie algebras by the corresponding small Gothic letters $\mathfrak{g}, \dots$. We assume that the pair $(\mathfrak{g}, \mathfrak{h})$ is effective, i.e., $\mathfrak{h}$ contains no nontrivial ideals of $\mathfrak{g}$. The automorphisms of $\mathfrak{g}$ induced by σ, τ are denoted by the same letters σ, τ. There is a decomposition of $\mathfrak{g}$ into direct sums of ± 1-eigenspaces of σ and τ: $\mathfrak{g} = \mathfrak{h} + \mathfrak{q}$ and $\mathfrak{g} = \mathfrak{k} + \mathfrak{p}$, as well as the combined decomposition,

$$\mathfrak{g} = \mathfrak{k} \cap \mathfrak{h} + \mathfrak{k} \cap \mathfrak{q} + \mathfrak{p} \cap \mathfrak{h} + \mathfrak{p} \cap \mathfrak{q}.$$

The subspace $\mathfrak{q}$ can be identified with the tangent space of G/H at the point $x^0 = He$, it is invariant with respect to the adjoint representations $\mathrm{Ad}_{\mathfrak{g}}$ of H and $\mathrm{ad}_{\mathfrak{g}}$ of $\mathfrak{g}$.

A Cartan subspace of $\mathfrak{q}$ is a maximal Abelian subalgebra in $\mathfrak{q}$ consisting of semisimple elements. The rank of G/H is the dimension of a Cartan subspace of $\mathfrak{q}$. If the center of G is finite, then the discrete series for G/H exists if and only if

$$\operatorname{rank} G/H = \operatorname{rank} K/K \cap H, \tag{3.1}$$

i.e., if any Cartan subspace of $\mathfrak{k} \cap \mathfrak{q}$ is also a Cartan subspace of $\mathfrak{q}$.

§4. Para-Hermitian symmetric spaces of the first category

We continue with the assumptions of §3 and assume, in addition, that G/H is *symplectic*. Then $\mathfrak{h}$ has a nontrivial center $Z(\mathfrak{h})$. For simplicity we assume that G/H is an orbit $\operatorname{Ad} G \cdot Z_0$ of an element $Z_0 \in \mathfrak{g}$. We can assume that G is simple. Then the statement "G/H is para-Hermitian of the first category" means that the center $Z(\mathfrak{h})$ of $\mathfrak{h}$ is one-dimensional: $Z(\mathfrak{h}) = \mathbb{R}Z_0$, and Z_0 can be normalized so that the operator $I = (\operatorname{ad} Z_0)_{\mathfrak{q}}$ on $\mathfrak{q}$ has eigenvalues ± 1. Therefore, $Z_0 \in \mathfrak{p} \cap \mathfrak{h}$. A symplectic structure on G/H is defined by the bilinear form $\omega(X,Y) = B(X, IY)$ on $\mathfrak{q}$, where $B(X,Y)$ is the Killing form of $\mathfrak{g}$.

The ± 1-eigenspaces $\mathfrak{q}^{\pm} \subset \mathfrak{q}$ of I are Lagrangian, H-invariant, and irreducible. They are Abelian subalgebras of $\mathfrak{g}$. So $\mathfrak{g}$ becomes a graded Lie algebra:

$$\mathfrak{g} = \mathfrak{q}^- + \mathfrak{h} + \mathfrak{q}^+ (= \mathfrak{g}_{-1} + \mathfrak{g}_0 + \mathfrak{g}_{+1}).$$

The involution τ gives rise to the following isomorphisms:

$$\tau \colon \mathfrak{q}^{\pm} \to \mathfrak{q}^{\mp}, \quad 1+\tau \colon \mathfrak{q}^{\pm} \to \mathfrak{k} \cap \mathfrak{q}, \quad 1-\tau \colon \mathfrak{q}^{\pm} \to \mathfrak{p} \cap \mathfrak{q}.$$

So there exists an isomorphism

$$\psi \colon \mathfrak{k} \cap \mathfrak{q} \to \mathfrak{p} \cap \mathfrak{q}, \qquad \psi(X + \tau X) = X - \tau X, \quad X \in \mathfrak{q}^+.$$

The map ψ defines a duality between the two Riemannian symmetric Lie algebras $(\mathfrak{k}, \mathfrak{k} \cap \mathfrak{h}, \sigma)$ and $(\mathfrak{g}_0, \mathfrak{k} \cap \mathfrak{h}, \sigma)$, where $\mathfrak{g}_0 = \mathfrak{k} \cap \mathfrak{h} + \mathfrak{p} \cap \mathfrak{q}$. The former is of compact type; the latter, noncompact.

The pair $(\mathfrak{q}^+, \mathfrak{q}^-)$ is a Jordan pair [**16**] with multiplication $\{XYZ\} = \frac{1}{2}[[X,Y],Z]$ (we follow the notation of [**24**], in [**16**] the coefficient $1/2$ is omitted).

Let r and κ be the rank and the genus of this Jordan pair. In $\mathfrak{q}^+$ take a maximal system $e_1, \dots, e_r$ of pairwise orthogonal idempotents: $\{e_i e_j^* e_j\} = \delta_{ij} e_j$, where $e_i^* = -\tau e_i \in \mathfrak{q}^-$. Let $\mathfrak{a} \subset \mathfrak{p} \cap \mathfrak{q}$ and $\mathfrak{b} \subset \mathfrak{k} \cap \mathfrak{q}$ be the subspaces with the bases $X_i = e_i + e_i^* = e_i - \tau e_i$ and $Y_i = e_i - e_i^* = e_i + \tau e_i$, respectively. They are maximal Abelian subalgebras (Cartan subspaces) in $\mathfrak{p} \cap \mathfrak{q}$ and, moreover, even in $\mathfrak{q}$. We see that the ranks of $(\mathfrak{q}^+, \mathfrak{q}^-)$, G/H, and $K/K \cap H$ coincide, so that condition (3.1) is satisfied and G/H has a discrete series.

Set $Q^{\pm} = \exp \mathfrak{q}^{\pm}$. The subgroups $P^{\pm} = HQ^{\pm} = Q^{\pm}H$ are maximal parabolic subgroups of G, with H as a Levi subgroup. One has the following decompositions:

$$G = \overline{Q^+ H Q^-} \tag{4.1}$$

$$= \overline{Q^- H Q^+} \tag{4.2}$$

$$= Q^+ H K \tag{4.3}$$

$$= Q^- H K, \tag{4.4}$$

where bar means closure and the sets under the bar are open and dense in G. Let us call (4.1) the *Gauss decomposition* and (4.3) the *Iwasawa-type decomposition.* Allowing some slang, let us call (4.2) the *anti-Gauss decomposition* and (4.4) the *anti-Iwasawa-type decomposition.* For an element in G all three factors in (4.1), (4.2) and the first factors in (4.3), (4.4) are defined uniquely, whereas the second and the third factors in (4.3), (4.4) are defined up to an element of $K \cap H$.

For $g \in G$ we define the transformations $\xi \mapsto \tilde{\xi}$ of $\mathfrak{q}^-$ and $\eta \mapsto \hat{\eta}$ of $\mathfrak{q}^+$ taking $\tilde{\xi}$ and $\hat{\eta}$ from the Gauss and the anti-Gauss decompositions:

$$\exp \xi \cdot g = \exp Y \cdot \tilde{h} \cdot \exp \tilde{\xi}, \tag{4.5}$$

$$\exp \eta \cdot g = \exp X \cdot \hat{h} \cdot \exp \hat{\eta}. \tag{4.6}$$

These $\tilde{\xi}$ and $\hat{\eta}$ are defined on open and dense sets in $\mathfrak{q}^-$ and $\mathfrak{q}^+$ respectively, depending on g.

Therefore, G acts on $\mathfrak{q}^- \times \mathfrak{q}^+$: $(\xi, \eta) \mapsto (\tilde{\xi}, \hat{\eta})$. The stabilizer of (0.0) is $P^+ \cap P^- = H$, so that we obtain the embedding (defined on an open dense set)

$$\mathfrak{q}^- \times \mathfrak{q}^+ \hookrightarrow G/H. \tag{4.7}$$

We may regard (ξ, η) as coordinates in G/H.

The coset spaces $S^+ = G/P^-$, $S^- = G/P^+$, $S = K/K \cap H$ are compact manifolds, diffeomorphic to each other by the following correspondence:

$$s^0 k \leftrightarrow s^{\pm} k, \qquad k \in K, \tag{4.8}$$

where $s^+ = P^- e$, $s^- = P^+ e$, $s^0 = (K \cap H)e$ are the basic points. The natural action of G on $S^{\pm}$ yields to two actions of G on S: $s \mapsto \tilde{s}$ and $s \mapsto \hat{s}$, where $s = s^0 k$, $\tilde{s} = s^0 \tilde{k}$, $\hat{s} = s^0 \hat{k}$, and $\tilde{k}$, $\hat{k}$ are obtained from the Iwasawa and the anti-Iwasawa decompositions:

$$kg = \exp Y_1 \cdot \tilde{h}_1 \cdot \tilde{k}, \tag{4.9}$$

$$kg = \exp X_1 \cdot \hat{h}_1 \cdot \hat{k}. \tag{4.10}$$

Set

$$\tilde{s} = s \cdot g; \tag{4.11}$$

then

$$\hat{s} = s \cdot \tau(g). \tag{4.12}$$

The group G acts on $S^- \times S^+$ in a natural way. The stabilizer of the point (s^-, s^+) is H again, so that we obtain the following equivariant embedding

$$G/H \hookrightarrow S^- \times S^+. \tag{4.13}$$

The identification (4.8) gives rise to the equivariant embedding

$$G/H \hookrightarrow S \times S, \tag{4.14}$$

where G acts on $S \times S$ by $(s, t) \mapsto (\tilde{s}, \hat{t})$. The image of (4.14) is a single open dense orbit. Denote it by Ω. Thus, $S \times S$ is a compactification of G/H. For the G-orbit structure of $S \times S$, see [**14**]. Note that G/H can be represented as the tangent bundle of the manifold S.

The connection between the Gauss and the anti-Gauss decompositions gives us an operator and a function, both very important (see (4.16), (4.17) below). Let $\xi \in \mathfrak{q}^-$, $\eta \in \mathfrak{q}^+$. Decompose the anti-Gauss product $\exp \xi \cdot \exp(-\eta)$ according to the Gauss decomposition:

$$\exp \xi \cdot \exp(-\eta) = \exp Y \cdot h \cdot \exp X. \tag{4.15}$$

Denote this h by $h(\xi,\eta)$. On $\mathfrak{q}^+$ define the operator

$$K(\xi,\eta)=\operatorname{Ad}h(\xi,\eta)^{-1}|_{\mathfrak{q}^+} \tag{4.16}$$

which is the analog of the Bergman transform for Hermitian symmetric spaces. In terms of Jordan pairs it becomes:

$$K(\xi,\eta)Y=Y-2\{\eta\xi Y\}+\{\eta\{\xi Y\xi\}\eta\}.$$

Under the action of G the operator $K(\xi,\eta)$ is transformed as follows:

$$K(\tilde{\xi},\hat{\eta})=(\operatorname{Ad}\hat{h}^{-1})_{\mathfrak{q}^+}K(\xi,\eta)(\operatorname{Ad}\tilde{h})_{\mathfrak{q}^+},$$

where $\tilde{h}$ and $\hat{h}$ are taken from (4.5) and (4.6).

The function $\det K(\xi,\eta)$ is a polynomial in ξ, η. Moreover, $\det K(\xi,\eta)=N(\xi,\eta)^\kappa$, where $N(\xi,\eta)$ is an irreducible polynomial in ξ and η of degree r in ξ and η separately [**16**]. In view of (4.7), the function

$$b(\xi,\eta)=[\det K(\xi,\eta)]^{-1} \tag{4.17}$$

can be regarded as a function on G/H, becoming an the analog of the Bergman kernel. It is invariant with respect to H.

Let us introduce a metric and a measure on G/H, both G-invariant. Take a basis $E_1,\dots,E_m$ of $\mathfrak{q}^-$ in such a way that $B(E_i,\tau E_j)=\delta_{ij}$. Then $F_i=\tau E_i$ form a basis of $\mathfrak{q}^+$. Let ξ_i and η_i be the coordinates of $\xi\in\mathfrak{q}^-$ and $\eta\in\mathfrak{q}^+$ in these bases. Then the desired metric and measure are:

$$ds^2=2\sum k^{ij}(\xi,\eta)\,d\xi_i\,d\eta_j,$$

$$dx=|b(\xi,\eta)|\,d\xi\,d\eta, \tag{4.18}$$

where k^{ij} are the entries of $K(\xi,\eta)^{-1}$ and $d\xi=d\xi_1\cdots d\xi_m$, $d\eta=d\eta_1\cdots d\eta_m$. The function $F(\xi,\eta)=\ln b(\xi,\eta)$ is the potential of the metric:

$$2k^{ij}=\frac{\partial^2F}{\partial\xi_i\,\partial\eta_j}.$$

The spaces $\mathfrak{q}^-$ and $\mathfrak{q}^+$ can be embedded in S:

$$\xi\mapsto s^0\cdot\exp\xi,\qquad \eta\mapsto s^0\cdot\exp\tau(\eta),$$

where $\xi\in\mathfrak{q}^-$, $\eta\in\mathfrak{q}^+$, see (4.11), (4.12), with open dense images; thus either ξ or η can be considered as a coordinate system on S. In these coordinates let us write a K-invariant measure ds on S:

$$ds=\sqrt{b(\xi,\tau\xi)}\,d\xi \tag{4.19}$$

$$=\sqrt{b(\tau\eta,\eta)}\,d\eta. \tag{4.20}$$

We now define an important function $\|s,t\|$ on $S\times S$. For $s,t\in S$ take k_s, k_t so that $s=s^0k_s$, $t=s^0k_t$, and apply to $k_sk_t^{-1}$ the Gauss decomposition,

$$k_sk_t^{-1}=\exp Y\cdot h\cdot\exp X. \tag{4.21}$$

It turns out that $\det(\operatorname{Ad}h)_{\mathfrak{q}^+}$ depends only on s, t, but not on the choice of k_s, k_t. We set

$$\|s,t\|=|\det(\operatorname{Ad}h)_{\mathfrak{q}^+}|^{-1/\kappa}, \tag{4.22}$$

where h is taken from (4.21). Formula (4.22) defines $\|s,t\|$ on an open dense subset of $S \times S$. This function is continuous, symmetric, and invariant with respect to the diagonal action of K. It can be expanded on the whole $S \times S$, keeping all these properties. Let us write out its explicit expression. By K-invariance, it is enough to do this for points from $(s^0, s^0 \exp \mathfrak{b})$ with $\mathfrak{b} \subset \mathfrak{k} \cap \mathfrak{q}$ as above in the beginning of this section. Using the basis Y_i of $\mathfrak{b}$ we have:

$$\left\|s^0, s^0 \exp \sum u_i Y_i\right\| = |\cos u_1| \cdots |\cos u_r| . \tag{4.23}$$

In terms of this function, we can rewrite (4.18) as follows:

$$dx = dx(s,t) = \|s,t\|^{-\kappa}\, ds\, dt,$$

where $x \mapsto (s,t)$ by (4.14). The orbit Ω is characterized by the condition

$$\|s,t\| \neq 0 .$$

The following table contains the list of simple symmetric Lie algebras $\mathfrak{g}/\mathfrak{h}$ that correspond to para-Hermitian symmetric spaces G/H with G simple. Here $G_{pq}(\mathbb{F})$ denotes the Grassmann manifold of p-planes in $\mathbb{F}^n$, where $\mathbb{F} = \mathbb{R}$ or $\mathbb{H}$; S^{m-1} is the sphere in $\mathbb{R}^m$; $P_2(\mathbb{O})$ denotes the octonion projective plane; $n = p + q$. For aesthetic reasons we denote Lie algebras by capital Latin letters instead of small Gothic ones.

$\mathfrak{g}$	$\mathfrak{h}$	S
$SL(n,\mathbb{R})$	$SL(p,\mathbb{R}) + SL(q,\mathbb{R}) + \mathbb{R}$	$G_{pq}(\mathbb{R})$
$SU^*(2n)$	$SU^*(2p) + SU^*(2q) + \mathbb{R}$	$G_{pq}(\mathbb{H})$
$SU(n,n)$	$SL(n,\mathbb{C}) + \mathbb{R}$	$U(n)$
$SO^*(4n)$	$SU^*(2n)$	$U(2n)/Sp(n)$
$SO(n,n)$	$SL(n,\mathbb{R})$	$SO(n)$
$SO(p,q)$	$SO(p-1,q-1) + \mathbb{R}$	$(S^{p-1} \times S^{q-1})/\mathbb{Z}_2$
$Sp(n,\mathbb{R})$	$SL(n,\mathbb{R}) + \mathbb{R}$	$U(n)/O(n)$
$Sp(n,n)$	$SU^*(2n) + \mathbb{R}$	$Sp(n)$
$E_{6(6)}$	$SO(5,5) + \mathbb{R}$	$G_{22}(\mathbb{H})/\mathbb{Z}_2$
$E_{6(-26)}$	$SO(1,9) + \mathbb{R}$	$P_2(\mathbb{O})$
$E_{7(7)}$	$E_{6(6)} + \mathbb{R}$	$SU(8)/Sp(4) \cdot \mathbb{Z}_2$
$E_{7(-25)}$	$E_{6(-26)} + \mathbb{R}$	$S^1 \cdot E_6/F_4$

§5. Representations induced from $P^{\pm}$

For $\mu \in \mathbb{C}$, let ω_μ be the character of H:

$$\omega_\mu(h) = |\det(\operatorname{Ad} h)_{\mathfrak{q}^+}|^{-\mu/\kappa}. \tag{5.1}$$

We restrict ourselves to such characters of H, for simplicity. Extend ω_μ to the character of $P^{\pm}$, setting it equal to 1 on $Q^{\pm}$. Consider the representations of G acting on $C^\infty(S)$:

$$T_\mu^{\mp} = \operatorname{Ind}_{P^\pm}^G \omega_{\pm\mu} .$$

In more detail,

$$(T_\mu^-(g)\varphi)(s) = \omega_\mu(\tilde{h}_1)\,\varphi(\tilde{s}), \qquad (T_\mu^+(g)\varphi)(s) = \omega_\mu(\hat{h}_1^{-1})\,\varphi(\hat{s}),$$

we use (4.9), (4.10) and put $s = s^0 k$, $\tilde{s} = s^0 \tilde{k}$, $\hat{s} = s^0 \hat{k}$; note that $\omega_\mu(\tilde{h}_1)$ and $\omega_\mu(\hat{h}_1^{-1})$ are well defined because $\omega_\mu(l) = 1$ for $l \in K \cap H$. For the same μ, the representations $T_\mu^\pm$ are connected by τ: $T_\mu^- = T_\mu^+ \circ \tau$, so that if τ is an inner automorphism, then T_μ^+ and T_μ^- are equivalent.

Let (φ, ψ) be the inner product in $L^2(S)$,

$$(\varphi, \psi) = \int_S \varphi(s) \overline{\psi(s)} \, ds, \tag{5.2}$$

for ds, see (4.19), (4.20). This Hermitian form is G-invariant for the pair $(T_\mu^+, T_{-\bar{\mu}-\kappa}^+)$ and $(T_\mu^-, T_{-\bar{\mu}-\kappa}^-)$. Therefore, for $\operatorname{Re}\mu = -\kappa/2$, the representations $T_\mu^\pm$ are unitarizable, and we obtain two *continuous series* of unitary representations.

In the generic case, $T_\mu^\pm$ are irreducible: reducibility is possible only for real μ satisfying some integrality conditions. Therefore the representations of the continuous series are irreducible for $\operatorname{Im}\mu \neq 0$.

On $C^\infty(S)$ define the operator A_μ:

$$(A_\mu \varphi)(s) = \int_S \|s, t\|^{-\mu-\kappa} \varphi(t) \, dt; \tag{5.3}$$

the integral converges absolutely for $\operatorname{Re}\mu < -\kappa + 1$ and extends to the μ-plane as a meromorphic function. This operator intertwines $T_\mu^\pm$ with $T_{-\mu-\kappa}^\mp$:

$$A_\mu T_\mu^\pm = T_{-\mu-\kappa}^\mp A_\mu .$$

Moreover,

$$A_{-\mu-\kappa} A_\mu = \frac{1}{c(\mu)} E, \tag{5.4}$$

where E is the identity operator and $c(\mu)$ is a meromorphic function.

For the study of the $T_\mu^\pm$, it should be useful to consider their K-types. Let us mention something in this direction.

The restriction of $T_\mu^\pm$ on K is the quasiregular representation T of K on S:

$$(T(k)\varphi)(s) = \varphi(sk).$$

Since S is a compact symmetric space, T decomposes into the direct multiplicity-free sum of irreducible representations π_ν of K with the highest weights ν acting on subspaces $L_\nu \subset C^\infty(S)$. The L_ν is an eigenspace of the operator A_μ with the eigenvalue $a_\nu(\mu)$. For any ν we have $a_\nu(\mu) a_\nu(-\mu-\kappa) = 1/c(\mu)$, so that $c(\mu)$ is invariant under $\mu \mapsto -\mu - \kappa$.

For $a_\nu(\mu)$ we can write an integral representation. Take $\mathfrak{a}$, $\mathfrak{b}$, X_i, Y_i, $\mathfrak{g}_0$ as in the beginning of §4. For $u = (u_1, \dots, u_r)$, put $X(u) = \sum u_i X_i$, $Y(u) = \sum u_i Y_i$. The group $B = \exp \mathfrak{b}$ is a subgroup of K, hence compact. In B introduce coordinates u by $u \mapsto \exp Y(u)$. Let Σ^+ be the system of positive roots of the pair $(\mathfrak{g}_0, \mathfrak{a})$, r_α being the multiplicity of the root α. Let $\psi_\nu \in L_\nu$ be $(K \cap H)$-spherical function, $\psi_\nu(s^0) = 1$. Then, by (5.3), (4.23) and [**12,** Chapter X, §1], we have

$$a_\nu(\mu) = C \int_B \prod_{i=1}^r |\cos u_i|^{-\mu-\kappa} \prod_{\alpha \in \Sigma^+} |\sin \alpha(X(u))|^{r_\alpha} \psi_\nu(s^0 \exp Y(u)) \, du,$$

where $du = du_1 \cdots du_r$ and C is a constant not depending on ν, μ.

In the study of irreducibility, a crucial role should be played by the operator $T_\mu^+(Z_0)$ (the operator $T_\mu^-(Z_0)$ differs by the sign only). It suffices to observe how this operator acts on the spherical functions ψ_ν. Its $K\cap H$-radial part L_μ is a differential operator in the coordinates $u_1,\dots,u_r$. Here the expression is:

$$L_\mu = \frac{1}{2}\sum_{i=1}^{r} \sin 2u_i \frac{\partial}{\partial u_i} - \mu\left(\sum_{i=1}^{r}\cos 2u_i - r + \frac{2m}{\kappa}\right),$$

where $2m = \dim G/H$.

The representations $T_\mu^\pm$ (degenerate series representations) were studied for specific spaces and with different degree of completeness; see, for instance, [**1, 7, 9, 13, 18, 20, 23, 25**].

§6. Supercomplete systems and symbols

Let us give an outline for the construction of a quantization in the spirit of Berezin. The main role belongs to the kernel of the intertwining operator from §5, i.e., to the function

$$\Phi_t(s) = \Phi(s,t) = \|s,t\|^\mu;$$

this function is an analog of Berezin's supercomplete system. The function Φ_t has the reproducing property (which is formula (5.4) written in another form):

$$\varphi(s) = c(\mu)\int_{S\times S}\varphi(\tilde s)\,\frac{\Phi(s,\tilde t)}{\Phi(\tilde s,\tilde t)}\,dx(\tilde s,\tilde t).$$

Let $\widehat{A}$ be an operator acting on functions on S. Define the (covariant) *symbol* $A(s,t)$ of $\widehat{A}$ as follows:

$$A(s,t) = \frac{(\widehat{A}\Phi_t)(s)}{\Phi(s,t)}.$$

We can regard it as a function $A(x)$ on G/H, using (4.14). The operator is recovered from its symbol:

$$(\widehat{A}\varphi)(s) = c(\mu)\int_{S\times S} A(s,\tilde t)\,\frac{\Phi(s,\tilde t)}{\Phi(\tilde s,\tilde t)}\,\varphi(\tilde s)\,dx(\tilde s,\tilde t).$$

The identity operator has 1 as its symbol. The multiplication of operators gives rise to the multiplication of the symbols $\widehat{A}_1\widehat{A}_2 = \widehat{(A_1 * A_2)}$, where

$$(6.1)\qquad \begin{aligned} (A_1 * A_2)(s,t) &= \int_{S\times S} A_1(s,\tilde t)\,A_2(\tilde s,t)\,\mathcal{B}(s,t;\tilde s,\tilde t)\,dx(\tilde s,\tilde t),\\ \mathcal{B}(s,t;\tilde s,\tilde t) &= c(\mu)\,\frac{\Phi(s,\tilde t)\,\Phi(\tilde s,t)}{\Phi(s,t)\,\Phi(\tilde s,\tilde t)}. \end{aligned}$$

Let us call this kernel the *Berezin kernel.*

Thus, we have a method for constructing a family of algebras $\mathcal{A}_h$: they consist of the symbols $A(s,t) = A(x)$ of operators from some class, the multiplication $*$ in $\mathcal{A}_h$ is given by (6.1), the representations are $A \mapsto \widehat{A}$. For the Planck constant we take $h = -d/\mu$, where d depends on normalizations of measures, metrics, etc.

Define the bilinear form $F_\mu(\varphi,\psi)$ on $C^\infty(S)$ by setting

$$F_\mu(\varphi,\psi) = (A_\mu\varphi,\overline{\psi}) = \int_S \|s,t\|^{-\mu-\kappa}\varphi(s)\,\psi(t)\,ds\,dt.$$

Let $\hat{A}'$ be the operator conjugated to the operator $\widehat{A}$ with respect to this form: $F_\mu(\widehat{A}\varphi, \psi) = F_\mu(\varphi, \widehat{A}'\psi)$. Then their symbols are connected by the transposition of the arguments: $A'(s,t) = A(t,s)$. The map $A \mapsto A'$ changes the order of the factors in the product (6.1): $(A_1 * A_2)' = A_2' * A_1'$, so it is an anti-involution of any $\mathcal{A}_h$. In order that CP be in agreement with this anti-involution, we must omit the factor $i = \sqrt{-1}$ in formula (1.1).

By (4.14), the Berezin kernel can be regarded as a function $\mathcal{B}(x, \tilde{x})$ on $G/H \times G/H$. In the coordinates ξ, η, it can be written in terms of the function (4.17):

$$\mathcal{B}(x, \tilde{x}) = c(\mu) \left| \frac{b(\xi, \tilde{\eta})\, b(\tilde{\xi}, \eta)}{b(\xi, \eta)\, b(\tilde{\xi}, \tilde{\eta})} \right|^{-\mu/\kappa},$$

where $(\xi, \eta) \mapsto x$, $(\tilde{\xi}, \tilde{\eta}) \mapsto \tilde{x}$ according to (4.7). In particular (recall that $x^0 = He$ is the basic point of G/H):

$$\mathcal{B}(x, x^0) = c(\mu)\, |b(\xi, \eta)|^{\mu/\kappa}.$$

The kernel of an intertwining operator depends on a realization of a representation. If we use the coordinates ξ, η, then we must take the function $\Phi(\xi, \eta) = |b(\xi, \eta)|^{-\mu/\kappa}$, in direct analogy with the Hermitian case.

§7. Tensor products

We have already seen that the representation $T_\mu^\pm$ for $\operatorname{Re}\mu = -\kappa/2$ has an invariant Hermitian form, namely (5.2). In the general case, there is no other invariant Hermitian form for $T_\mu^\pm$. But such forms exist for tensor products. Indeed, let $\mu \in \mathbb{R}$, then the representation $T_{-\mu-\kappa}^- \otimes T_{-\mu-\kappa}^+$ acting on $C^\infty(S \times S)$ has the following invariant Hermitian form:

$$E_\mu'(\varphi_1, \varphi_2) = c(\mu) \int \varphi_1(s,t)\, \overline{\varphi_2(\tilde{s}, \tilde{t})} (\|s, \tilde{t}\| \cdot \|\tilde{s}, t\|)^\mu \, ds\, dt\, d\tilde{s}\, d\tilde{t}.$$

Let us restrict this representation to the space $\mathcal{D}(\Omega)$ (the space of C^∞-functions on Ω with compact support). Recall that $\Omega = \{(s,t) \mid \|s,t\| \neq 0\}$ is a G-orbit. An operator $\varphi \mapsto f$ on $\mathcal{D}(\Omega)$ defined by

$$f(s,t) = \varphi(s,t)\, \|s,t\|^{\mu+\kappa}$$

takes the representation $T_{-\mu-\kappa}^- \otimes T_{-\mu-\kappa}^+$ of G to the representation U of G in $\mathcal{D}(\Omega)$ by translations (see (4.11), (4.12)):

$$U(g) f(s,t) = f(s \cdot g, t \cdot \tau(g)),$$

and the Hermitian form E_μ' to the Hermitian form E_μ with the Berezin kernel (let us call E_μ the *Berezin form*):

$$E_\mu(f_1, f_2) = \int f_1(s,t)\, \overline{f_2(\tilde{s}, \tilde{t})}\, \mathcal{B}(s, t; \tilde{s}, \tilde{t})\, dx(s,t)\, dx(\tilde{s}, \tilde{t}), \tag{7.1}$$

or, in terms of G/H:

$$(U(g) f)(x) = f(xg),$$
$$E_\mu(f_1, f_2) = \int f_1(x)\, \overline{f_2(\tilde{x})}\, \mathcal{B}(x, \tilde{x})\, dx\, d\tilde{x}. \tag{7.2}$$

Thus, we obtain a densely defined G-invariant Hermitian form E_μ on $L^2(G/H)$ (with $\mathcal{D}(G/H)$ as the domain). The integral (7.1), or (7.2), converges absolutely for $\operatorname{Re}\mu > -1$ and is understood as the analytic continuation for other μ's. The representation U of G in $\mathcal{D}(G/H)$ by translations together with the Berezin form E_μ can be considered as the analog of the canonical representation from [**27**].

Let us call the transform of functions on G/H defined by the Berezin kernel the *Berezin transform*. In any case, it is well defined on $\mathcal{D}(G/H)$. Perhaps, here the Berezin form is somewhat preferable to deal with than the Berezin transform.

We can regard $\mathcal{B}(x, x^0)$ as a H-invariant distribution on G/H. Suppose that we succeed expanding $\mathcal{B}(x, x^0)$ in terms of spherical functions (distributions) on G/H. This is equivalent to writing a Plancherel formula for E_μ. Then we can write expressions of E_μ in terms of Laplace operators $\Delta_1, \dots, \Delta_r$ on every single series of representations occurring in $L^2(G/H)$. This gives us information about the behavior of E_μ on this series as $\mu \to -\infty$, and we can say whether CP is true on this series.

§8. Examples

(a) The hyperboloid of one sheet (the imaginary Lobachevskiĭ plane) G/H, where $G = SL(2, \mathbb{R})$, $H = GL(1, \mathbb{R})$. The Lie algebra $\mathfrak{g}$ consists of real 2×2 matrices with zero trace. Let $Z_0 = \operatorname{diag}\{1/2, -1/2\}$. Then H consists of diagonal matrices, $\mathfrak{h} = Z(\mathfrak{h}) = \mathbb{R}Z_0$,

$$\mathfrak{q}^- = \left\{ \begin{pmatrix} 0 & 0 \\ \xi & 0 \end{pmatrix} \right\}, \quad \mathfrak{q}^+ = \left\{ \begin{pmatrix} 0 & \eta \\ 0 & 0 \end{pmatrix} \right\}. \tag{8.1}$$

The space G/H consists of matrices

$$x = \frac{1}{2} \begin{pmatrix} x_3 & x_1 - x_2 \\ -x_1 - x_2 & -x_3 \end{pmatrix}$$

satisfying the condition $\det x = -1/4$. In $\mathbb{R}^3$, define the bilinear form $[x, y] = -x_1 y_1 + x_2 y_2 + x_3 y_3$. Then the condition $\det x = -1/4$ is $[x, x] = 1$, i.e., exactly the equation of the hyperboloid of one sheet.

The group G acts on G/H by $x \mapsto g^{-1} x g$ and on $\mathfrak{q}^-$ and $\mathfrak{q}^+$ by fractional linear transformations:

$$\xi \mapsto \frac{\alpha\xi + \gamma}{\beta\xi + \delta}, \quad \eta \mapsto \frac{\delta\eta + \beta}{\gamma\eta + \alpha}, \quad g = \begin{pmatrix} \alpha & \beta \\ \gamma & \delta \end{pmatrix} \in G.$$

The embedding (4.7) is

$$x_1 = \frac{\xi + \eta}{1 - \xi\eta}, \quad x_2 = \frac{\xi - \eta}{1 - \xi\eta}, \quad x_3 = \frac{1 + \xi\eta}{1 - \xi\eta}.$$

The manifold S is the unit circle $|u| = 1$ in $\mathbb{C}$. For this example it is convenient to take the embedding (4.14) as follows: $x \mapsto (u, v)$, $|u| = |v| = 1$, where

$$u = e^{i\alpha} = \frac{x_3 + i x_2}{x_1 + i}, \qquad v = e^{i\beta} = \frac{x_3 + i x_2}{x_1 - i}$$

(now α, β are not entries of g). The action of an element of G on u, v is a fractional linear function from $SU(1,1)$, the same for both u, v.

Let us take the measure dx and the Laplace-Beltrami operator Δ on G/H as follows

$$dx = \frac{dx_1 dx_2}{|x_3|} = \frac{2\, d\xi\, d\eta}{(1-\xi\eta)^2} = \frac{d\alpha\, d\beta}{1-\cos(\alpha-\beta)},$$
$$\Delta = (1-\xi\eta)^2 \frac{\partial^2}{\partial\xi\,\partial\eta} = -2\,(1-\cos(\alpha-\beta))\,\frac{\partial^2}{\partial\alpha\,\partial\beta}.$$

Let U be the unitary representation of G on $L^2(G/H)$ by translations (the quasiregular representation). It decomposes into irreducible unitary representations of three series: the continuous series representations T_σ, $\sigma = -1/2 + iu$, $u > 0$, with multiplicity 2, and the discrete series representations $T_n^{\pm}$, $n = 0, 1, 2, \dots$, with multiplicity 1, see, for instance, [**17**]. Correspondingly, $L^2(G/H)$ decomposes into the direct sum of four subspaces:

$$L^2(G/H) = L_c^{(0)} + L_c^{(1)} + L_d^{+} + L_d^{-}.$$

Let us write out the expressions for the Berezin transform on these subspaces in terms of Δ:

$$\mathcal{B} = \frac{\Gamma(-\mu+\sigma)\Gamma(-\mu-\sigma-1)}{\Gamma(-\mu)\Gamma(-\mu-1)}\,\frac{\sin\mu\pi + (-1)^{\varepsilon}\sin\sigma\pi}{\sin\mu\pi} \qquad \text{on } L_c^{(\varepsilon)},$$
$$\mathcal{B} = \frac{\Gamma(-\mu+\sigma)\Gamma(-\mu-\sigma-1)}{\Gamma(-\mu)\Gamma(-\mu-1)} \qquad \text{on } L_d^{\pm},$$

where the right-hand sides should be regarded as functions of $\Delta = \sigma(\sigma+1)$. As in §2 for L_d^+ and L_d^- we obtain:

$$\mathcal{B} \sim E - \frac{1}{\mu}\Delta \qquad (\mu \to -\infty).$$

Thus, CP holds for the discrete spectrum and does not hold for the continuous spectrum. As to algebras with the multiplication (6.1), we can take as such the subspaces of L_d^+ or L_d^- consisting of K-finite vectors. They have no identity element.

(b) The space G/H, where $G = SL(n,\mathbb{R})$, $H = GL(n-1,\mathbb{R})$, $n \geqslant 3$. Here it is more convenient to consider G/H as the orbit of the matrix $x^0 = \operatorname{diag}\{0,\dots,0,1\}$ under the action $x \mapsto g^{-1}xg$ of G. Then G/H consists of matrices x of rank one and trace one. This space has rank $r = 1$ and genus $\varkappa = n$. The spaces of examples (a) and (b) exhaust all para-Hermitian symmetric spaces of rank one up to the covering.

The stabilizer H of x^0 consists of matrices $\operatorname{diag}\{a, b\}$, where $a \in GL(n-1,\mathbb{R})$, $b = (\det a)^{-1}$.

The subalgebras $\mathfrak{q}^-$ and $\mathfrak{q}^+$ consist of matrices of the form (8.1), where ξ is the row $(\xi_1,\dots,\xi_{n-1})$ and η is the column $(\eta_1,\dots,\eta_{n-1})$ from $\mathbb{R}^{n-1}$. The embedding (4.7) is

$$x = \frac{1}{1-\xi\eta}\begin{pmatrix} -\xi\eta & -\eta \\ \xi & 1 \end{pmatrix}.$$

In these coordinates on G/H, the Laplace–Beltrami operator is:

$$\Delta = (1-\xi\eta)\sum(\delta_{ij} - \xi_i\eta_j)\,\frac{\partial^2}{\partial\xi_i\,\partial\eta_j}.$$

For $x, y \in \mathbb{R}^n$ we write $\langle x, y\rangle = x_1y_1 + \cdots + x_ny_n$ and $|x| = \sqrt{\langle x, x\rangle}$. The manifold S is the unit sphere S^{n-1}: $|s| = 1$ in $\mathbb{R}^n$ with the identification of points s and $-s$, i.e., S is the $(n-1)$-dimensional real projective space. We have $\|s, t\| = |\langle s, t\rangle|$. Any matrix $x \in G/H$ can be written as

$$x = \frac{t's}{\langle t, s\rangle},$$

where $t, s \in S^{n-1}$, $\langle t, s\rangle \neq 0$, the prime denotes matrix transposition. Let ds be the Euclidean measure on S^{n-1}. The following measure dx on G/H is G-invariant:

$$dx = |\langle t, s\rangle|^{-n} dt\, ds.$$

The supercomplete system is $\Phi(s, t) = |\langle s, t\rangle|^{\mu}$. In terms of G/H, the Berezin kernel is

$$\mathcal{B}(x, \tilde{x}) = c(\mu)\,|\operatorname{tr}(x\tilde{x})|^{\mu},$$

where

$$c(\mu) = \left\{2^{n+1}\pi^{n-2}\Gamma(-\mu-n+1)\Gamma(\mu+1)\left[\cos\left(\mu+\frac{n}{2}\right)\pi - \cos\frac{n\pi}{2}\right]\right\}^{-1}.$$

The quasiregular representation U of G on G/H decomposes into irreducible unitary representations of two series: the continuous series representations $T_{\sigma,\varepsilon}$, $\sigma = \frac{1}{2}(1-n) + iu$, $u > 0$, $\varepsilon = 0, 1$, and the discrete series representations $T_{\sigma(m)}$, $\sigma(m) = \frac{1}{2}(2-n) + m$, $m = 0, 1, 2, \ldots$; all with multiplicity 1, see **[20, 21, 8]**. Let us write the expressions of the Berezin form ($\mu < (1-n)/2$) in terms of Δ:

$$\mathcal{B} = \frac{\Gamma(-\mu+\sigma)\Gamma(-\mu-\sigma-n+1)}{\Gamma(-\mu)\Gamma(-\mu-n+1)}\,\frac{\cos\mu\pi + (-1)^{\varepsilon}\cos\sigma\pi}{\cos\mu\pi + 1} \qquad (n \text{ odd}),$$

$$\mathcal{B} = \frac{\Gamma(-\mu+\sigma)\Gamma(-\mu-\sigma-n+1)}{\Gamma(-\mu)\Gamma(-\mu-n+1)}\,\frac{\sin\mu\pi + (-1)^{\varepsilon}\sin\sigma\pi}{\sin\mu\pi} \qquad (n \text{ even}).$$

The right-hand sides should be regarded as functions of $\Delta = \sigma(\sigma+n-1)$. In both formulas the first fraction behaves like $1 - \mu^{-1}\Delta$ as $\mu \to -\infty$. It is just what we need for CP. In the second fractions, the term with $(-1)^{\varepsilon}$ disappears on the discrete spectrum. So we have CP on the discrete spectrum for n even.

References

1. D. Barbash, S. Sahi, and B. Speh, *Degenerate series representations for $GL(2n, \mathbb{R})$ and Fourier analysis*, Sympos. Math. **31** (1989), 45–69.
2. F. A. Berezin, *Quantization on complex bounded domains*, Dokl. Akad. Nauk SSSR **211** (1973), no. 6, 1263–1266; English transl., Soviet Math. Dokl. **14** (1973), 1209–1213.
3. ______, *Quantization*, Izv. Akad. Nauk SSSR Ser. Mat. **38** (1974), no. 5, 1116–1175; English transl., Math. USSR-Izv. **8** (1974), 1109–1165.
4. ______, *Quantization on complex symmetric spaces*, Izv. Akad. Nauk SSSR Ser. Mat. **39** (1975), no. 2, 363–402; English transl., Math. USSR-Izv. **9** (1975), 341–379.
5. ______, *A connection between the co- and the contravariant symbols of operators on classical complex symmetric spaces*, Dokl. Akad. Nauk SSSR **19** (1978), no. 1, 15–17; English transl., Soviet Math. Dokl. **19** (1978), no. 4, 786–789.
6. M. Berger, *Les espaces symétrique non compacts*, Ann. Sci. École Norm. Super. **74** (1957), 85–177.
7. J. Cailliez and J. Oberdoerffer, *Série complémentaire pour le groupe $SU(n, n, \mathbb{F})$*, C. R. Acad. Sci. Paris Sér. I Math. **297** (1983), no. 5, 279–281.

8. G. van Dijk and M. Poel, *The Plancherel formula for the pseudo-Riemannian space* $SL(n,\mathbb{R})/GL(n-1,\mathbb{R})$, Compositio Math. **58** (1986), 371–397.
9. ———, *The irreducible unitary* $GL(n-1,\mathbb{R})$*-spherical representations of* $SL(n,\mathbb{R})$, Compositio Math. **73** (1990), 1–30.
10. A. Erdelyi, W. Magnus, F. Oberhettinger, and F. Tricomi, *Higher transcendental functions*, vol. I, McGraw-Hill, New York, 1953.
11. I. M. Gelfand, M. I. Graev, and N. Ya. Vilenkin, *Integral geometry and representation theory. Generalized functions*, vol. 5, Fizmatgiz, Moscow, 1962; English transl., Academic Press, New York, 1966.
12. S. Helgason, *Differential geometry and symmetric spaces*, Academic Press, New York, 1962.
13. K. D. Johnson, *Degenerate principal series on tube domains*, Contemp. Math., vol. 138, Amer. Math. Soc., Providence, RI, 1992, pp. 175–187.
14. S. Kaneyuki, *On orbit structure of compactifications of para-Hermitian symmetric spaces*, Japan. J. Math. **13** (1987), no. 2, 333–370.
15. S. Kaneyuki and M. Kozai, *Paracomplex structures and affine symmetric spaces*, Tokyo J. Math. **8** (1985), no. 1, 81–98.
16. O. Loos, *Jordan pairs*, Lecture Notes in Math., vol. 460, Springer-Verlag, 1975.
17. V. F. Molchanov, *Harmonic analysis on the hyperboloid of one sheet*, Dokl. Akad. Nauk SSSR **171** (1966), no. 4, 794–797; English transl., Soviet Math. Dokl. **7** (1966), no. 6, 1553–1556.
18. ———, *Representations of a pseudo-orthogonal group associated with a cone*, Mat. Sb. **81** (1970), no. 3, 358–375; English transl., Math. USSR-Sb. **10** (1970), no. 3, 333–347.
19. ———, *Quantization on the imaginary Lobachevskiĭ plane*, Funktsional. Anal. i Prilozhen. **14** (1980), no. 2, 73–74; English transl., Functional Anal. Appl. **14** (1980), no. 2, 142–144.
20. ———, *The Plancherel formula for the tangent bundle of a projective space*, Dokl. Akad. Nauk SSSR **260** (1981), no. 5, 1067–1070; English transl., Soviet Math. Dokl. **24** (1981), no. 2, 393–396.
21. ———, *The Plancherel formula for the pseudo-Riemannian space* $SL(3,\mathbb{R})/GL(2,\mathbb{R})$, Sibirsk. Mat. Zh. **23** (1982), no. 5, 142–151; English transl., Siberian Math. J. **23** (1982), no. 5, 703–711.
22. G. Olafsson, *Causal symmetric spaces*, Math. Goettingensis **15** (1990).
23. B. Orsted, *Composition series for analytic continuation of holomorphic discrete series of* $SU(n,n)$, Trans. Amer. Math. Soc. **260** (1980), no. 2, 563–573.
24. I. Satake, *Algebraic structures of symmetric domains*, Iwanami Shoten Publishers and Princeton Univ. Press, 1980.
25. B. Speh, *Degenerate series representations of the universal covering group of* $SU(2,2)$, J. Funct. Anal. **33** (1979), 95–118.
26. A. Unterberger and H. Upmeier, *The Berezin transform and invariant differential operators*, Preprint.
27. A. M. Vershik, I. M. Gelfand, and M. I. Graev, *Representations of the group* $SL(2,R)$ *where* R *is a ring of functions*, Uspekhi Mat. Nauk **28** (1973), no. 5, 83–128; English transl., London Math. Soc. Lecture Note Ser. **69** (1982), 15–110.

Translated by THE AUTHOR

Amer. Math. Soc. Transl.
(2) Vol. **175**, 1996

Integral Operators with Gaussian Kernels and Symmetries of Canonical Commutation Relations

Yu. A. Neretin

To the memory of F. A. Berezin

The first chapter of the book *The method of second quantization* by F. Berezin (Moscow, 1965) contains the description of the fundamentals of analysis in boson and fermion Fock spaces and the basic techniques for dealing with the symbol of operators. The greater part of Berezin's book (Chapters II–III) is devoted to automorphisms of canonical commutation and autocommutation relations. The theory of symbols of operators was developed afterwards by many authors, including Berezin himself (see [**2–5**]). The problem of the automorphism of commutation relations formulated by Friedrichs in 1953 (see [**6**]) attracted considerable attention in the fifties and the beginning of the sixties (see [**7–10**]), and the book by Berezin [**1**], summarizing the results of that period, seemed to contain exhaustive answers to the issues under discussion. It became clear only much later that this is not quite the case. The aim of this paper is to describe both the Berezin construction and (mainly) later constructions that appeared in the course of its development (see [**11–15**]). We restrict ourselves to the boson case (for the fermion case, see [**16**]).

§1. Boson Fock space with a finite number of degrees of freedom

1.1. Fock space with n degrees of freedom. Let $n = 0, 1, 2, \dots$. A boson Fock space F_n is the space of holomorphic functions $f(z) = f(z_1, \dots, z_n)$ on $\mathbb{C}^n$ that satisfy the condition

$$\iint_{\mathbb{C}^n} |f(z)|^2 \, d\mu(z) < \infty, \tag{1.1}$$

where $d\mu$ is a Gaussian measure on $\mathbb{C}^n$ with density $(1/\pi^n)e^{-|z|^2}$ (so that the measure of the whole $\mathbb{C}^n$ is 1). In F_n the inner product is defined by

$$\langle f, g\rangle = \iint_{\mathbb{C}^n} f(z)\overline{g(z)} \, d\mu(z). \tag{1.2}$$

It turns out that the space F_n with this inner product is a Hilbert space.

1991 *Mathematics Subject Classification.* Primary 81R99.

REMARK. The space F_0 is a one-dimensional space consisting of functions on the singleton $\mathbb{C}^0$.

1.2. Standard basis. It is easy to see by a straightforward calculation that the set of functions

$$e_{\alpha_1,\dots,\alpha_n}(z_1,\dots,z_n)=\prod_{j=1}^{n}\frac{z_j^{\alpha_j}}{\sqrt{\alpha_j!}},\tag{1.3}$$

where α_j are nonnegative integers, forms an orthonormal basis in F_n.

1.3. Overfilled basis. Let $h=(h_1,\dots,h_n)\in\mathbb{C}^n$. In F_n we consider the function

$$\varphi_h(z)=\exp(z,h)=\exp\left(\sum z_j\overline{h}_j\right).\tag{1.4}$$

A simple calculation shows that

$$\langle\varphi_h(z),\varphi_r(z)\rangle=\exp(r,h).\tag{1.5}$$

Moreover, it is easy to demonstrate that for any function the following remarkable equality (the "reproducing property") holds

$$\langle f(z),\varphi_h(z)\rangle=f(h).\tag{1.6}$$

1.4. Gaussian vectors. Let P be a symmetric matrix of size $n\times n$ and $\|P\|<1$ (where $\|\cdot\|$ here and in what follows denotes the *Euclidean norm* of a matrix; for any matrix $\|A\|$ is the greatest of the eigenvalues of the matrix $\sqrt{A^*A}$). Now consider the function

$$b[P]=\exp\left\{\frac{1}{2}\sum p_{ij}z_iz_j\right\},$$

where p_{ij} are matrix elements of the matrix P. It will be more convenient to write this expression in the form

$$b[P]=\exp\left\{\frac{1}{2}zPz^t\right\},\tag{1.7}$$

where z is the row matrix $z=(z_1\ z_2\ \dots\ z_n)$, and the index t denotes matrix transposition.

A simple calculation shows that

$$\langle b[P],b[Q]\rangle=\det[(1-P\overline{Q})^{-1/2}],\tag{1.8}$$

where raising to the power $-1/2$ is defined by the equation

$$(1+x)^{-1/2}=1-\frac{1}{2}x+\frac{(-\frac{1}{2})(-\frac{1}{2}-1)}{2!}x^2+\cdots.$$

1.5. The canonical isomorphism $L^2(\mathbb{R}^n) \leftrightarrow F_n$. Consider the space $L^2(\mathbb{R}^n)$ with the inner product

$$\langle f_1, f_2\rangle = \frac{1}{\pi^{n/2}} \int_{\mathbb{R}^n} f(x)\overline{g(x)}\, dx .$$

In $L^2(\mathbb{R}^n)$ consider the set of functions

$$\widetilde{\varphi}_h(x) = \exp\Big\{ -\frac{1}{2}\Big(\sum \overline{h}_j^2 - 2\sqrt{2}\sum x_j\overline{h}_j + \sum x_j^2\Big)\Big\}, \tag{1.9}$$

where $h = (h_1, \dots, h_n) \in \mathbb{C}^n$. It is easy to verify that

$$\langle \widetilde{\varphi}_h, \widetilde{\varphi}_r\rangle = \exp(r, h) . \tag{1.10}$$

It is rather evident that linear combinations of the functions φ_h are dense in $L^2(\mathbb{R}^n)$. Therefore (see equations (1.5) and (1.10)), the correspondence

$$\{\widetilde{\varphi}_h \in L^2(\mathbb{R}^n)\} \longleftrightarrow \{\varphi_h \in F_n\}$$

extends to the linear unitary operator $\mathcal{BB}\colon L^2(\mathbb{R}^n) \to F_n$.

It is easy to write out an explicit expression for this operator

$$\begin{aligned} \mathcal{BB}f(z) &= \langle f(x), \widetilde{\varphi}_h(x)\rangle_{L^2(\mathbb{R})} \\ &= \int_{\mathbb{R}^n} f(x) \exp\Big\{ -\frac{1}{2}\Big(\sum z_j^2 - 2\sqrt{2}\sum z_j x_j + \sum x_j^2\Big)\Big\}\, dx . \end{aligned} \tag{1.11}$$

1.6. Correspondence of the functions. In $L^2(\mathbb{R}^n)$ consider the set of Hermitian functions

$$H_n(x) = \sigma_n(x)e^{-x^2/2}.$$

Recall that it is determined by the following conditions: σ_k is a polynomial of degree kth and

$$\langle H_k(x), H_l(x)\rangle_{L^2(\mathbb{R})} = \delta_{k,l} .$$

Then the set of functions

$$\tilde{e}_{\alpha_1,\dots,\alpha_n}(x) = H_{\alpha_1}(x_1)\cdot \dots \cdot H_{\alpha_n}(x_n)$$

forms an orthonormal basis in $L^2(\mathbb{R}^n)$. A simple calculation shows that

$$\mathcal{BB}\tilde{e}_{\alpha_1,\dots,\alpha_n} = e_{\alpha_1,\dots,\alpha_n} .$$

Now consider a symmetric matrix of size $n \times n$ with a strictly positive definite real part. Consider the vector

$$\tilde{b}[R] = \exp\Big\{ -\frac{1}{2}xRx^t\Big\} \in L^2(\mathbb{R}^n).$$

One can easily see that

$$\langle \tilde{b}[R_1], \tilde{b}[R_2]\rangle = \det(R_1 + R_2)^{-1/2},$$

where $Q^{-1/2}$ is the analytic continuation of the function $A^{-1/2}$ defined on positive definite matrices. The existence of such an analytic continuation is seen from the following formula. Let $Q = A + iB$, where $A = A^t$ is positive definite and real, and $B = B^t$ is real. Then

$$Q = A^{1/2}(E + iA^{1/2}BA^{1/2})A^{1/2},$$

and the matrix $E + iA^{-1/2}BA^{-1/2}$ has eigenvalues of the form $1 + is$, where $s \in \mathbb{R}$.

A straightforward calculation shows that

$$\mathcal{BB}\cdot\tilde{b}[R]=\det(1+R)^{-1/2}b[(R-E)(E+R)^{-1}].$$

1.7. Symbols of operators (kernels). Let A be a bounded operator $F_n\to F_m$. With it we associate the function

$$K_A(u,\overline{v})=\langle A\varphi_v,\varphi_u\rangle_{F_m},\tag{1.12}$$

which is called the *kernel* or *symbol* of the operator A. It is clear that the function K_A is holomorphic in the variables $u_1,\dots,u_m$ and antiholomorphic in the variables $v_1,\dots,v_n$.

Consider the matrix elements

$$s^{\alpha_1\alpha_2\dots}_{\beta_1\beta_2\dots}=\langle Ae_{\beta_1\beta_2\dots},e_{\alpha_1\alpha_2\dots}\rangle$$

of the operator A in the standard basis $e^{\cdots}_{\cdots}$. It is easily verified that

$$K_A(u,\overline{v})=\sum s^{\alpha_1\dots\alpha_m}_{\beta_1\dots\beta_n}e_{\alpha_1\dots\alpha_m}(u)\,e_{\beta_1\dots\beta_n}(\overline{v}).\tag{1.13}$$

It can be seen from the latter expression that the operator A can be reconstructed from its symbol A: the function K_A turns out to be the generating function for matrix elements of the operator A (for this reason in the book by Berezin the symbols K_A are called *generating functionals*).

The operator A can be reconstructed from its symbol K_A by using a very simple and elegant formula, namely

$$Af(z)=\int K(z,\overline{u})\,f(u)\,d\mu(u).\tag{1.14}$$

Thus all bounded operators $F_n\to F_m$ are integral operators in the strict sense of the word.

Note that if the kernel of the operator A is defined by the expression $K(z,\overline{u})$, then the kernel of the operator A^* is $K(u,\overline{z})$.

1.8. Correspondence of the kernels. We have seen that any operator in F_n has a kernel. In $L^2(\mathbb{R}^n)$ this is certainly not the case. Consider the integral operator

$$Af(x)=\int_{\mathbb{R}^n}\widetilde{K}(x,y)f(y)\,dy$$

in $L^2(\mathbb{R}^n)$. Then the kernel K of the corresponding operator in F_n is obviously given by the expression

$$\begin{aligned}K(z,\overline{u})&=\langle A\varphi_u,\varphi_z\rangle\\&=\frac{1}{\pi^n}\iint\widetilde{K}(x,y)\exp\Big\{-\frac{1}{2}\Big(\sum(z_j^2+\overline{u}_j^2)\\&\qquad-2\sqrt{2}\sum(z_j+\overline{u}_j)x+2\sum x_j^2\Big)\Big\}\,dx\,dy.\end{aligned}$$

§2. Symmetries of commutation relations

2.1. Creation-annihilation operators. Consider the linear space $V_{2n} = V_{2n}^+ \oplus V_{2n}^- = \mathbb{C}^n \oplus \mathbb{C}^n$. We write its elements in the form

$$v = (v^+; v^-) = (v_1^+, \dots, v_n^+; v_1^-, \dots, v_n^-).$$

With each $v \in V_{2n}$ we associate the operator $\hat{a}(v)$ in F_n defined by the equation

$$\hat{a}(v) f(z) = \left(\sum v_j^+ z_j + v_j^- \frac{\partial}{\partial z_j} \right) f(z). \tag{2.1}$$

These operators are usually called *creation-annihilation operators*. It is easy to see that

$$[\hat{a}(v), \hat{a}(w)] = \sum (v_j^- w_j^+ - v_j^+ w_j^-) E, \tag{2.2}$$

where $[\,\cdot\,,\cdot\,]$ denotes the commutator of operators. It can also be easily seen that

$$\hat{a}(v_+; v_-)^* = \hat{a}(\overline{v}_-; \overline{v}_+). \tag{2.3}$$

We regard equations (2.3)–(2.4) as formal, not being interested in the domain of definition of the operators.

The equations (2.2) can be represented in a simpler way. We introduce in F_n the *creation operators*

$$\hat{a}_j^+ f(z) = z_j f(z)$$

and the *annihilation operators*

$$\hat{a}_j^- f(z) = \frac{\partial}{\partial z_j} f(z).$$

Then equations (2.2)–(2.3) can be written in the form

$$[\hat{a}_k^+, \hat{a}_m^+] = [\hat{a}_k^-, \hat{a}_m^-] = 0, \qquad [\hat{a}_k^+, \hat{a}_m^-] = -\delta_{km} E, \tag{2.4}$$

$$(\hat{a}_k^+)^* = \hat{a}_k^-. \tag{2.5}$$

The set of formal equalities (2.4)–(2.5) is called the *canonical commutation relations*.

It would be more pleasant for us to speak about creation-annihilation operators $\hat{a}(v)$. So now on the space V_{2n} we introduce simple geometrical structures, which will be important in what follows. First, we define a skew-symmetric bilinear form Λ on V_{2n} by the formula

$$\Lambda(v, w) = \sum (v_j^- w_j^+ - v_j^+ w_j^-). \tag{2.6}$$

Secondly, we fix the antilinear operator C in V_{2n}:

$$C(v_+; v_-) = (\overline{v}_-; \overline{v}_+).$$

It can be easily seen that $C^2 = E$ and the set of fixed vectors of the operator C is a real subspace $V_{2n}^{\mathbb{R}}$ in V_{2n} consisting of vectors of the form $v = (h; \overline{h})$. We can regard the space V_{2n} itself as a complexification of the linear space $V_{2n}^{\mathbb{R}}$.

Finally, in V_{2n} define the indefinite Hermitian form

$$M(v, w) = \frac{1}{i} \Lambda(v, Cw) = \frac{1}{i} \sum (v_j^+ \overline{w}_j^+ - v_j^- \overline{w}_j^-). \tag{2.7}$$

Then equations (2.2)–(2.3) can be written as

$$[\hat{a}(v), \hat{a}(w)] = \Lambda(v, w)E, \qquad \hat{a}(v)^* = \hat{a}(Cv).$$

2.2. Correspondence of the creation-annihilation operators. Linear combinations of the operators

$$f \mapsto x_j f, \qquad f \mapsto \frac{\partial}{\partial x_j} f$$

in $L^2(\mathbb{R}^n)$ correspond to the creation-annihilation operators in F_n. More precisely, the correspondence is given by the relations

$$\frac{1}{\sqrt{2}}\left(x_j - \frac{\partial}{\partial x_j}\right) \leftrightarrow z_j, \qquad \frac{1}{\sqrt{2}}\left(x_j + \frac{\partial}{\partial x_j}\right) \leftrightarrow \frac{\partial}{\partial z_j}. \tag{2.8}$$

Now consider the space $V_{2n} = \widetilde{V}_{2n}^+ \oplus \widetilde{V}_{2n}^- = \mathbb{C}^n \oplus \mathbb{C}^n$. For any $v \in \widetilde{V}_{2n}$ define the operator $\tilde{a}(v)$ in $L^2(\mathbb{R}^n)$ by the formula

$$\tilde{a}(v)f(x) = \left(\sum v_k^+ x_k + i \sum v_k^- \frac{\partial}{\partial x_k}\right) f(x).$$

One can easily see that

$$\begin{aligned} &[\tilde{a}(v), \tilde{a}(w)] = i \sum (v_k^- w_k^+ - v_k^+ w_k^-)E, \\ &\tilde{a}(v^+; v^-)^* = \tilde{a}(\overline{v}^+; \overline{v}^-). \end{aligned} \tag{2.9}$$

Here a simpler language is of course possible. Further, we introduce the coordinate operators

$$Q_j f(x) = x_j f(x)$$

and the "momentum operators"

$$P_j f(x) = i \frac{\partial}{\partial x_j} f(x).$$

Then

$$\begin{gathered} [P_k, P_m] = [Q_k, Q_m] = 0, \qquad [Q_k, P_m] = -i\delta_{k,m} E, \\ P_k^* = P_k, \qquad Q_k^* = Q_k. \end{gathered}$$

In this case, however, we also prefer to deal with the operators $\tilde{a}(v)$ and now introduce additional geometrical structures on $\widetilde{V}_{2n}$.

We define the skew-symmetric bilinear form $\tilde{\Lambda}$ on $\widetilde{V}_{2n}$ by the equation

$$\tilde{\Lambda}(v, w) = i \sum (v_j^- w_j^+ - v_j^+ w_j^-). \tag{2.10}$$

We define the complex conjugation operator $\widetilde{C}$ in the obvious way: $\widetilde{C}v = \overline{v}$. Finally, we introduce the Hermitian indefinite form

$$\widetilde{M}(v, w) = -\Lambda(v, Cw) = -i \sum (v_j^- \overline{w}_j^+ - v_j^+ \overline{w}_j^-). \tag{2.11}$$

Then equations (2.9) can be written in the form

$$[\tilde{a}(v), \tilde{a}(w)] = \Lambda(v, w)E, \qquad \tilde{a}(v)^* = \tilde{a}(Cv).$$

Observe now that the correspondence relation (2.8) identifies the linear spaces $\widetilde{V}_{2n}$ and V_{2n}. To be more precise, the canonical isomorphism $\widetilde{V}_{2n} \to V_{2n}$ is a linear operator with matrix

$$K = \frac{1}{\sqrt{2}} \begin{pmatrix} E & iE \\ iE & E \end{pmatrix}. \tag{2.12}$$

It is easy to verify that the isomorphism K carries the form $\tilde{\Lambda}$ into the form Λ, the form $\widetilde{M}$ into the form M, and the complex conjugation operator $\widetilde{C}$ into the operator C.

2.3. Two realizations of the symplectic group. Consider the group $\mathrm{Sp}(2n, \mathbb{R})$ consisting of real matrices

$$\begin{pmatrix} A & B \\ C & D \end{pmatrix}$$

which preserve the skew-symmetric bilinear form with matrix $\begin{pmatrix} 0 & E \\ -E & 0 \end{pmatrix}$.

Thus, real symplectic matrices $T = \begin{pmatrix} A & B \\ C & D \end{pmatrix}$ act in the complex space $\widetilde{V}_{2n}$ preserving the form $\tilde{\Lambda}$. The operator T obviously commutes with the complex conjugation operator $\widetilde{C}$, and, therefore, T also preserves the Hermitian form $\widetilde{M}$. It can be easily deduced that the group $\mathrm{Sp}(2n, \mathbb{R})$ consists of exactly all the operators in $\widetilde{V}_{2n}$ that preserve the two forms $\tilde{\Lambda}$ and $\widetilde{M}$.

Since $\widetilde{V}_{2n}$ and V_{2n} can be canonically identified, the group $\mathrm{Sp}(2n, \mathbb{R})$ can be defined as a group of matrices in V_{2n}. These matrices have a bloc structure of the type

$$\begin{pmatrix} \Phi & \Psi \\ \overline{\Psi} & \overline{\Phi} \end{pmatrix} \tag{2.13}$$

and preserve both forms Λ and M.

REMARK. If a matrix of type (2.13) preserves one of the forms Λ or M, then it preserves the other one.

2.4. Symmetry theorem.

THEOREM 2.1 (complex version). (a) *For any symplectic matrix*

$$Q = \begin{pmatrix} \Phi & \Psi \\ \overline{\Psi} & \overline{\Phi} \end{pmatrix} : V_{2n} \to V_{2n}$$

there exists a unique (up to a multiplicative factor) bounded operator $W(Q)\colon F_n \to F_n$ *that satisfies the condition*

$$\hat{a}(Qv) = W(Q)\,\hat{a}(v)\,W(Q)^{-1} \tag{2.14}$$

for any $v \in V_{2n}$.

(b) *The operator* $W(Q)$ *is unitary up to a multiplicative factor.*

(c) *For all symplectic matrices* P *and* Q *of the type* (2.13), *the following relation holds*:

$$W(PQ) = \lambda(P, Q)\,W(P)\,W(Q), \tag{2.15}$$

where $\lambda(P, Q) \in \mathbb{C}$.

THEOREM 2.1′ (real version). (a) *For any matrix* $S = \begin{pmatrix} A & B \\ C & D \end{pmatrix} \in \mathrm{Sp}(2n, \mathbb{R})$ *there exists a unique (up to a multiplicative factor) bounded operator* $\widetilde{W}(S)$ *in* $L^2(\mathbb{R}^n)$ *which satisfies the equation*

$$\tilde{a}(Sv) = \widetilde{W}(S)\,\tilde{a}(v)\,\widetilde{W}(S)^{-1} \tag{2.16}$$

for any $v \in \widetilde{V}_{2n}$.

(b) *The operator* $\widetilde{W}(S)$ *is unitary up to multiplication by a constant.*

(c) *For all* $S_1, S_2 \in \mathrm{Sp}(2n, \mathbb{R})$ *the following equation is satisfied:*

$$\widetilde{W}(S_1 S_2) = \lambda(S_1 S_2)\,\widetilde{W}(S_1)\widetilde{W}(S_2), \tag{2.17}$$

where $\lambda(S_1, S_2) \in \mathbb{C}$.

It is clear that both these theorems are equivalent.

2.5. Proof of the symmetry theorem. We now prove the real version of the theorem.

1. Any bounded operator commuting with all $\hat{a}(v)$ is a scalar multiple of the identity operator scalar. This is more or less obvious.

2. Assume that there are two solutions $\widetilde{W}_1(S)$ and $\widetilde{W}_2(S)$ of equation (2.16). Then

$$\widetilde{W}_1(S)\,\tilde{a}(v)\,W_1(S)^{-1} = \tilde{a}(Sv) = \widetilde{W}_2(S)\,\tilde{a}(v)\,W_2(S)^{-1}.$$

It follows that $W_1(S)^{-1}W_2(S)$ commutes with all $\hat{a}(v)$. Therefore, $\widetilde{W}_1(S)$ and $\widetilde{W}_2(S)$ are proportional to each other.

3. Let the solutions $W(P)$ and $W(Q)$ of equation (2.16) exist for the operators P and Q. Then

$$\tilde{a}(PQv) = \widetilde{W}(P)\,\tilde{a}(Qv)\,\widetilde{W}(P)^{-1} = \widetilde{W}(P)\,\widetilde{W}(Q)\,\tilde{a}(v)\,\widetilde{W}(Q)^{-1}\widetilde{W}(P)^{-1},$$

i.e., the operator $\widetilde{W}(P)\,\widetilde{W}(Q)$ satisfies equation (2.16) that must be satisfied by the operator $\widetilde{W}(PQ)$. Taking into account the uniqueness of the solution of equation (2.16), we obtain equation (2.17).

4. By virtue of the above statement, it is sufficient to prove the existence of the operators $\widetilde{W}(P)$ for some set of generators of the group $\mathrm{Sp}(2n, \mathbb{R})$.

5. Matrices of the type

$$\begin{pmatrix} 0 & E \\ -E & 0 \end{pmatrix}, \quad \begin{pmatrix} E & T \\ 0 & E \end{pmatrix}, \quad \begin{pmatrix} A & 0 \\ 0 & A^{t-1} \end{pmatrix},$$

where $T = T^t$, generate the whole group $\mathrm{Sp}(2n, \mathbb{R})$.

6. The operator $\widetilde{W}\begin{pmatrix} 0 & E \\ -E & 0 \end{pmatrix}$ is the Fourier transform

$$\widetilde{W}\begin{pmatrix} 0 & E \\ -E & 0 \end{pmatrix} f(x) = \frac{1}{(2\pi)^{n/2}} \int_{\mathbb{R}^n} f(y)\, e^{-i\sum x_k y_k}\, dy. \tag{2.18}$$

Further,

$$\widetilde{W}\begin{pmatrix} E & T \\ 0 & E \end{pmatrix} f(x) = \exp\Big\{ -\frac{i}{2}\, xTx^t \Big\} f(x), \tag{2.19}$$

$$\widetilde{W}\begin{pmatrix} A & \\ & A^{t-1} \end{pmatrix} f(x) = f(xA). \tag{2.20}$$

7. The operators (2.18) and (2.19) are unitary and the operator (2.20) is unitary up to a multiplication by the constant $\det(A)^{1/2}$.

This ends the proof of the theorem.

REMARK. Note the useful formula

$$\widetilde{W}\begin{pmatrix} E & 0 \\ H & E \end{pmatrix} f(x) = \exp\Big\{ i \sum h_{kl} \frac{\partial^2}{\partial x_k \partial x_l} \Big\} f(x). \tag{2.21}$$

REMARK. Almost any symplectic matrix can be represented in the form

$$\begin{pmatrix} A & B \\ C & D \end{pmatrix} = \begin{pmatrix} E & 0 \\ H & E \end{pmatrix} \begin{pmatrix} R & 0 \\ 0 & R^{t-1} \end{pmatrix} \begin{pmatrix} E & T \\ 0 & E \end{pmatrix}.$$

For this reason equations (2.19)–(2.21) provide a more or less explicit expression for the operator $\widetilde{W}(Q)$ for almost all $Q \in \mathrm{Sp}(2n, \mathbb{R})$.

2.6. The Berezin formula. In fact, the simplest and most natural way to prove Theorem 2.1 is to produce a formula for the operator $W(Q)$.

THEOREM 2.2. *Let* $Q = \begin{pmatrix} \Phi & \Psi \\ \overline{\Psi} & \overline{\Phi} \end{pmatrix}$ *be the symplectic operator* $V_{2n} \to V_{2n}$. *Then*

$$W(Q) f(z) = \int \exp\Big\{ \frac{1}{2} (z \overline{u}) \begin{pmatrix} \overline{\Psi}\Phi^{-1} & \Phi^{t-1} \\ \Phi^{-1} & -\Phi^{-1}\Psi \end{pmatrix} \begin{pmatrix} z^t \\ \overline{u}^t \end{pmatrix} \Big\} f(u)\, d\mu(u). \tag{2.22}$$

PROOF. This can be proved by a straightforward verification.

PROPOSITION 2.1. *The operators*

$$W^0 \begin{pmatrix} \Phi & \Psi \\ \overline{\Psi} & \overline{\Phi} \end{pmatrix} = \pm (\det \Phi)^{-1/2} W \begin{pmatrix} \Phi & \Psi \\ \overline{\Psi} & \overline{\Phi} \end{pmatrix} \tag{2.23}$$

are unitary and satisfy the equation

$$W^0(P) W^0(Q) = \pm W^0(PQ).$$

2.7. The real version of the Berezin formula. Let $\begin{pmatrix} A & B \\ C & D \end{pmatrix}$ be a real symplectic matrix, the matrix C being invertible. Then

$$W\begin{pmatrix} A & B \\ C & D \end{pmatrix} f(x) = \int \exp\Big\{ -i\,(xy) \begin{pmatrix} AC^{-1} & -C^{t-1} \\ -C^{-1} & C^{-1}D \end{pmatrix} \begin{pmatrix} x^t \\ y^t \end{pmatrix} \Big\} f(y)\, dy.$$

For an accurate definition of this operator one has to do some work. Thus one can define this transformation on $L^1 \cap L^2$ and then continuously extend it onto L^2.

2.8. Gaussian densities and operators $\hat{a}(v)$. Let T be a symmetric matrix of size $n \times n$ and $\|T\| < 1$. Consider the set $L = L_T$ of all $v \in V_{2n}$ such that

$$\hat{a}(v)\, b[T] = 0 \tag{2.24}$$

(cf. equation (2.14)). Equation (2.24) can be written in the form $v_+ = Tv_-$.

The matrix T is symmetric, and, therefore, L is a Lagrangian subspace in V_{2n} (see 4.2 below). The condition $\|T\| < 1$ implies that $\|v_+\|^2 < \|v_-\|^2$, and this means that the form M is positive definite on L_T.

Conversely, with each Lagrangian subspace L in V_{2n} on which the form M is negative definite we associate a Gaussian vector $b[T]$ such that $L = L_T$.

The similarity between the operators $W(Q)$ and the Gaussian vectors $b[T]$ cannot but look suspicious (see (1.7) and (2.22), (2.14) and (2.24)). If one carries out the necessary calculations explicitly, the suspicion of similarity is strengthened. Below in §4, we shall understand that the vectors $W(Q)$ and the operators $b[T]$ are actually objects of equal status.

§3. Integral operators with Gaussian kernels and strange multiplication of matrices

3.1. Operators with Gaussian kernels. Consider a symmetric bloc matrix $S = \begin{pmatrix} K & L \\ L^t & M \end{pmatrix}$ of size $(m+n) \times (m+n)$. With each such matrix we associate the operator

$$B[S] \colon F_n \to F_m$$

defined by the expression

$$B\begin{bmatrix} K & L \\ L^t & M \end{bmatrix} f(z) = \int \exp\left\{\frac{1}{2}(z\overline{u})\begin{pmatrix} K & L \\ L^t & M \end{pmatrix}\begin{pmatrix} z^t \\ \overline{u}^t \end{pmatrix}\right\} f(u)\, d\mu(u). \tag{3.1}$$

In this formula z and $\overline{u}$ are row matrices,

$$z = (z_1\; z_2\; \dots\; z_m), \qquad \overline{u} = (\overline{u}_1\; \overline{u}_2\; \dots\; \overline{u}_n).$$

THEOREM 3.1. *The operator $B[S]$ is bounded if and only if the matrix*

$$S = \begin{pmatrix} K & L \\ L^t & M \end{pmatrix}$$

satisfies the conditions

1^+. $\|S\| \leqslant 1$;

2^+. $\|K\| < 1$ *and* $\|M\| < 1$.

EXAMPLE. The operator $B\begin{bmatrix} 0 & L \\ L^t & 0 \end{bmatrix}$ is the change of variable operator $f(z) \mapsto f(zL)$.

3.2. Comments to Theorem 2.1. We now prove those statements of Theorem 2.1 that are trivial.

1. For any bounded operator A its kernel $K(u, \overline{v})$ satisfies the condition

$$K(u, \overline{v}) = \langle A\varphi_v, \varphi_u \rangle \leqslant \|A\| \cdot \|\varphi_u\| \cdot \|\varphi_v\| = \|A\| e^{\|u\|^2/2} e^{\|v\|^2/2}.$$

Hence it immediately follows that $\|S\| \leqslant 1$.

2. $B[S] \cdot 1 = b[K]$, $B[S]^* \cdot 1 = b[\overline{M}]$. Hence $\|K\| < 1$ and $\|M\| < 1$.

3. If $\|S\| < 1$, the kernel of the operator $B[S]$ is square-integrable, i.e., $B[S]$ is a Hilbert–Schmidt operator; in particular, it is bounded.

3.3. Multiplication of the operators $B[S]$. Let

$$S_1 = \begin{pmatrix} K & L \\ L^t & M \end{pmatrix} \quad \text{and} \quad S_2 = \begin{pmatrix} P & Q \\ R & T \end{pmatrix}$$

be symmetric bloc matrices of size $(m+n) \times (m+n)$ and $(n+k) \times (n+k)$, respectively, satisfying conditions 1^+ and 2^+ of Theorem 3.1. Then the bounded operators

$$B[S_2]\colon F_k \to F_n, \qquad B[S_1]\colon F_n \to F_m$$

are defined.

THEOREM 3.2. *$B[S_1]B[S_2] = \lambda(S_1, S_2) B[S_1 * S_2]$, where*

$$\lambda(S_1, S_2) = \det[(1 - MP)^{-1/2}], \tag{3.2}$$

$$S_1 * S_2 = \begin{pmatrix} K + LP(1-MP)^{-1}L^t & L(1-PM)^{-1}Q \\ Q^t(1-MP)^{-1}L^t & R + Q^t(1-MP)^{-1}MQ \end{pmatrix}. \tag{3.3}$$

REMARK. It is useful to note the following formula:

$$P(1-MP)^{-1} = (1-PM)^{-1}P.$$

PROOF. This can be verified by a direct calculation.

3.4. The category $\mathcal{B}$. We now introduce a certain category $\mathcal{B}$. Its objects are the numbers $n = 1, 2, \ldots$, and the morphisms $n \to m$ are the matrices

$$S = \begin{pmatrix} K & L \\ L^t & M \end{pmatrix}$$

of size $(m+n) \times (m+n)$ that satisfy the following conditions:

1^+. S is symmetric;

2^+. $\|S\| \leqslant 1$;

3^+. $\|K\| < 1$, $\|M\| < 1$.

Let the morphisms

$$S_1 = \begin{pmatrix} K & L \\ L^t & M \end{pmatrix} : m \to n, \qquad S_2 = \begin{pmatrix} P & Q \\ R & T \end{pmatrix} : k \to n$$

be given. Then their product

$$S_1 * S_2 = \begin{pmatrix} K & L \\ L^t & M \end{pmatrix} * \begin{pmatrix} P & Q \\ R & T \end{pmatrix}$$

is, by definition, given by equation (3.3):

$$S_1 * S_2 = \begin{pmatrix} K + LP(1-MP)^{-1}L^t & L(1-PM)^{-1}Q \\ Q^t(1-MP)^{-1}L^t & R + Q^t(1-MP)^{-1}MQ \end{pmatrix}.$$

The definition of the category $\mathcal{B}$ implicitly contains the following theorem.

THEOREM 3.3. (a) *If* S_1 *and* S_2 *satisfy conditions* 1^+–3^+, *then* $S_1 * S_2$ *also satisfies these conditions.*

(b) $S_1 * (S_2 * S_3) = (S_1 * S_2) * S_3$.

Statement (b) can be verified by a direct calculation, although it turns out to be somewhat more complicated than one would expect. On the other hand, the theorem is an obvious logical consequence of the difficult Theorem 3.1 and the easy Theorem 3.2. Note that Theorem 3.3 can also be derived from the easy part of Theorem 3.1 (see 3.2).

3.5. Category notation. We shall use the usual category notation: $\mathrm{Ob}(\mathcal{K})$ for the set of objects of the category $\mathcal{K}$, $\mathrm{Mor}_{\mathcal{K}}(V, W)$ for the set of morphisms $V \to W$, $\mathrm{End}_{\mathcal{K}}(V) = \mathrm{Mor}_{\mathcal{K}}(V, V)$ for the semigroup of endomorphisms of the object V, and $\mathrm{Aut}_{\mathcal{K}}(V)$ for the group of automorphisms of the object V.

3.6. The semigroup $\Gamma_n = \mathrm{End}_{\mathcal{B}}(n)$. Let us first discuss the semigroup Γ_n of endomorphisms of the object n of the category $\mathcal{B}$ (or, which is almost the same, the semigroup of operators in F_n with Gaussian kernels).

Consider the subset Δ_n consisting of unitary matrices S in Γ_n.

PROPOSITION 3.1. (a) *If* $S_1, S_2 \in \Delta_n$, *then* $S_1 * S_2 \in \Delta_n$.

(b) *The set* Δ_n *is a group with respect to the operation* $*$. *Moreover,* Δ_n *is isomorphic to the group* $\mathrm{Sp}(2n, \mathbb{R})$, *and the isomorphism* $\tau\colon \mathrm{Sp}(2n, \mathbb{R}) \to \Delta_n$ *is given by the formula*

$$\tau \begin{pmatrix} \Phi & \Psi \\ \overline{\Psi} & \overline{\Phi} \end{pmatrix} = \begin{pmatrix} \overline{\Psi\Phi}^{-1} & \Phi^{t-1} \\ \Phi^{-1} & -\Phi^{-1}\Psi \end{pmatrix}.$$

REMARK. Obviously,

$$W \begin{pmatrix} \Phi & \Psi \\ \overline{\Psi} & \overline{\Phi} \end{pmatrix} = \sigma B \left[\tau \begin{pmatrix} \Phi & \Psi \\ \overline{\Psi} & \overline{\Phi} \end{pmatrix} \right],$$

where $\sigma \in \mathbb{C}$ (see equation (2.22)).

Further, in Γ_n consider the subset Π_n consisting of matrices $S = \begin{pmatrix} K & L \\ L^t & M \end{pmatrix}$ such that the matrix L is invertible.

PROPOSITION 3.2. (a) *If* $S_1, S_2 \in \Pi_n$, *then* $S_1 * S_2 \in \Pi_n$.

(b) Π_n *is isomorphic to a subsemigroup* Σ_n *in the group* $\mathrm{Sp}(2n, \mathbb{C})$.

We now explicitly describe the semigroup $\Sigma_n \subset \mathrm{Sp}(2n, \mathbb{C})$. This semigroup consists of operators Q in the space V_{2n} (see 2.1) that preserve the skew-symmetric form Λ and "contract" the Hermitian form M. The expression "an operator Q contracts a form M" means that

$$M(Qv, Qv) \leqslant M(v, v)$$

for all $v \in V_n$.

The isomorphism $\Sigma_n \to \Pi_n$ is defined by

$$\begin{pmatrix} A & B \\ C & D \end{pmatrix} \mapsto \begin{pmatrix} CA^{-1} & A^{t-1} \\ A^{-1} & -A^{-1}B \end{pmatrix}.$$

Thus, we have the following pattern:

$$\Gamma_n \supset \{\Pi_n \subset \mathrm{Sp}(2n,\mathbb{C})\} \supset \Delta_n = \mathrm{Sp}(2n,\mathbb{R}).$$

The question which naturally arises is whether Γ_n is embedded in $\mathrm{Sp}(2n,\mathbb{C})$. The answer to this question proves to be negative.

PROPOSITION 3.3. *The semigroup Γ_n is not embedded in any other group.*

PROOF. $S_1 * O = S_2 * O$ does not imply $S_1 = S_2$.

Thus, we have partially reduced the multiplication $*$ to the ordinary matrix multiplication but we have not succeeded in doing this completely.

3.7. Real version of the same question. Consider a symmetric matrix $S = \begin{pmatrix} K & L \\ L^t & M \end{pmatrix}$ of size $(m+n)\times(m+n)$. Take the integral operator $\widetilde{B}[S]\colon L^2(\mathbb{R}^n) \to L^2(\mathbb{R}^m)$:

$$\widetilde{B}[S]f(x) = \frac{1}{\pi^{n/2}}\int_{\mathbb{R}^n} \exp\left\{-\frac{1}{2}(xy)\begin{pmatrix} K & L \\ L^t & M \end{pmatrix}\begin{pmatrix} x^t \\ y^t \end{pmatrix}\right\} f(y)\,dy. \tag{3.4}$$

PROPOSITION 3.4. *The operator $\widetilde{B}[S]$ is a Hilbert–Schmidt operator if and only if the real part of the matrix is strictly positive definite.*

PROOF. Notice that the kernel is square-integrable precisely when the real part of the matrix is strictly positive definite.

REMARK. Of course, in this case the operator $\widetilde{B}[S]$ is smoothing, i.e., it carries $L^2(\mathbb{R}^n)$ into $C^\infty(\mathbb{R}^m)$.

Let $S_1 = \begin{pmatrix} K & L \\ L^t & M \end{pmatrix}$ and $S_2 = \begin{pmatrix} P & Q \\ Q^t & R \end{pmatrix}$ be matrices of size $(m+n)\times(m+n)$ and $(n+k)\times(n+k)$, respectively, with positive definite real parts. Then the operators $\widetilde{B}[S_1]\colon L^2(\mathbb{R}^n)\to L^2(\mathbb{R}^m)$ and $\widetilde{B}[S_2]\colon L^2(\mathbb{R}^k)\to L^2(\mathbb{R}^n)$ are defined.

THEOREM 3.3. $\widetilde{B}[S_1]\widetilde{B}[S_2] = \lambda(S_1,S_2)\widetilde{B}[S_1\circ S_2]$, *where*

$$\lambda(S_1,S_2) = \det[(M+P)^{-1/2}]$$

$$S_1\circ S_2 = \begin{pmatrix} K - L(M+P)^{-1}L^t & -L(M+P)^{-1}Q \\ -Q^t(M+P)^{-1}L^t & R - Q^t(M+P^{-1}Q) \end{pmatrix}. \tag{3.5}$$

Note that equation (3.5) gives rise to the same thoughts as equation (3.3). Consider two examples of the operators $\widetilde{B}[S]$.

EXAMPLE 1. The solution $\psi(x,t)$ of the heat equation

$$\left(\frac{\partial}{\partial t} - \frac{\partial^2}{\partial x^2}\right)\psi(x,t) = 0$$

with initial condition $\psi(x,t) = q(x)$ in $L^2(\mathbb{R})$, is given by the operator

$$\psi(x,t) = \frac{1}{\sqrt{2\pi t}}\int_{\mathbb{R}} \exp\left\{-\frac{(x-y)^2}{2t}\right\} q(y)\,dy.$$

EXAMPLE 2. The evolution of the harmonic oscillator, i.e., the solution of the equation

$$\frac{\partial}{\partial t} f(x,t) = i\Big(- \frac{\partial^2}{\partial x^2} + x^2\Big) f(x,t)$$

with initial condition $f(x,t) = q(x)$ is given by

$$f(x,t) = (i \sin t)^{-1/2} \widetilde{B}\left[i \begin{pmatrix} \cot t & (\sin t)^{-1} \\ -(\sin t)^{-1} & \cot t \end{pmatrix} \right] q(x),$$

where $\operatorname{Im} t \geqslant 0$ and $t \neq \pi k$.

§4. The symplectic category

4.1. Linear relations. Let V and W be linear spaces. A linear relation $P\colon V \rightrightarrows W$ is an arbitrary linear subspace P in $V \oplus W$.

EXAMPLE. A graph gr_A of any linear operator $A\colon V \to W$ is a linear relation.

If $P\colon V \rightrightarrows W$ and $Q\colon W \rightrightarrows Y$ are linear subspaces, their product $QP\colon V \rightrightarrows Y$ is defined; it consists of all $(v,y) \in V \oplus Y$ such that there exists $w \in W$ such that $(v,w) \in P$ and $(w,y) \in Q$.

4.2. The symplectic category. Objects of the category Sp are spaces V_{2n} introduced in 2.1 (recall that $n = 0, 1, 2, \dots$). Recall that these spaces are equipped with two forms $\Lambda = \Lambda_{2n}$ and $M = M_{2n}$. On $V_{2n} \oplus V_{2m}$ we introduce the following two forms:

$$\Lambda_{2n,2m}((v,w),(v',w')) = \Lambda_{2n}(v,v') - \Lambda_{2m}(w,w'), \tag{4.1}$$

$$M_{2n,2m}((v,w),(v',w')) = M_{2n}(w,w') - M_{2m}(w,w'). \tag{4.2}$$

We define the set $\operatorname{Mor}_{\mathrm{Sp}}(V_{2n}, V_{2m})$ as the set of all linear relations $P\colon V_{2n} \rightrightarrows V_{2m}$ that satisfy the following three conditions:

(a) P is a Lagrangian (with respect to $\Lambda_{2n,2m}$) subspace in $V_{2n} \oplus V_{2m}$;

(b) the form M is nonnegative definite on P;

(c) if $(v,0) \in P$, $v \neq 0$, then $M_{2n}(v,v) > 0$, whereas $(0,w) \in P$, $w \neq 0$ implies $M_{2m}(w,w) < 0$.

REMARK. Recall that the subspace P is said to be *Lagrangian* if $\Lambda(v,w) = 0$ for all $v, w \in P$ and the dimension of subspace P is half the dimension of space (in our case it is equal to $m + n$).

REMARK. Condition (c) is only slightly stronger than (b): (b) would imply that $M_{2n}(v,v) \geqslant 0$, $M_{2m}(w,w) \leqslant 0$.

The morphisms of the category Sp are multiplied as linear relations.

PROPOSITION 4.1. Sp *is indeed a category.*

The difficulty arises only in verifying that the product of linear relations has the required dimension.

4.3. Several remarks. A) Let $A \in \mathrm{Sp}(2n, \mathbb{R})$, i.e., let the operator A in V_{2n} preserve both forms Λ and M. Then its graph gr_A obviously satisfies conditions (a), (b), and (c), and, consequently, gr_A is a morphism $V_{2n} \to V_{2n}$.

Indeed, the condition $\Lambda(Av, Av') = \Lambda(v, v')$ implies that gr_A is Lagrangian, and the condition $M(Av, Av') = M(v, v')$ implies that the form $M_{2n,2n}$ vanishes on gr_A and is, in particular, nonnegative definite.

It can be easily seen that the group of automorphisms of the object V_{2n} of the category Sp coincides exactly with $\mathrm{Sp}(2n, \mathbb{R})$.

B) Let $P \in \mathrm{Mor}(V_{2n}, V_{2n})$ be a graph of some operator A. Then condition (a) implies that

$$\Lambda(Av, Av') - \Lambda(v, v') = 0,$$

i.e., $A \in \mathrm{Sp}(2n, \mathbb{C})$, and condition (b) implies that

$$M(Av, Av) \leqslant M(v, v),$$

i.e., we again encounter the semigroup Σ from 3.6.

C) If $m \neq n$, no $P \in \mathrm{Mor}(V_{2n}, V_{2m})$ can be either a graph of the operator $V_{2n} \to V_{2m}$ or a graph of the operator $V_{2m} \to V_{2n}$.

4.4. Angular operator. Our next aim is to choose convenient coordinates on the set $\mathrm{Mor}(V_{2n}, V_{2m})$. But first let us introduce an auxiliary definition.

Let W_1 and W_2 be linear subspaces of dimensions k and l, respectively. Let $L \subset W_1 \oplus W_2$ be a subspace of dimension k. Assume that $L \cup W_2 = 0$. Then L is a graph of some linear operator $A\colon W_1 \to W_2$. This operator is called the *angular operator* of the subspace L.

Further, let $\dim W_1 = \dim W_2 = k$. Let a skew-symmetric bilinear form with the matrix $\begin{pmatrix} 0 & E \\ -E & 0 \end{pmatrix}$ be introduced in $W_1 \oplus W_2$.

PROPOSITION 4.2. *A k-dimensional subspace L in $W_1 \oplus W_2$ such that $L \cap W_2 = 0$ is Lagrangian if and only if its angular operator has a symmetric matrix.*

PROOF. Proving the proposition is an easy exercise.

Let $\dim W_1 = k$ and $\dim W_2 = l$. Introduce a Hermitian sign-indefinite form M with the matrix $\begin{pmatrix} E & 0 \\ 0 & -E \end{pmatrix}$ in $W_1 \oplus W_2$.

PROPOSITION 4.3. *Let $L \in W_1 \oplus W_2$ be a k-dimensional subspace. Then the following conditions are equivalent:*

1. *The form M is positive (nonnegative) definite on L.*
2. *L has an angular operator $A = A(L)$, and $\|A\| < 1$ ($\|A\| \leqslant 1$, respectively).*

PROOF. Proving the proposition is an exercise.

4.5. Coordinates on $\mathrm{Mor}_{\mathrm{Sp}}(V_{2n}, V_{2m})$.

PROPOSITION 4.4. (a) *Let $P \in \mathrm{Mor}_{\mathrm{Sp}}(V_{2n}, V_{2m})$. Then P is a graph of the operator*

$$S = S(P) = \begin{pmatrix} K & L \\ L^t & M \end{pmatrix} : V_{2m}^- \oplus V_{2n}^+ \to V_{2m}^+ \oplus V_{2n}^-,$$

where S satisfies the following conditions:

1^+. *S is symmetric ($S = S^t$);*

2^+. $\|S\| \leqslant 1$;

3^+. $\|K\| < 1$, $\|N\| < 1$.

(b) *Conversely, a graph of any operator $S\colon V_{2m}^- \oplus V_{2n}^+ \to V_{2m}^+ \oplus V_{2n}^-$ satisfying conditions 1^+–3^+, is a morphism $V_{2n} \to V_{2m}$ in the category* Sp.

PROOF. The statement follows from Propositions 4.2–4.3, which should be applied to $W_1 \oplus W_2 = (V_{2m}^- \oplus V_{2n}^+) \oplus (V_{2m}^+ \oplus V_{2n}^-)$.

The matrix $S = S(P)$ is called the *Potapov transformation of the linear relation* P.

PROPOSITION 4.5. *Let* $P \in \mathrm{Mor}_{\mathrm{Sp}}(V_{2n}, V_{2m})$ *and* $Q \in \mathrm{Mor}_{\mathrm{Sp}}(V_{2k}, V_{2n})$. *Then*

$$S(PQ) = S(P) * S(Q),$$

where the matrix multiplication $*$ *is defined by equation* (3.3).

PROOF. The above statement is proved by a calculation.

Thus, the category $\mathcal{B}$ of §3 is isomorphic to the category Sp.

4.6. Basic Theorem. Suppose that $P \in \mathrm{Mor}_{\mathrm{Sp}}(V_{2n}, V_{2m})$. Let $S(P)$ be its Potapov transformation. Define the operator

$$W(P) = B[S(P)]\colon F_n \to F_m. \tag{4.3}$$

THEOREM 4.1. (a) *Let* $P \in \mathrm{Mor}_{\mathrm{Sp}}(V_{2n}, V_{2m})$ *and* $Q \in \mathrm{Mor}_{\mathrm{Sp}}(V_{2k}, V_{2n})$. *Then*

$$W(P)\,W(Q) = \lambda(P,Q)\,W(PQ),$$

where $\lambda(P,Q) \in \mathbb{C}$.

(b) *The operators* $W(P)$ *satisfy the relation*

$$\hat{a}(w)\,W(P) = W(P)\,\hat{a}(v) \tag{4.4}$$

for all $v \in V_{2n}$ *and* $w \in V_{2m}$ *such that* $(v, w) \in P$.

(c) *Equation* (4.4) *defines the operator* $W(P)$ *in a unique way up to a multiplicative factor.*

(d) *Let* $P \in \mathrm{Sp}(2n, \mathbb{R}) \simeq \mathrm{Aut}(V_{2n})$. *Then the operator* $W(P)$ *coincides, up to a factor, with the operator* $W(P)$ *from Theorem* 2.1.

Thus, we have obtained what is called the *projective representation* of a category. To each object V_{2n} of the category Sp we associate a boson Fock space, and to each $P \in \mathrm{Mor}_{\mathrm{Sp}}(V_{2n}, V_{2m})$ a linear operator $W(P)\colon F_n \to F_m$, in such a way that identity (4.3) is satisfied. In this case for any $n = 0, 1, 2, \dots$ the group $\mathrm{Sp}(2n, \mathbb{R}) \simeq \mathrm{Aut}_{\mathrm{Sp}}(V_{2n})$ acts in F_n, and this operation coincides with the representation of the group $\mathrm{Sp}(2n, \mathbb{R})$ from Theorem 2.1.

4.7. Proof of Theorem 4.1. Statement (a) follows from Propositions 4.4–4.5 and Theorem 3.2.

Further, let $K(z, \overline{u})$ be the kernel of the operator $W(P)$. Then the kernel of the operator $\hat{a}(w)\,W(P)$ is

$$\Bigl(\sum w_j^+ z_j + \sum w_j^- \frac{\partial}{\partial z_j}\Bigr) K(z, \overline{u}),$$

and the kernel of $W(P)\,\hat{a}(w)$ is

$$\Bigl(\sum v_j^+ \frac{\partial}{\partial \overline{u}_j} + \sum v_j^- \overline{u}_j\Bigr) K(z, \overline{u});$$

consequently we obtain the system of equations for the function $K(z,\overline{u})$

$$\left(\sum w_j^+ z_j + \sum w_j^- \frac{\partial}{\partial z_j} - \sum v_k^+ \frac{\partial}{\partial \overline{u}_k} - \sum v_k^- \overline{u}_k\right) K(z,\overline{u}) = 0. \tag{4.5}$$

It can easily be verified that, on the one hand, the kernel of the operator satisfies the system (4.5) (this yields statement (b)) and, on the other hand, the solution of this set is unique (this yields statement (c)).

Finally, statement (d) follows from (b).

4.8. The real version of Theorem 4.1. Recall that V_{2n} is canonically identified with $\widetilde{V}_{2n}$ (see 2.2).

THEOREM 4.1′. (a) *For any* $P \in \operatorname{Mor}(\widetilde{V}_{2n}, \widetilde{V}_{2m})$ *there exists a unique (up to a multiplicative factor) bounded operator* $\widetilde{W}(P)\colon L^2(\mathbb{R}^n) \to L^2(\mathbb{R}^m)$ *such that*

$$\tilde{a}(w)\widetilde{W}(P) = \widetilde{W}(P)\tilde{a}(v)$$

for all $(v,w) \in P$.

(b) *For all* $P \in \operatorname{Mor}(\widetilde{V}_{2n}, \widetilde{V}_{2m})$ *and* $Q \in \operatorname{Mor}(\widetilde{V}_{2k}, \widetilde{V}_{2n})$ *the following equation is satisfied*

$$\widetilde{W}(QP) = \lambda(Q,P)\widetilde{W}(Q)\widetilde{W}(P).$$

It is clear that this theorem is a corollary of Theorem 4.1. The situation with the expression for the operator $\widetilde{W}(P)$ is somewhat worse than in the complex case.

Consider $P \in \operatorname{Mor}(\widetilde{V}_{2n}, \widetilde{V}_{2m})$. Assume that P is given by an equation of the type

$$\begin{pmatrix} w^+ \\ v^+ \end{pmatrix} = \begin{pmatrix} A & B \\ -B^t & C \end{pmatrix} \begin{pmatrix} w^- \\ v^- \end{pmatrix}.$$

This is not always satisfied, but is valid at least in the case when for any $(v,w) \in P$ the condition $M(v,v) = M(w,w)$ implies $v = 0$ and $w = 0$ (the inequality in condition (b) in 4.2 is strict). If this is the case, the operator $\widetilde{W}(P)\colon L^2(\mathbb{R}^n) \to L^2(\mathbb{R}^m)$ is given by the expression

$$\widetilde{W}(P) = B\left[i\begin{pmatrix} A & B \\ B^t & -C \end{pmatrix}\right].$$

4.9. Involution in the category Sp. With each $P \in \operatorname{Mor}(V_{2n}, V_{2m})$ we associate the linear relation $P^{(*)}\colon V_{2m} \rightrightarrows V_{2n}$ which is the orthogonal complement of P in $V_{2n} \oplus V_{2m}$ with respect to $M_{2n,2m}$. It is easily seen that $P^{(*)} \in \operatorname{Mor}(V_{2m}, V_{2n})$. Moreover,

$$P^{(*)(*)} = P, \qquad (PQ)^{(*)} = Q^{(*)}P^{(*)}.$$

Finally,

$$W(P^{(*)}) = W(P)^*.$$

§5. The Krein–Shmulyan functor

5.1. Cartan matrix balls. Let $n = 0, 1, 2, \ldots$. A set of all symmetric complex matrices of size $n \times n$ with norm < 1 will be called the *matrix ball* $\mathcal{Z}_n$.

EXAMPLE. For $n = 1$ the ball $\mathcal{Z}_n$ is the circle $|z| < 1$ on $\mathbb{C}$, i.e., the Lobachevskiĭ plane.

The group $\mathrm{Sp}(2n, \mathbb{R})$ acts on $\mathcal{Z}_n$ via linear-fractional transformations of the type

$$\begin{pmatrix} \Phi & \Psi \\ \overline{\Psi} & \overline{\Phi} \end{pmatrix} : T \mapsto (\Phi T + \Psi)(\overline{\Psi} T + \overline{\Phi})^{-1}. \tag{5.1}$$

The stabilizer of the point $0 \in \mathcal{Z}_n$ consists of matrices of the form $\begin{pmatrix} B & 0 \\ 0 & \overline{B} \end{pmatrix}$, where B is a unitary matrix of size $n \times n$. It is easily verified that a matrix ball $\mathcal{Z}_n$ is a $\mathrm{Sp}(2n, \mathbb{R})$-homogeneous space, and thus $\mathcal{Z}_n$ is a homogeneous symmetric space $\mathrm{Sp}(2n, \mathbb{R})/U(n)$.

5.2. The Siegel upper half-plane. Let H_n be a space of symmetric matrices of size $n \times n$ with a positive definite imaginary part. The group $\mathrm{Sp}(2n, \mathbb{R})$ is defined as a group of real symplectic matrices which acts on H_n via linear-fractional transformations of the type

$$\begin{pmatrix} A & B \\ C & D \end{pmatrix} : S \mapsto (AS + B)(CS + D)^{-1}.$$

The stabilizer of the point $S = iE$ consists of matrices of the form $\begin{pmatrix} A & B \\ -B & A \end{pmatrix} \in \mathrm{Sp}(2n, \mathbb{R})$. This stabilizer is isomorphic to the group $U(n)$. It is easily verified that $\mathrm{Sp}(2n, \mathbb{R})$ acts on H_n transitively, and, consequently, H_n is the symmetric homogeneous subspace $\mathrm{Sp}(2n, \mathbb{R})/U(n)$.

We see that $\mathcal{Z}_n$ and H_n is the same homogeneous space. The isomorphism $\mathcal{Z}_n \to H_n$ is established by the linear-fractional transformation

$$S = (T + iE)(iT + E)^{-1} \tag{5.2}$$

(the "Cayley transformation").

5.3. Embedding of $\mathcal{Z}_n$ and H_n into the Grassmannian. In the two previous subsections there were some statements that are quite difficult to prove directly. Everything becomes much more transparent if one turns to a different realization of the regions considered.

Let $T \in \mathcal{Z}_n$. Consider a linear subspace L_T in V_{2n} with the angular operator $T\colon V_{2n}^- \to V_{2n}^+$, i.e., a set of vectors in V_{2n} of the form $(Tv^-; v^-) \in V_{2n}^+ \oplus V_{2n}^-$. As noted above (see 4.4), the space L_T satisfies the following conditions:

(a) L_T is Lagrangian with respect to Λ.

(b) The form M is strictly negative definite on L_T.

Conversely, any subspace L satisfying conditions (a) and (b) has an angular operator $T \in \mathcal{Z}_n$.

Thus, we have established an isomorphism of $\mathcal{Z}_n$ with a certain open region $\mathcal{R}_n$ in a Lagrangian Grassmannian of the space V_{2n}.

Let $T \in \mathcal{Z}_n$ and $g = \begin{pmatrix} \Phi & \Psi \\ \overline{\Psi} & \overline{\Phi} \end{pmatrix} \in \mathrm{Sp}(2n, \mathbb{R})$. Then $L_T \in \mathcal{R}_n$ consists of vectors of the form (Tv_-, v_-). Acting by the operators g, we get a vector space of the form

$$((\Phi T + \Psi) v_-; (\overline{\Psi} T + \overline{\Phi}) v_-).$$

Denoting by w_- the vector $(\overline{\Psi}T + \overline{\Phi})v_-$, we see that the space gL_T consists of vectors

$$((\Phi T + \Psi)(\overline{\Psi}T + \overline{\Phi})^{-1}w_-; w_-),$$

i.e., the angular operator of the subspace gL_T is $(\Phi T + \Psi)(\overline{\Psi}T + \overline{\Phi})^{-1}$. Thus, we come to equation (5.1).

The situation is just as simple for the upper half-plane H_n. With any $S \in H_n$ we associate the subspace with the angular operator $S\colon \widetilde{V}_{2n}^- \to \widetilde{V}_{2n}^+$ this subspace satisfies conditions (a) and (b).

Recalling that V_{2n} and $\widetilde{V}_{2n}$ are the same (see the isomorphism (2.12)), we get the isomorphism $\mathcal{Z}_n \to H_n$ defined by equation (5.2).

Below in this section we shall deal with matrix balls $\mathcal{Z}_n$ only.

5.4. Krein–Shmulyan functor. Consider $P \in \mathrm{Mor}_{\mathrm{Sp}}(V_0, V_{2n})$. The morphism P is a subspace in $V_0 \oplus V_{2n} = V_{2n}$, and this subspace satisfies conditions (a) and (b) from the previous subsection. Thus we get the canonical isomorphism

$$\mathcal{Z}_n \simeq \mathrm{Mor}_{\mathrm{Sp}}(V_0, V_{2n}).$$

Now consider the following trivial abstract construction. Let $\mathcal{K}$ be some category and R_0 be one of its objects. Let us construct a functor F from the category $\mathcal{K}$ to the category of sets and mappings using the following rule: for any $R \in \mathrm{Ob}(\mathcal{K})$ the set $F(R)$ is

$$F(R) = \mathrm{Mor}(R_0, R),$$

and for any $\psi \in \mathrm{Mor}(R_1, R_2)$ the mapping $F(\psi)\colon F(R_1) \to F(R_2)$ is defined by the equation $F(\psi)q = \psi q$.

Let us apply this unsophisticated construction to the category Sp and to the object V_0. A matrix ball $\mathcal{Z}_n$ is associated with each object V_{2n}. Suppose that $P \in \mathrm{Mor}_{\mathrm{Sp}}(V_{2n}, V_{2m})$, and $S(P) = \begin{pmatrix} K & L \\ L^t & M \end{pmatrix}$ is its Potapov transform. Then the transformation

$$\tau(P)T = K + LT(1 - MT)^{-1}L^t \tag{5.3}$$

acting from $\mathcal{Z}_n$ to $\mathcal{Z}_m$ is associated with the morphism P. These transformations are called *generalized linear-fractional Krein–Shmulyan transformations* (it is worth noting that even in the case $m = n$ they cannot generally be written either in the form $(AT + B)(CT + D)^{-1}$ or in the form $(C'T + D')^{-1}(A'T + B')$).

Naturally, the equation

$$\tau(PQ) = \tau(P)\tau(Q) \tag{5.4}$$

is satisfied.

5.5. Equivariant embeddings of matrix balls into Fock spaces. Let $T \in \mathcal{Z}_n$, $P \in \mathrm{Mor}(V_{2n}, V_{2m})$, and $S(P) = \begin{pmatrix} K & L \\ L^t & M \end{pmatrix}$ be the Potapov transform of the morphism P. Then

$$\begin{aligned} W(P)b[T] &= B\begin{bmatrix} K & L \\ L^t & M \end{bmatrix} b[T] \\ &= \det[(1 - MT)^{-1/2}]\, b[K + LT(1 - MT)^{-1}L^t]. \end{aligned} \tag{5.5}$$

Now consider a projectivized Fock space $\mathbb{P}F_n$ and the embedding $\sigma\colon \mathcal{Z}_n \to \mathbb{P}F_n$ defined by the expression $\sigma(T) = b[T]$. Then equation (5.5) implies

$$W(P)\sigma(T) = \sigma(\tau(T)). \tag{5.6}$$

REMARK. Let $Q \in \mathrm{Mor}_{\mathrm{Sp}}(V_0, V_{2n})$ and let $T \in \mathcal{Z}_n$ be its Potapov transformation (i.e., the angular operator). Then the operator $W(Q)\colon F^0 \to F^n$ is defined by the equation $s \mapsto s \cdot b[T]$, where $s \in F_0 \simeq \mathbb{C}$. Therefore, equation (5.5) is just a special case of equations (3.2)–(3.3).

5.6. Fixed points and eigenvectors.

PROPOSITION 5.1. *Let $P \in \mathrm{End}(V_{2n})$ and $T \in \mathcal{Z}_n$. The following conditions are equivalent:*

(a) *T is a fixed point of the transformation $\tau(P)\colon \mathcal{Z}_n \to \mathcal{Z}_n$.*

(b) *The vector $b[T] \in F_n$ is an eigenvector for the operator $W(P)\colon F_n \to F_n$.*

PROOF. See equation (5.5).

Now let $P \in \mathrm{End}(V_{2n})$ satisfy the additional condition

$$(v, w) \in P \implies M(v, v) > M(w, w)$$

(for $v \neq 0$ and $w \neq 0$), i.e., P is an interior point of the semigroup $\mathrm{End}(V_{2n})$. Denote by $\mathrm{End}^0(V_{2n})$ the set of all such linear relations. It is easy to see that $P \in \mathrm{End}^0(V_{2n})$ is equivalent to $\|S(P)\| < 1$.

It is easy to verify that in the case $P \in \mathrm{End}^0(V_{2n})$ the mapping $\tau(P)$ takes the matrix ball $\mathcal{Z}_n$ strictly inside $\mathcal{Z}_n$, i.e., there exists $\varepsilon > 0$ such that $\|\tau(P)T\| \leqslant 1-\varepsilon$ for all $T \in \mathcal{Z}_n$. Since $\mathcal{Z}_n$ is a ball from the topological point of view, we see that $\tau(P)$ necessarily has a fixed point.

5.7. Estimate for the norm of $B[S]$. Let the operator $B[S]$ be self-adjoint. This is equivalent to the statement that the matrix $S = \begin{pmatrix} K & L \\ L^t & M \end{pmatrix}$ satisfies the conditions $K = \overline{M}$ and $L = L^*$.

Let $\|S\| < 1$. Then, as we have seen in the previous subsection, the operator $B[S]$ has an eigenvector of the form $b[T]$. The group $\mathrm{Aut}(V_{2n}) \simeq \mathrm{Sp}(2n, \mathbb{R})$ acts transitively on $\mathcal{Z}_n$, and, consequently, there exists $P \in \mathrm{Aut}(V_{2n})$ such that $\tau(P)0 = T$. Consider the operator $W(P)B[S]W(P)^{-1}$. This self-adjoint operator has the form $\varkappa \cdot B\begin{bmatrix} K & L \\ L^t & M \end{bmatrix}$, and the vector $b[0] = 1$ is its eigenvector. Hence, $K = 0$ and so $M = \overline{K} = 0$.

The operator $B\begin{bmatrix} 0 & Q \\ Q^t & 0 \end{bmatrix}$, however, is nothing but the change of variables operator $f(z) \mapsto f(zQ)$. Without loss of generality, one can assume that the matrix Q is diagonal:

$$Q = \begin{pmatrix} \lambda_1 & & \\ & \ddots & \\ & & \lambda_n \end{pmatrix}$$

(with $\lambda_j \in \mathbb{R}$ and $|\lambda_j| < 1$). The vectors $e_{\alpha_1 \dots \alpha_n}$ (see equation (1.3)) are eigenvectors for $B\begin{bmatrix} 0 & Q \\ Q^t & 0 \end{bmatrix}$ with eigenvalues $\lambda_1^{\alpha_1} \cdots \lambda_n^{\alpha_n}$. The largest of these numbers corresponds to $\alpha_1 = \cdots = \alpha_n = 0$, and, consequently, the norm of the operator $B[S]$ is reached at the vector $b[0] = 1$. Therefore, the norm of the operator $b[S]$ is reached at the vector $b[T]$. Thus (see equation (5.5)), the following theorem is proved.

THEOREM 5.1. *Let $P \in \operatorname{End}(V_{2n})$ be self-adjoint (i.e., $P^{(*)} = P$). Suppose that $S = S(P) = \begin{pmatrix} K & L \\ L^t & M \end{pmatrix}$ and $T \in \mathcal{Z}_n$ is a fixed point of the transformation $\tau(P)$. Then we have*

$$\|B[S]\| = \det[(1 - MT)^{-1/2}]. \tag{5.7}$$

One can prove (this is not quite obvious, see [**56,** §3]) that if $\|T\| < 1$, then

$$|\det(1 - MT)| \geqslant \det(1 - |M|),$$

where $|M| = \sqrt{M^*M}$. This gives

COROLLARY. *For a self-adjoint operator $B[S]$ the following inequality holds:*

$$\|B[S]\| \leqslant \det(1 - |M|)^{-1/2}. \tag{5.8}$$

Note that Theorem 3.1 follows from this inequality.

REMARK. The above arguments give the following additional information: if $\|S\| < 1$, then

(a) All eigenvectors of the operator $B[S]$ have the form $p(z)\exp\{\frac{1}{2}zTz^t\}$, where $p(z)$ is a polynomial.

(b) The set of eigenvalues of the operator $\|B[S]\|^{-1}B[S]$ has the form $\lambda_1^{\alpha_1}\cdots\lambda_n^{\alpha_n}$, where $-1 < \lambda_j < 1$ and $\alpha_1, \ldots, \alpha_n$ are nonnegative integers.

§6. Geometry of matrix balls and the Compression Theorem

The subject of this section stands somewhat apart from the main contents of the paper; the Compression Theorem will be used later, in 7.6.

6.1. Composite distance. Let $T_1, T_2 \in \mathcal{Z}_n$. Consider the matrix

$$L(T_1, T_2) = (1 - T_1^*T_1)^{-1/2}(1 - T_1^*T_2)(1 - T_2^*T_2)^{-1/2}. \tag{6.1}$$

Suppose that

$$\lambda_1(T_1, T_2) \geqslant \lambda_2(T_1, T_2) \geqslant \cdots$$

are its *singular values*. Recall that by singular values of a matrix M we mean eigenvalues of the matrix $\|M\| = \sqrt{M^*M}$.

PROPOSITION 6.1. (a) *For all j, we have $\lambda_j \geqslant 1$.*

(b) *Let $T_1, T_2, S_1, S_2 \in \mathcal{Z}_n$. An element $g \in \operatorname{Aut}(V_{2n}) \simeq \operatorname{Sp}(2n, \mathbb{R})$ such that $gT_1 = S_1$ and $gT_2 = S_2$ exists if and only if*

$$\lambda_j(T_1, T_2) = \lambda_j(S_1, S_2)$$

for all j.

The set of numbers

$$\varphi_j = \cosh^{-1}\lambda_j(T_1, T_2) \tag{6.2}$$

is called the *composite distance* between T_1 and T_2 in $\mathcal{Z}_n$. It will be more convenient for us to consider the set of numbers

$$\varphi_1(T_1,T_2) \geqslant \varphi_2(T_1,T_2) \geqslant \cdots \tag{6.3}$$

as unbounded, setting $\varphi_j(T_1,T_2) = 0$ for $j > n$.

6.2. Another description of the values λ_j. Recall that with each element $T \in \mathcal{Z}_n$ a subspace $P = P_T \subset V_{2n}$, the graph of the operator $T\colon V_{2n}^- \to V_{2n}^+$, is associated. Consider the orthogonal (with respect to the form M) projection operator $L\colon P_{T_1} \to P_{T_2}$.

It turns out that λ_j are precisely the singular values of the operator L.

We shall state this in a slightly different manner. The form $(-M)$ is positive definite on P_{T_1} and P_{T_2}. Therefore, P_{T_1} and P_{T_2} are Euclidean spaces. Consider the Hermitian form

$$B(v_1,v_2) = -M(v_1,v_2)\colon P_{T_1} \times P_{T_2} \to \mathbb{C}.$$

Then λ_j are the singular values of the form B.

Now we write out the matrix of the form B. Note that the operator

$$Q_T v_- = ((1-T^*T)^{-1/2}v_-; T(1-T^*T)^{-1/2}v_-)$$

is a unitary (with respect to the form $(-M)$) operator $V_- \to P_T$, and thus one can regard v_- as a coordinate on P_T; now it is easily seen that

$$-M(Q_{T_1}v_-, Q_{T_2}v'_-) = -M(v_-, L(T_1,T_2)v_-),$$

where $L(T_1,T_2)v_-$ is given by equation (6.1). Thus, we have demonstrated the equivalence of the two definitions of the values λ_j.

6.3. Minimax description of the values λ_j. Matrix eigenvalues and singular values have natural minimax descriptions (see, for example, [**17, 57**]). In our case this gives

$$\lambda_j = \max_{\substack{R\subset P_{T_2}\\ \dim R=j}} \left(\min_{\substack{v\in R\\ M(v,v)=-1}} \left(\max_{\substack{w\in P_{T_1}\\ M(w,w)=-1}} |M(v,w)| \right)\right), \tag{6.4}$$

where R runs over all j-dimensional spaces in P_{T_2}.

REMARK. Recall that the *angles* $\psi_1 \geqslant \psi_2 \geqslant \cdots$ between the subspaces P_1 and P_2 in standard Euclidean space $\mathbb{R}^n$ are given by the expression

$$\cos\Psi_j = \max_{\substack{R\subset P_2\\ \dim R=j}} \left(\min_{\substack{v\in R\\ \|v\|=1}} \left(\max_{\substack{w\in P_1\\ \|w\|=1}} \langle v,w\rangle \right)\right).$$

Thus it turns out that the composite distance is the hyperbolic analog of angles.

6.4. Infinitesimal metric structure. Suppose that $T \in \mathcal{Z}_n$ and ∂T is a tangent vector at the point T. We construct a matrix

$$\Sigma(T,\partial T) = (1-T^*T)^{-1/2}\partial T^*(1-TT^*)^{-1/2}.$$

Let

$$\sigma_1(T,\partial T) \geqslant \sigma_2(T,\partial T) \geqslant \cdots$$

be its singular values.

PROPOSITION 6.2. *Let $T, S \in \mathcal{Z}_n$, and suppose that ∂T and ∂S are tangent vectors at the points T and S, respectively. The following conditions are equivalent:*

1. *There exists $g \in \mathrm{Aut}_{\mathrm{Sp}}(V_{2n}) \simeq \mathrm{Sp}(2n, \mathbb{R})$ that takes T to S and ∂T to ∂S.*
2. *$\sigma_j(T, \partial T) = \sigma_j(S, \partial S)$ for all j.*

6.5. Metric in $\mathcal{Z}_n$. Note that the expression

$$ds^2 = \sum \sigma_j^2(T, \partial T)$$

is a $\mathrm{Sp}(2n, \mathbb{R})$-invariant Riemann metric in $\mathcal{Z}_n$.

THEOREM 6.1. *The distance in $\mathcal{Z}_n$ that corresponds to the Riemann metric ds^2 is given by the formula*

$$\rho(T_1, T_2) = \left[\sum (\varphi_j(T_1, T_2))^2\right]^{1/2}.$$

6.6. Invariant metrics in $\mathcal{Z}_n$. Let $p(x_1, \dots, x_n)$ be a norm in $\mathbb{R}^n$ invariant under coordinate permutations and under all possible changes of direction of the axes. Consider an invariant Finsler metric in $\mathcal{Z}_n$,

$$K_p(T, \partial T) = p(\sigma_1(T, \partial T), \dots, \sigma_n(T, \partial T)).$$

THEOREM 6.2. *A distance in $\mathcal{Z}_n$ which is defined by the Finsler metric K_p is given by*

$$\rho_p(T_1, T_2) = p(\varphi_1(T_1, T_2), \dots, \varphi_n(T_1, T_2)).$$

6.7. The Compression Theorem.

THEOREM 6.3. *Let $P \in \mathrm{Mor}_{\mathrm{Sp}}(V_{2n}, V_{2m})$. Then for all $T_1, T_2 \in \mathcal{Z}_n$ and for all j one has*

$$\varphi_j(\tau(p)T_1, \tau(p)T_2) \leqslant \varphi_j(T_1, T_2). \tag{6.5}$$

PROOF. The proof of this fact is rather simple: for $n = m = 1$ it is verified by a straightforward calculation, and the general case is reduced to the one-dimensional case by using equality (6.4).

COROLLARY. *For any metric ρ_p the following inequality holds:*

$$\rho_p(\tau(P)T_1, \tau(P)T_2) \leqslant \rho_p(T_1, T_2). \tag{6.6}$$

COROLLARY. *Let $\tau(P)$ take the pair $(T, \partial T)$ to $(S, \partial S)$. Then for all j the following inequality holds:*

$$\sigma_j(S, \partial S) \leqslant \sigma_j(T, \partial T).$$

6.8. Integer distance. In $\mathcal{Z}_n$ we introduce one more $\mathrm{Sp}(2n, \mathbb{R})$-invariant metric

$$n(T_1, T_2) = \mathrm{rk}\,(T_1 - T_2).$$

It is easily seen that $n(T_1, T_2)$ is exactly the number of nonzero numbers in the set (6.3). Obviously, for any $P \in \mathrm{Mor}_{\mathrm{Sp}}(V_{2n}, V_{2m})$ the following inequality holds:

$$n(\tau(p)T_1, \tau(p)T_2) \leqslant n(T_1, T_2). \tag{6.7}$$

6.9. Inverse theorems.

THEOREM 6.4. *Let* $q\colon \mathcal{Z}_n \to \mathcal{Z}_m$ *be a holomorphic mapping that satisfies condition* (6.7), *and let there exist* $T_1, T_2 \in \mathcal{Z}_n$ *such that* $n(q(T_1), q(T_2)) \geqslant 2$. *Then* q *is of the form* $\tau(P)$ *for some* $P \in \mathrm{Mor}_{\mathrm{Sp}}(V_{2n}, V_{2m})$.

THEOREM 6.5. *Let* $q\colon \mathcal{Z}_n \to \mathcal{Z}_m$ *be a holomorphic mapping that satisfies the following conditions*:

1°. *Let the mapping* q *take the pair (point* T; *tangent vector* ∂T *at the point* T) *to the pair* $(S; \partial S)$. *Then the condition* $\mathrm{rk}(\partial T) \leqslant 1$ *implies* $\mathrm{rk}(\partial S) \leqslant 1$.

2°. $\dim q(\mathcal{Z}_n) > 1$.

3°. *There exist* $T_1, T_2 \in \mathcal{Z}_n$ *such that* $n(q(T_1), q(T_2)) \geqslant 2$. *Then* q *is of the form* $\tau(P)$ *for some* $P \in \mathrm{Mor}_{\mathrm{Sp}}(V_{2n}, V_{2m})$.

COROLLARY. *Let* $q\colon \mathcal{Z}_n \to \mathcal{Z}_m$ *be a holomorphic mapping that satisfies condition* (6.5) *and condition* 3° *of Theorem* 6.5. *Then* q *is of the form* $\tau(P)$ *for some* $P \in \mathrm{Mor}_{\mathrm{Sp}}(V_{2n}, V_{2m})$.

COROLLARY. *Let* $q\colon \mathcal{Z}_n \to \mathcal{Z}_m$ *be a holomorphic mapping that satisfies condition* (6.6) *and condition* 3° *of Theorem* 6.5. *Then* q *is of the form* $\tau(P)$ *for some* $P \in \mathrm{Mor}_{\mathrm{Sp}}(V_{2n}, V_{2m})$.

§7. The infinite-dimensional case

7.1. Boson Fock space F_∞ with an infinite number of degrees of freedom. Note that the space F_n can be canonically isometrically embedded into F_{n+1} according to the relation

$$f(z_1, \dots, z_n) \mapsto f(z_1, \dots, z_n, 0).$$

Therefore, we can identify F_n with a subspace in F_{n+1}. Let $\mathcal{F} = \bigcup_{j=1}^{\infty} F_j$. We denote by F_∞ the completion of $\mathcal{F}$ with respect to the norm defined by the inner product.

Let $f_j \in F_j$, and let the sequence $f_1, f_2, \dots$ be fundamental in $\mathcal{F}$. It is easily demonstrated that for any vector $u = (u_1, u_2, \dots) \in l_2$ the sequence of numbers $f_j(u_1, u_2, \dots)$ converges. Denote by $f(u)$ the limit of this sequence. It is easily verified that $f(u)$ is a holomorphic function on l_2.

Thus, the space F_∞ is identified with some space of holomorphic functions on l_2.

Further, the canonical embedding $F_n \to F_{n+1}$ identifies the basis vectors $e_{\alpha_1 \dots \alpha_n} \in F_n$ and $e_{\alpha_1 \dots \alpha_n 0} \in F_{n+1}$. Let $\alpha_1, \alpha_2, \dots$ be an infinite sequence of nonnegative integers, all the α_j's from some place on being equal to 0. Define the function

$$e_{\alpha_1 \alpha_2 \dots}(z_1, z_2, \dots) = \prod_{j=1}^{\infty} \frac{z_j^{\alpha_j}}{\sqrt{\alpha_j!}}.$$

It is easily seen that this set of functions forms an orthonormal basis in F_∞ (by the way, one can see from this that elements of F_∞ are entire functions on l_2).

For the inner product in F_∞ we keep the notation

$$\langle f, g \rangle = \int f(z) \overline{g(z)} \exp(-|z|^2)\, dz\, d\overline{z},$$

but we consider this relation as formal (the left-hand side being the definition of the right-hand side). We denote by F_ω the Fock spaces with $\omega = 0, 1, 2, \dots, \infty$ degrees of freedom.

7.2. Overfilled basis. Let $a = (a_1, a_2, \dots) \in l_2$. Define the function $\varphi_a(z)$ as before:

$$\varphi_a(z) = \exp(z, a) = \exp\Big(\sum z_j \overline{a}_j\Big).$$

As before,

$$\langle f(z), \varphi_a(z)\rangle = f(a).$$

7.3. Gaussian vectors. Let T be a Hilbert–Schmidt operator in l_2; moreover, assume that $T = T^t$, $\|T\| < 1$, and $c = (c_1, c_2, \dots) \in l_2$. Define the vector

$$b[T \,|\, c^t] = \exp\Big\{\frac{1}{2}\, zTz^t + zc^t\Big\}.$$

PROPOSITION 7.1. *The vectors $b[T \,|\, c^t]$ lie in F_∞, and*

$$\langle b[T \,|\, c^t], b[Q \,|\, d^t]\rangle = \det[(1 - T\overline{Q})^{1/2}] \exp\left\{\frac{1}{2}(c\overline{d}) \begin{pmatrix} -T & E \\ E & -\overline{Q} \end{pmatrix}^{-1} \begin{pmatrix} c^t \\ \overline{d}^t \end{pmatrix}\right\}. \tag{7.1}$$

We denote by F_ω^0 the subspace in F_ω, consisting of finite linear combinations of the Gaussian vectors $b[T \,|\, c^t]$.

7.4. Operators. Let A be a bounded operator $F_\omega \to F_\nu$. With it we associate the function $K(z, \overline{u})$ (the *kernel*) defined by the same equation (1.12) as before. Equation (1.13) remains valid.

As before, we write

$$Af(z) = \int K(z, \overline{u}) f(u) \exp(-|u|^2)\, du,$$

but now by this we mean that

$$Af(z) = \langle f(u), \overline{k_z(\overline{u})}\rangle,$$

where $k_z(\overline{u}) = K(z, \overline{u})$ is the function obtained from $K(z, \overline{u})$ for a fixed value of the argument z.

If $K(z, \overline{u})$ is the kernel of the operator $A\colon F_\omega \to F_\nu$, and $L(u, \overline{v})$ is the kernel of the operator $B\colon F_\varkappa \to F_\omega$, then the kernel $M(z, \overline{v})$ of the operator AB can be written in the form

$$M(z, \overline{v}) = \int K(z, \overline{u}) L(u, \overline{v}) \exp\{-|u|^2\}\, du = \langle l_{\overline{v}}(u), \overline{k_z(\overline{u})}\rangle,$$

where $l_{\overline{v}}(u) = L(u, \overline{v})$ and $k_z(\overline{u}) = K(z, \overline{u})$.

7.5. Operators $B[S]$. Let $S = \begin{pmatrix} K & L \\ L^t & M \end{pmatrix}$ be a matrix of size $(\varkappa + \omega) \times (\varkappa + \omega)$. Define the operator $B[S]\colon F_\omega \to F_\varkappa$ by equation (3.1).

THEOREM 7.1. *If the operator $B[S]$ is bounded, then*
1. $\|S\| \leqslant 1$;
2. $\|K\| < 1$, $\|M\| < 1$;
3. *K and M are Hilbert–Schmidt operators.*

From now on we assume that the matrix S satisfies conditions 1–3.

THEOREM 7.2. *The operator $B[S]$ is a well-defined (possibly unbounded) operator that takes F^0_ω to $F^0_\varkappa$.*

THEOREM 7.3. *The operators $B[S]$ are multiplied according to the same formulas* (3.2), (3.3) *as in the finite-dimensional case.*

THEOREM 7.4. *If K and M are the nuclear operators, then the operator $B[S]$ is bounded.*

THEOREM 7.5. *If $\|S\| < 1$, the operator $B[S]$ is bounded.*

THEOREM 7.6. *If the operator $B[S]$ is not bounded, it still has a unique maximal closed extension.*

7.6. Sketches of the proofs. *Theorem* 7.1. See 3.2.

Theorem 7.2. See equation (7.1), which enables us to calculate $B[S]\, b[T \,|\, c^t]$.

Theorem 7.4. The estimate (5.8) survives at infinity.

Theorem 7.5. The infinite-dimensional analog $\mathcal{Z}_\infty$ of the space $\mathcal{Z}_n$ is the set of symmetric Hilbert–Schmidt matrices. The Compression Theorem remains valid also for $\mathcal{Z}_\infty$, and one can easily deduce from it that the mapping (5.3) is contractive if $\left\| \begin{pmatrix} K & L \\ L^t & M \end{pmatrix} \right\| < 1$. Further, the argument is similar to that of 5.6.

Theorem 7.6. Let $0 < \sigma < 1$. Then the operator $B \begin{bmatrix} 0 & \sigma E \\ \sigma E & 0 \end{bmatrix}$ is injective, and the operator

$$H = B \begin{bmatrix} 0 & \sigma E \\ \sigma E & 0 \end{bmatrix} B[S] = B \begin{bmatrix} \sigma^2 K & \sigma L \\ \sigma L^t & M \end{bmatrix}$$

satisfies the assumption of Theorem 7.5. If $B[S]$ had different closed extensions, the bounded operator H would also have different closed extensions, and if $B[S]$ did not have a closed extension, H would also have none.

7.7. Linear relations. Consider the Hilbert space $V_{2\infty} = V^+_{2\infty} \oplus V^-_{2\infty} = l_2 \oplus l_2$ equipped with the two forms Λ and M defined by equations (2.6), (2.7). Just as in 2.1, with each $v \in V_{2\infty}$ we associate an operator $\hat{a}(v)$ according to equation (2.1).

Let $S = \begin{pmatrix} K & L \\ L^t & M \end{pmatrix}$ be a symmetric matrix of size $(\mu+\nu) \times (\mu+\nu)$ satisfying the conditions of Theorem 7.1. We associate with it the linear relation $P = P_S$ consisting of the vectors $(v, w) \in V_{2\nu} \oplus V_{2\mu}$ satisfying the conditions

$$\begin{pmatrix} w_+ \\ v_- \end{pmatrix} = \begin{pmatrix} K & L \\ L^t & M \end{pmatrix} \begin{pmatrix} w_- \\ v_+ \end{pmatrix}.$$

It is clear (see Theorem 7.3) that the multiplication of linear relations corresponds to the multiplication of operators $B[S]$.

7.8. Commutation relations. Let $Q = \begin{pmatrix} \Phi & \Psi \\ \overline{\Psi} & \overline{\Phi} \end{pmatrix}$ be a bounded invertible operator in $V_{2\infty}$ preserving the form Λ.

THEOREM 7.7. (a) *A bounded operator $W(Q)$ such that the relation*

$$\hat{a}(Qv) = W(Q)\, \hat{a}(v)\, W(Q)^{-1}$$

holds for any $v \in V_{2\infty}$, *exists if and only if* Ψ *is a Hilbert–Schmidt operator. If the operator* $W(Q)$ *exists, it is unique up to a multiplicative factor.*

(b) *The operator* $W(Q)$ *is given by equation* (2.22).

(c) *The operators*

$$\det[(\Phi^*\Phi)^{-1/4}]\,W(Q)$$

are unitary.

We denote by Sp_∞ the group of all symplectic matrices $\begin{pmatrix} \Phi & \Psi \\ \overline{\Psi} & \overline{\Phi} \end{pmatrix}$ such that Ψ is a Hilbert–Schmidt operator.

Equation (4.4) also remains valid, but since all the operators appearing in it are, in general, unbounded, the exact formulation of the corresponding theorem is rather long, and so we omit it.

7.9. Fixed point. Let $\overline{\mathcal{Z}}_\infty$ be the set of all symmetric ($T = T^t$) operators in l_2 with norm < 1 equipped with the weak operator topology. It follows from the Schauder–Tikhonov theorem that for any matrix S satisfying the conditions of Theorem 7.1 the map

$$T \mapsto K + LT(1 - MT)^{-1}L^t \tag{7.2}$$

has a fixed point $T_0 \in \overline{\mathcal{Z}}_\infty$.

Formula (5.7) for the norm of a self-adjoint operator remains valid (although the proof becomes more complicated).

7.10. A counterexample. Let N be a diagonal matrix with eigenvalues $\nu_1, \nu_2, \dots$, where $0 < \nu_j < 1$. Consider the operator $B\begin{bmatrix} N & E - N \\ E - N & N \end{bmatrix}$. The fixed point of the transformation (7.2) is $T_0 = E$, and we see that $\|Q\| = \prod(1 - \nu_j)^{-1/2}$. If $\sum \nu_j^2 < \infty$ and $\sum \nu_j = \infty$, then the operator Q turns out to be unbounded.

§8. Lie algebra

Since the symplectic group $\mathrm{Sp}(2n, \mathbb{R})$ acts in $L^2(\mathbb{R}^n) \simeq F_n$, the symplectic Lie algebra $\mathrm{sp}(2n, \mathbb{R})$ must also act there.

8.1. Action in $L^2(\mathbb{R}^n)$. The Lie algebra of the group of real symplectic matrices consists of real matrices of the form

$$\begin{pmatrix} A & B \\ C & -A^t \end{pmatrix}, \qquad C = C^t, \quad B = B^t.$$

The corresponding operator in $L^2(\mathbb{R}^n)$ is defined by the equation

$$\widetilde{W}\begin{pmatrix} A & B \\ C & -A^t \end{pmatrix} = \sum a_{kj} x_k \frac{\partial}{\partial x_j} + \frac{i}{2}\sum b_{kj} x_k x_j - \frac{i}{2}\sum c_{kj} \frac{\partial^2}{\partial x_k\, \partial x_j}.$$

If $Q_1 = \begin{pmatrix} A_1 & B_1 \\ C_1 & -A_1^t \end{pmatrix}$ and $Q_2 = \begin{pmatrix} A_2 & B_2 \\ C_2 & -A_2^t \end{pmatrix}$, then

$$[\widetilde{W}(Q_1), \widetilde{W}(Q_2)] = \widetilde{W}([Q_1, Q_2]) + \frac{1}{2}\operatorname{tr}(B_2 C_1 - B_1 C_2).$$

Hence $\widetilde{W}$ is a projective representation of the Lie algebra $\mathrm{sp}(2n,\mathbb{R})$. The substitution

$$\widetilde{W}'\begin{pmatrix} A & B \\ C & -A^t \end{pmatrix} = \widetilde{W}\begin{pmatrix} A & B \\ C & -A^t \end{pmatrix} + \frac{1}{2}\operatorname{tr} A$$

yields a linear representation W' of the algebra $\mathrm{sp}(2n,\mathbb{R})$.

8.2. Action in F_n. In this case the symplectic Lie algebra is realized as the algebra of matrices of the form

$$R = i\begin{pmatrix} -P & -Q \\ \overline{Q} & \overline{P} \end{pmatrix}, \qquad P = P^*, \quad Q = Q^t.$$

The corresponding operator in F_n is given by the equation

$$W(R) = i\left[\sum p_{kj} z_k \frac{\partial}{\partial z_j} + \frac{1}{2}\sum q_{kj} z_k z_j + \frac{1}{2}\sum \overline{q}_{kj} \frac{\partial^2}{\partial z_k\, \partial z_j}\right].$$

8.3. Generators of one-parameter semigroups. In 3.6 and 4.3 we discussed the semigroup $\Xi_n = \mathrm{End}_{\mathrm{Sp}}(V_{2n}) \cap \mathrm{Sp}(2n,\mathbb{C})$. It consists of the operators A in V_{2n} satisfying the conditions

$$\Lambda(Av, Av') = \Lambda(v, v'), \qquad M(Av, Av) \leqslant M(v,v).$$

Consider an operator $\mathcal{D}$ in V_{2n} such that

$$\Lambda(\mathcal{D}v, v') + \Lambda(v, \mathcal{D}v') = 0, \tag{8.1}$$

$$M(\mathcal{D}v, v) + M(v, \mathcal{D}v) = 2\operatorname{Re} M(\mathcal{D}v, v) \leqslant 0. \tag{8.2}$$

The first condition means that $\mathcal{D}$ lies in a complex symplectic algebra, and the second one is called the *M-dissipativity.*

It is easily seen that the following conditions are equivalent:

1. $\mathcal{D}$ satisfies equations (8.1), (8.2);
2. $\exp(t\mathcal{D}) \in \Xi_n$.

Denote by $\mathcal{M}$ the set of all operators $\mathcal{D}$ satisfying equations (8.1)–(8.2). It is easily seen that $\mathcal{M}$ is a convex cone in a complex symplectic algebra, invariant with respect to the associated action of the real symplectic group $\mathrm{Sp}(2n,\mathbb{R})$. In addition, it contains the cone C that consists of the operators $\mathcal{D}$ satisfying the conditions

$$M(\mathcal{D}v, w) = M(v, \mathcal{D}w) \qquad (M\text{-self-adjointness}),$$
$$M(\mathcal{D}v, v) \leqslant 0 \qquad (M\text{-dissipativity}).$$

Obviously, $C \subset ig$, $\mathcal{M} = g \oplus iC$.

THEOREM 8.1. *If* $\mathcal{D} = \begin{pmatrix} -P & -Q \\ \overline{Q} & \overline{P} \end{pmatrix} \in C$, *then the operator*

$$W'(\mathcal{D}) = \sum p_{kj} z_k \frac{\partial}{\partial z_j} + \frac{1}{2}\sum q_{kj} z_k z_j + \frac{1}{2}\sum \overline{q}_{kj} \frac{\partial^2}{\partial z_k \partial z_j} + \frac{1}{2}\operatorname{tr} P$$

is negative definite.

In particular, $W'(\mathcal{D})$ generates a contraction semigroup of self-adjoint operators in F_n.

8.4. One-parameter subgroups in Sp_∞. Consider a matrix of the form $H = \begin{pmatrix} -P & -Q \\ \overline{Q} & \overline{P} \end{pmatrix}$, where $Q = Q^t$ is a Hilbert–Schmidt operator, and P is, in general, an unbounded self-adjoint operator. Standard statements on one-parameter operator semigroups (see [**18,** X.8]) demonstrate that for all $t \in \mathbb{R}$ the exponential $\exp(itH)$ exists.

THEOREM 8.2. *Let* $\begin{pmatrix} \Phi & \Psi \\ \overline{\Psi} & \overline{\Phi} \end{pmatrix} = \exp(itH)$. *With the constraints stated above, we have* $\begin{pmatrix} \Phi & \Psi \\ \overline{\Psi} & \overline{\Phi} \end{pmatrix} \in \mathrm{Sp}_\infty$, *the operator* $(\Phi e^{-iPt} - E)$ *being a nuclear operator.*

This theorem does not describe all one-parameter groups in Sp_∞.

COUNTEREXAMPLE. Let matrices P and Q be diagonal with $Q = E$, and the eigenvalues of P increase sufficiently rapidly (e.g., let them be equal to j^2). Then $\exp\left(it\begin{pmatrix} -P & -Q \\ \overline{Q} & \overline{P} \end{pmatrix}\right)$ is a one-parameter subgroup in Sp_∞.

8.5. Self-adjointness of quadratic operators. The set $F_\infty^k \subset F_\infty$ of all homogeneous functions $f \in F_\infty$ of degree k is naturally identified with the kth symmetric power $S^k l_2$ of the space l_2. For this reason, the Fock space F_∞ is identified with $\bigoplus_{k=0}^\infty S^k l_2$.

Let P be an essentially self-adjoint operator in l_2 with domain of definition L, and Q be the Hilbert–Schmidt operator for which $Q = Q^t$. Denote by $s(Q)$ an operator in F_∞ acting in each F_∞^k as the kth symmetric power of the operator Q. In F_∞ consider the operator

$$\mathcal{H} = \frac{1}{2}\sum q_{kj} z_k z_j + s(Q) + \frac{1}{2}\sum \overline{q}_{kj} \frac{\partial^2}{\partial z_k \partial z_j}. \tag{8.3}$$

THEOREM 8.3. *The operator $\mathcal{H}$ is essentially self-adjoint on the subspace $\mathcal{L}$ consisting of finite linear combinations of vectors of the type $x_1 \cdot x_2 \cdot \dots \cdot x_k$, where $x_k \in L$.*

8.6. Correspondence of one-parameter subgroups. Let $\begin{pmatrix} -P & -Q \\ \overline{Q} & \overline{P} \end{pmatrix}$ satisfy the conditions of 8.4, and $\mathcal{H}$ is given by equation (8.3).

THEOREM 8.4. *We have*

$$\exp(it\mathcal{H}) = \det(\Phi e^{iPt})^{1/2} W\left[\exp\left\{it\begin{pmatrix} -P & -Q \\ \overline{Q} & \overline{P} \end{pmatrix}\right\}\right]. \tag{8.4}$$

§9. Canonical forms

Here we consider only the finite-dimensional case.

9.1. Canonical decomposition. Consider the semigroup $\mathrm{End}_{\mathrm{Sp}}(V_{2n})$ and inside it the subsemigroup Ξ_n consisting of actual operators. Recall that we have the conjugation operation $P \mapsto P^{(*)}$ in $\mathrm{End}_{\mathrm{Sp}}(V_{2n})$ (see 4.9) and the *unitary elements* $\mathrm{End}(V_{2n})$ (i.e., elements satisfying the condition $P^{(*)} = P^{-1}$) form the group $\mathrm{Sp}(2n, \mathbb{R}) \simeq \mathrm{Aut}(V_{2n})$.

THEOREM 9.1. *Any element $g \in \Xi_n$ can be uniquely represented as a product $g = UR$, where U is a unitary element, and $R \in \Xi_n$ is self-adjoint ($R^{(*)} = R$).*

THEOREM 9.1′. *Any element $P \in \mathrm{End}(V_{2n})$ can be represented in the form of a product $P = UR$, where U is a unitary element, and $R \in \mathrm{End}(V_{2n})$ is self-adjoint.*

9.2. THEOREM 9.2. *Any self-adjoint element $R \in \Xi_n$ can be uniquely represented in the form $R = \exp(H)$, where H satisfies the conditions*

$$M(Hv, w) = M(v, Hw), \tag{9.1}$$

$$M(Hv, v) \leqslant 0 \tag{9.2}$$

for all $v, w \in V_{2n}$.

Recall that the set of all H satisfying (9.1)–(9.2) forms a convex cone invariant with respect to the associated operation $H \mapsto g^{-1}Hg$ of the group $\mathrm{Sp}(2n, \mathbb{R})$; in 8.3 we have denoted this cone by C.

9.3. Canonical forms of the elements of the cone C.

THEOREM 9.3. *By the action of the group $\mathrm{Sp}(2n, \mathbb{R})$ any element of the cone C can be reduced to the canonical form $K\colon V_{2n}^{+} \oplus V_{2n}^{-} \to V_{2n}^{+} \oplus V_{2n}^{-}$,*

$$K = \left(\begin{array}{ccc:ccc:ccc:ccc}
-\lambda_1 & & & & & & & & & & & \\
 & \ddots & & & & & & & & & & \\
 & & -\lambda_k & & & & & & & & & \\
\hdashline
 & & & -1 & & & & & & 1 & & \\
 & & & & \ddots & & & & & & \ddots & \\
 & & & & & -1 & & & & & & 1 \\
\hdashline
 & & & & & & -\lambda_1 & & & & & \\
 & & & & & & & \ddots & & & & \\
 & & & & & & & & -\lambda_k & & & \\
\hdashline
 & & & -1 & & & & & & 1 & & \\
 & & & & \ddots & & & & & & \ddots & \\
 & & & & & -1 & & & & & & 1
\end{array}\right),$$

where $\lambda_j \geqslant 0$.

REMARK. The Jordan form of the matrix K contains only Jordan boxes of size 1×1 and Jordan boxes of size 2×2 with zero eigenvalues. In this case the interior of the cone C (i.e., the set of all H such that $M(Hv, v) < 0$ for $v \neq 0$) consists of operators with nonzero eigenvalues.

REMARK. Obviously, this theorem makes it also possible to construct canonical forms for the self-adjoint elements of the semigroup Ξ_n.

9.4. Canonical forms of quadratic operators. Theorem 9.3 has the following corollary. Let

$$\mathcal{H} = a + \frac{1}{2}\sum q_{kj} z_k z_j + \sum p_{kj} z_k \frac{\partial}{\partial z_j} + \frac{1}{2}\sum \overline{q}_{kj} \frac{\partial^2}{\partial z_k \partial z_j}$$

be a positive definite quadratic operator (in particular, $\begin{pmatrix} P & Q \\ -\overline{Q} & -\overline{P} \end{pmatrix} \in C$). Then, using the transformation $\mathcal{H} \mapsto W(P)\mathcal{H}W(P)^{-1}$, where $P \in \mathrm{Sp}(2n, \mathbb{R})$, it can be reduced to the form

$$\tau + \sum_{j=1}^{k} \lambda_j z_j \frac{\partial}{\partial z_j} + \frac{1}{2} \sum_{\mu > k} \left(z_\mu + \frac{\partial}{\partial z_\mu} \right)^2,$$

where $\tau \in \mathbb{R}$ and $0 \leqslant \lambda_j$.

Similarly, using the transformation $\mathcal{R} \mapsto \widetilde{W}(P)^{-1} \mathcal{R} \widetilde{W}(P)$, where $P \in \mathrm{Sp}(2n, \mathbb{R})$, we can reduce any positive operator

$$\mathcal{R} = s + i \sum a_{kj} x_k \frac{\partial}{\partial x_j} + \frac{1}{2} \sum b_{kj} x_k x_j + \frac{1}{2} \sum c_{kj} \frac{\partial^2}{\partial x_k \partial x_j}$$

in $L^2(\mathcal{R}_n)$ to the form

$$\sigma + \sum_{j=1}^{k} \mu_j \left(- \frac{\partial^2}{\partial x_j^2} + x_j^2 \right) - \sum_{m > k} \frac{\partial^2}{\partial x_m^2},$$

where $\sigma \in \mathbb{R}$ and $\mu_j \geqslant 0$ (i.e., $\mathcal{R}$ is split into the sum of the harmonic oscillator and the Laplace operator).

9.5. Canonical forms of the elements $P \in \mathrm{Mor}_{\mathrm{Sp}}(V_{2n}, V_{2m})$**.** The group $\mathrm{Sp}(2n, \mathbb{R}) \times \mathrm{Sp}(2m, \mathbb{R})$ acts in an obvious manner on the set $\mathrm{Mor}_{\mathrm{Sp}}(V_{2n}, V_{2m})$. Now we describe the orbits of this group.

Consider the following four types of canonical morphisms:

(a) The morphism $\alpha\colon V_0 \to V_2$ consists of elements of the form

$$(0; v^-) \in V_2^+ \oplus V_2^+ = V_2 = V_0 \oplus V_2 .$$

(b) The morphism $\beta\colon V_2 \to V_0$ consists of elements of the form

$$(v^+; 0) \in V_2^+ \oplus V_2 = V_2 = V_2 \oplus V_0 .$$

(c) The morphisms $\gamma_s\colon V_2 \to V_2$, where $0 < s \leqslant 1$, consist of the elements

$$(x, y; sx, s^{-1}y) \in V_2 \oplus V_2 .$$

(d) The morphisms $\delta_t\colon V_2 \to V_2$, where $0 < t < 1$, consist of the elements

$$(x, y + t(x - y); x + t(y - x), y) \in V_2 \oplus V_2$$

(for different t these morphisms are equivalent).

Further, note that direct sums of the form $V_{2n_1} \oplus \cdots \oplus V_{2n_\mu}$ are obviously identified with $V_{2(n_1 + \cdots + n_\mu)}$.

THEOREM 9.4. *Any morphism* $P\colon V_{2n} \to V_{2m}$ *is equivalent to a direct sum of morphisms of the form* (a)–(d).

SKETCH OF THE PROOF. The morphism $P^{(*)}P$ is self-adjoint, and we can make use of Theorem 9.3.

9.6. Canonical forms of the operators $B[S]$**.** Now we reformulate Theorem 9.4 in the language of the operators $B[S]$.

THEOREM 9.4$'$. *The transformation*

$$B[S] \mapsto \lambda W(P)\, B[S]\, W(Q)^{-1},$$

where $P \in \mathrm{Sp}(2n, \mathbb{R})$, $Q \in \mathrm{Sp}(2m, \mathbb{R})$, *and* $\lambda \in \mathbb{C}$, *reduces any operator* $B[S]\colon F_n \to F_m$ *to the form*

$$B\left[\begin{array}{c|c|c|c|c|c}
 & & & \begin{smallmatrix} s_1 & & \\ & \ddots & \\ & & s_k \end{smallmatrix} & & \\ \hline
 & \begin{smallmatrix} t_1 & & \\ & \ddots & \\ & & t_\alpha \end{smallmatrix} & & & \begin{smallmatrix} 1-t_1 & & \\ & \ddots & \\ & & 1-t_\alpha \end{smallmatrix} & \\ \hline
 & & & & & \\ \hline
\begin{smallmatrix} s_1 & & \\ & \ddots & \\ & & s_k \end{smallmatrix} & & & & & \\ \hline
 & \begin{smallmatrix} 1-t_1 & & \\ & \ddots & \\ & & 1-t_\alpha \end{smallmatrix} & & & \begin{smallmatrix} t_1 & & \\ & \ddots & \\ & & t_\alpha \end{smallmatrix} & \\ \hline
 & & & & &
\end{array}\right], \tag{9.3}$$

where $0 < s \leqslant 1$ *and* $0 < t < 1$.

THEOREM 9.4$''$. *The transformation*

$$B[S] \mapsto \lambda W(P)\, B[S]\, W(P)^{-1},$$

where $\lambda \in \mathbb{C}$ *and* $P \in \mathrm{Sp}(2n, \mathbb{R})$, *reduces any self-adjoint operator* $B[S]\colon F_n \to F_n$ *to the form* (9.3).

In both cases the numbers s_j are invariants of the operator $B[S]$, whereas the set t_μ may be replaced by any other set $t'_1, \dots, t'_\alpha$, where $0 < t'_\nu < 1$.

9.7. The invariants s_j. For simplicity, consider the case $n = m$. With each operator $B[S]\colon F_n \to F_n$ we can associate the set of numbers $s_1, \dots, s_n$ consisting of $s_1, \dots, s_\alpha$ (see equation (9.3)) and also of μ ones and $n - \alpha - \mu$ zeros.

THEOREM 9.5. *The roots* λ *of the equation*

$$\det\left((1 - S^*S) - \lambda\left(\begin{pmatrix} -1 & 0 \\ 0 & 1 \end{pmatrix} - S^* \begin{pmatrix} -1 & 0 \\ 0 & 1 \end{pmatrix} S\right)\right) = 0 \tag{9.4}$$

are $\mathrm{Sp}(2n, \mathbb{R}) \times \mathrm{Sp}(2n, \mathbb{R})$*-invariants of the operator* $B[S]$. *In addition*

$$\lambda = \pm(1 - s_j^2)/(1 + s_j^2). \tag{9.5}$$

Note that the roots of equation (9.4) coincide with the eigenvalues of the matrix

$$X(S) = (1 - S^*S)^{1/2}\left(\begin{pmatrix} -1 & 0 \\ 0 & 1 \end{pmatrix} - S^* \begin{pmatrix} -1 & 0 \\ 0 & 1 \end{pmatrix} S\right)^{-1} (1 - S^*S)^{1/2}. \tag{9.6}$$

REMARK. The invariants s_j do not separate the orbits of the group $\mathrm{Sp}(2n, \mathbb{R}) \times \mathrm{Sp}(2n, \mathbb{R})$ that lie on the boundary of the ball $\|S\| \leqslant 1$, i.e., on the "surface" $\|S\| = 1$.

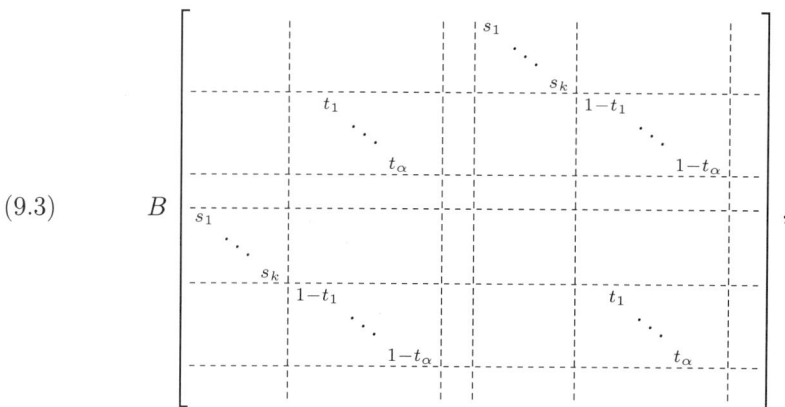

§10. Formulas for the norm of the operators $B[S]$

We consider the case of the operators $B[S]\colon F_n \to F_n$ only. The general case is reduced to the following one: instead of the operator $B[S]\colon F_n \to F_n$ one can consider the operator $B[\widetilde{S}]\colon F_N \to F_N$, where $N = \max(m, n)$, and $\widetilde{S}$ is a matrix S filled to a required size by zeros.

10.1. Formulas for the norm of a self-adjoint operator. Let $\|S\| < 1$, and let $B[S]$ be self-adjoint, where $S = \begin{pmatrix} K & L \\ L^t & \overline{K} \end{pmatrix}$. Let s_j be the invariants of the operator $B[S]$.

It is obvious (see 5.7) that $B[S]$ is a nuclear operator. Let us calculate its trace in two ways. First, the trace is equal to the integral of the kernel of the operator over the diagonal:

$$\operatorname{tr} B[S] = \int \exp\left\{\frac{1}{2}(z\overline{z})\begin{pmatrix} K & L \\ L^t & M \end{pmatrix}\begin{pmatrix} z \\ \overline{z} \end{pmatrix} \exp\{-|z|^2\}\right\} dz$$
$$= \det\left[\begin{pmatrix} 1-L & -K \\ -M & 1-L^t \end{pmatrix}^{-1/2}\right].$$

Secondly, the trace is equal to the sum of the eigenvalues (see 5.7):

$$\operatorname{tr} B[S] = \|B[S]\| \sum s_1^{k_1} \cdots s_n^{k_n} = \|B[S]\| \prod_{j=1}^{n} (1-s_j)^{-1}.$$

Hence

$$\|B[S]\| = \det\left[\begin{pmatrix} 1-L & -K \\ -M & 1-L^t \end{pmatrix}^{-1/2}\right] \prod_{j=1}^{n}(1-s_j). \tag{10.1}$$

Further, note that

$$B\begin{bmatrix} 0 & -E \\ -E & 0 \end{bmatrix} f(z) = f(-z),$$

and this operator commutes with any $B[S]$:

$$B\begin{bmatrix} 0 & -E \\ -E & 0 \end{bmatrix} B[S] = B[S]\, B\begin{bmatrix} 0 & -E \\ -E & 0 \end{bmatrix} = B[JSJ],$$

where $J = \begin{pmatrix} -E & 0 \\ 0 & E \end{pmatrix}$. Calculating $\operatorname{tr}\left(B\begin{bmatrix} 0 & -E \\ -E & 0 \end{bmatrix} B[S]\right)$ in the same two ways, we obtain

$$\|B[S]\| = \det\left[\begin{pmatrix} 1+L & -K \\ -M & 1+L^t \end{pmatrix}^{-1/2}\right] \prod_{j=1}^{n}(1+s_j). \tag{10.2}$$

10.2. Norm of an arbitrary operator $B[S]$. Let $\|S\| < 1$. Calculating $\operatorname{tr}(B[S]^* B[S])$ in two ways

$$\operatorname{tr}(B[S]^* B[S]) = \int_{\mathbb{C}^n}\int_{\mathbb{C}^n} |\exp\{1/2(z\,\overline{u})\, S(z\,\overline{u})^t\}|^2 d\mu(z)\, d\mu(u)$$
$$= \|B[S]\|^2 \prod_{j=1}^{n}(1-s_j^2),$$

we obtain

$$\|B[S]\| = \det(1 - S^*S)^{-1/4} \prod_{j=1}^{n} (1 - s_j^2)^{1/2}. \tag{10.3}$$

Calculating $\operatorname{tr}\left(B[S]^* B\begin{bmatrix} 0 & -E \\ -E & 0 \end{bmatrix} B[S]\right)$ in two ways, we get

$$\|B[S]\| = \det(J - S^*JSJ)^{-1/4} \prod_{j=1}^{n} (1 + s_j^2)^{1/2}. \tag{10.4}$$

Taking into account the fact that the eigenvalues of the matrix (9.6) are given by (9.5), we can write the latter expression in the form

$$\|B[S]\| = |\det Q|^{-1/4} \det((E + |R^{1/2}Q^{-1}R^{1/2}|)/2)^{-1/2}, \tag{10.5}$$

where $Q = J - S^*JS$, $R = 1 - S^*S$.

10.3. One more formula for the norm of $B[S]$. Let $S = \begin{pmatrix} K & L \\ L^t & M \end{pmatrix}$, the matrix L being invertible, and $\|S\| < 1$. By Theorem 9.4′, any matrix S such that $\|S\| < 1$ can be represented in the form

$$S = \begin{pmatrix} P & Q \\ Q^t & R \end{pmatrix} * \begin{pmatrix} 0 & \Sigma \\ \Sigma & 0 \end{pmatrix} * \begin{pmatrix} G & H \\ H^t & F \end{pmatrix},$$

where Σ is a diagonal matrix with eigenvalues $s_1, \dots, s_n$, and $\begin{pmatrix} P & Q \\ R & T \end{pmatrix}$, $\begin{pmatrix} G & H \\ H^t & F \end{pmatrix}$ are unitary matrices (see 3.6). Then (see Proposition 2.1) the operators

$$|\det(Q)|^{1/2} B\begin{bmatrix} P & Q \\ Q^t & R \end{bmatrix}, \qquad |\det(H)|^{1/2} B\begin{bmatrix} G & H \\ H^t & F \end{bmatrix}$$

are unitary. Taking into account the fact that $\left\| B\begin{bmatrix} 0 & \Sigma \\ \Sigma & 0 \end{bmatrix} \right\| = 1$, we find that the norm of the operator

$$X = |\det(Q)|^{1/2} |\det(H)|^{1/2} B\begin{bmatrix} P & Q \\ Q^t & R \end{bmatrix} B\begin{bmatrix} 0 & \Sigma \\ \Sigma & 0 \end{bmatrix} B\begin{bmatrix} G & H \\ H^t & F \end{bmatrix}$$

is equal to 1. Using equations (3.2)–(3.3) to calculate the right-hand side, we obtain

$$X = |\det(Q)|^{1/2} |\det(H)|^{1/2} \det(1 - \Sigma R \Sigma G)^{-1/2} B\begin{bmatrix} K & L \\ L^t & M \end{bmatrix}.$$

However,

$$L = Q\Sigma(1 - \Sigma R \Sigma G)^{-1} H,$$

and, comparing the two last expressions, we obtain

$$\left\| B\begin{bmatrix} K & L \\ L^t & M \end{bmatrix} \right\| = |\det(L)|^{-1/2} |\det(\Sigma)|^{1/2} = |\det(L)|^{-1/2} \prod_{j=1}^{n} |s_j|^{1/2}. \tag{10.6}$$

10.4. Remarks. The formulas just obtained are valid under somewhat more general conditions than those used in the derivation. Thus, equations (10.2), (10.4)–(10.6) remain valid also in the case $\|S\| = 1$.

§11. Remarks

11.1. Remarks on §1. *With regard to* 1.1. What we call the "boson Fock space" should be more precisely called the "holomorphic Bargmann–Segal–Berezin model" of the boson Fock space.

With regard to 1.3 *and* 1.7. The phenomena described here are related to the Bergman kernel function and are general for Hilbert spaces of holomorphic functions, see, for example, [**4**]. On the other hand, the sets of functions φ_u are special cases of the so-called "overfilled bases" (or "coherent states"), see [**5**].

With regard to 1.5. The isomorphism $\mathcal{B}B$ is called the Bargmann–Segal transform. See [**3**].

11.2. Remarks on §2. *With regard to* 2.5. Theorem 2.1 is a curious example of a theorem that has no author. It appeared in mathematics certainly between 1930 and 1958. On the one hand, it is a trivial consequence of the well-known Stone-von Neumann uniqueness theorem (1930) about the unitary equivalence of the set of self-adjoint operators $P_1, \dots, P_n$ and $Q_1, \dots, Q_n$, such that

$$[P_k, P_l] = [Q_k, Q_l] = 0, \qquad [P_k, Q_l] = i\delta_{kl}E$$

(see, for example, [**19,** §1.3]). In this case, however, the words "trivial consequence" only imply the triviality of the proof (after the theorem has already been formulated!). Theorem 2.1 itself is usually attributed to I. Segal [**7**] but this is not quite true, since the difficulties connected with the generalization of Theorem 2.1 to the infinite-dimensional case had been discussed earlier in the book by Friedrichs [**6**].

As soon as the theorem was formulated by Segal in a transparent manner, it evoked a great deal of interest (see A. Weil [**20**], Cartier [**21**]). After one of the first responses to this construction, the construction itself was then called the "Weil representation" instead of the historically justified but never used term "Friedrichs–Segal–Berezin representation". It seems that term "harmonic representation" proposed by I. Segal is better.

The representation $W(\cdot)$ of the group $\mathrm{Sp}(2n, \mathbb{R})$ is only one of the representations of only one of the series of semisimple groups, but its actual role in the theory of semisimple group representations is quite important (see, for example, [**19, 22, 58, 56**]). The role of the "Weil representation" turns out to be still more important in the theory of infinite-dimensional groups, where it plays the part of a sort of "universal object" (see [**23, 24, 26**]). Finally, the construction is important beyond representation theory itself, see [**1, 19–20, 53, 43**].

With regard to 2.6. The remarkable Berezin formula was obtained in [**8**] (see also [**1**]) but remained unnoticed and was rediscovered much later ([**25**]).

With regard to 2.7. In all probability, the operators $W(\cdot)$, which refer to a singular subset in $\mathrm{Sp}(2n, \mathbb{R})$, correspond to complex-valued Gaussian measures concentrated on subspaces in $\mathbb{R}^n \oplus \mathbb{R}^n$. I have not seen them being discussed anywhere else.

11.3. Remarks on §3. The basic question formulated in this section was posed many times in many different forms (we limit ourselves with references to [**11, 14, 27**]) and was solved only in [**12, 13, 16**].

With regard to 3.1–3.6. The semigroup $\mathrm{End}_{\mathcal{B}}(n)$ appeared in the unpublished paper [**11**].

With regard to 3.7. See [**14**].

11.4. Remarks on §4. The construction of this section appeared in [**12, 13**].

With regard to 4.4. See [**28, 29**].

With regard to 4.8. See the remarks on 2.7.

11.5. Remarks on §5. *With regard to* 5.1 and 5.2. See, for example, [**30**, §11.3].

With regard to 5.3. See, for example, [**31**].

With regard to 5.4. See [**32–34**].

With regard to 5.5–5.7. See [**13**]. Fixed points of generalized linear-fractional mappings were extensively studied in connection with the Krein theorem on the invariant subspace; see, for example, [**29**].

11.6. Remarks on §6. *With regard to* 6.1 *and* 6.5. See [**35, 30**]. The problem of the triangle inequality for composite distance does not seem to be solved and does not seem to have ever been formulated.

With regard to 6.6 *and* 6.8. The description of all invariant metrics on $\mathcal{Z}_n$ seems to be unknown.

With regard to 6.7. The Compression Theorem was announced in [**36**].

With regard to 6.9. In [**36**] an inexact assertion on the subject is formulated. Theorems of the type of the "basic theorem of the projective geometry" for symmetric subspaces were extensively discussed in the literature, see, for example, [**37–38**]. Theorem 6.5 is a strengthening of the theorem from [**39**].

11.7. Remarks on §7. *With regard to* 7.1. See the remarks on 1.1. Literally, the construction of the Fock space as $L^2(\mathbb{R}^n)$ collapses when passing to the limit $n \to \infty$. However, if the space L^2 over Gaussian measure is considered, then this passage can be realized; see [**54, 55, 40**]. Operators with Gaussian kernels in $L^2(\mathbb{R}^\infty)$ do not seem to have been studied systematically, see [**49**]

With regard to 7.5–7.10. See [**13**].

With regard to 7.7. Theorem 7.7 is usually attributed to Shale [**9**], but Berezin in [**1**] attributes it to Friedrichs [**6**]. This theorem is present in the book by Friedrichs, although it is not easy to find it. However, the proof by Friedrichs should be considered incorrect (although one can correct his proof by using the techniques of §5). Moreover, F. A. Berezin [**8**] proved Theorem 7.7a at the same time and independently from Shale and also derived the explicit formula for the operators $W(\cdot)$ (Theorem 7.7, (b), (c)). The latter circumstance remained unnoticed in mathematics, and this lent a pleasant but not very useful flavor of mysticism to the whole construction (which is basically very simple: one writes a formula and then verifies it). (To a still greater extent this also applies to the Berezin fermion construction.) It is worth noting that the books [**1**] and [**6**] are, in fact, written in a way that makes them hard to understand.

11.8. Remarks on §8. *With regard to* 8.1 *and* 8.2. These constructions are contemporary to Theorems 2.1 and 7.7.

With regard to 8.3. Invariant cones in Lie algebras were studied by Vinberg, Olshanskiĭ, Paneitz; see [**41, 48**].

With regard to 8.4–8.6. All the statements are taken from the book [**1**], they are very important in the applications of the construction to other infinite-dimensional groups. In [**1**] a statement stronger than Theorem 8.2 is formulated. A good proof of Theorem 8.3 is not known. The existing proof ([**1**]) is based on a calculation (both sides of equation (8.3) can be calculated explicitly), but the calculation is very laborious, and the justification that it is correct is much longer. It would be interesting to derive statements similar to those of 8.3.

11.9. Remarks on §9. *With regard to* 9.1 *and* 9.2. See [**28, 41**].
With regard to 9.5–9.7. See [**11, 15**].

11.10. Remarks on §10. See [**11, 15**]. For the L^p-theory of operators with Gauss kernels, see [**49–51**]; for more references see [**50–51**].

11.11. Remarks on §11. *With regard to* 11.3. The problem formulated in §3 arose in connection with the studies of the "hidden structures" related to the infinite-dimensional groups. In fact, formula (3.3) is only one example of "strange multiplication" appearing in the theory of infinite-dimensional groups. The work [**11**] was done in connection with the following question.

Take the "Weil representation" of the group Sp_∞. Consider the set of all operators of the form $\lambda W(g)$, where $g \in \mathrm{Sp}_\infty$ and $\lambda \in \mathbb{C}$; consider the closure of this set in the weak operator topology. One can understand this closure as a set of limit points of the group Sp_∞. It is easy to see that this weak closure is exactly the semigroup of bounded operators of the type $B[S]$.

In [**42**] (see also [**43**]) the complexification of the group of circle diffeomorphisms was constructed, and it has turned out that to understand this construction one needs to analyze the same semigroup of operators of the type of $B[S]$.

With regard to 11.4. As soon as the answer to the problem of §3 was obtained, the situation with "hidden structures" was clarified, and by now in most cases the description of "hidden structures" is realized (see [**43–47, 12–13, 16**]).

The category Sp is only one of the categories of linear relations related to infinite-dimensional classical groups. It appears that a large part of representation theory (including the theory of finite-dimensional representations of the classical groups) can be extended to the representation theory of categories (see [**36**]).

With regard to 11.5 *and* 11.6. Almost all the statements of §§5 and 6 can be extended to any Riemann symmetric space (see [**36**]).

With regard to 11.7. For the fermion analog of the operators B, see [**16**].

References

1. F. A. Berezin, *The method of second quantization*, "Nauka", Moscow, 1965; English transl., Academic Press, New York, 1966.
2. ———, *Some remarks on the associative hull of the Lie algebra*, Funktsional. Anal. i Prilozhen. **1** (1967), no. 2, 1–14; English transl. in Functional Anal. Appl. **1** (1968), no. 2.
3. ———, *Wick and anti-Wick operator symbols*, Mat. Sb. **86** (1971), no. 4, 578–610; English transl. in Math. USSR-Sb. **15** (1971).
4. ———, *Quantization on complex symmetric spaces*, Izv. Akad. Nauk SSSR Ser. Mat. **39** (1975), no. 2, 363–403; English transl., Math. USSR-Izv. **9** (1975), 341–379.
5. ———, *Covariant and contravariant symbols of operators*, Izv. Akad. Nauk SSSR Ser. Mat. **36** (1972), no. 5, 1134–1167; English transl. in Math. USSR-Izv. **6** (1972).
6. K. O. Friedrichs, *Mathematical aspects of quantum theory of fields*, Interscience, New York, 1953.
7. I. E. Segal, *Foundations of the theory of dynamical systems of infinitely many degrees of freedom*, Danske Vid. Selsk. Math. Phys. Medd. **31** (1959), no. 12, 1–39.
8. F. A. Berezin, *Canonical transformations in the secondary quantization representation*, Dokl. Akad. Nauk SSSR **137** (1961), no. 2, 311–314; English transl., Soviet Phys. Dokl. **6** (1961), 212–215.
9. D. Shale, *Linear symmetries of free boson fields*, Trans. Amer. Math. Soc. **103** (1962), 149–167.

10. D. Shale and W. Stinespring, *States of Clifford algebra*, Ann. Math. **80** (1964), 365–381.
11. G. I. Ol′shanskiĭ, *Subgroup of integral operators with Gaussian kernels*, Unpublished text (1984). (Russian)
12. M. L. Nazarov, Yu. A. Neretin, and G. I. Ol′shanskiĭ, *Semigroups engendrés par la représentation de Weil du group simplectique de dimension infinie*, C. R. Acad. Sci. Paris Sér. I Math. **309** (1989), no. 7, 443–446.
13. Yu. A. Neretin, *On a semigroup of operators in boson Fock space*, Funktsional. Anal. i Prilozhen. **24** (1990), no. 2, 63–73; English transl., Functional Anal. Appl **24** (1990), no. 2, 135–144.
14. R. Howe, *The oscillator semigroup*, Theta functions – Bowdois 1987, Proc. Sympos. Pure Math., vol. 48, Amer. Math. Soc., Providence, RI, 1988, pp. 61–131.
15. G. I. Ol′shanskiĭ, *The Weil representation and norms of Gaussian operators*, Funktsional. Anal. i Prilozhen. **28** (1994), no. 1, 51–67; English transl., Functional Anal. Appl. **28** (1994), no. 1, 42–54.
16. Yu. A. Neretin, *The spinor representation of an infinite-dimensional orthogonal semigroup and the Virasoro algebra*, Funktsional. Anal. i Prilozhen. **23** (1989), no. 3, 32–44; English transl., Functional Anal. Appl **23** (1990), no. 3, 196–207.
17. V. B. Lidskiĭ, *Inequalities for eigenvalues and singular values*, Appendix to the book: F. R. Gantmakher. The theory of matrices. 2nd–4th editions. (Russian)
18. M. Reed and B. Simon, *Methods of modern mathematical physics*, vol. 2, Academic Press, New York, 1975.
19. G. Lion and M. Vergne, *The Weil representation, Maslov index and the theta series*, Birkhäuser, Basel, 1980.
20. A. Weil, *Sur certain groupes d'opérateurs unitaires*, Acta Math. **111** (1964), 143–211.
21. P. Cartier, *Quantum mechanical commutation relations and theta-functions*, Algebraic Groups and Discontinuos Subgroups, Proc. Sympos. Pure Math., vol. 9, Amer. Math. Soc., Providence, RI, 1966, pp. 361–383.
22. R. Howe, *Remarks on classical invariant theory*, Trans. Amer. Math. Soc. **313** (1989), no. 2, 539–570.
23. Yu. A. Neretin, *Representations of the Virasoro algebra and of the affine algebras*, Itogi Nauki i Tekhniki. Sovremennye Problemy Matematiki. Fundamental′nye Napravleniya., vol. 22, VINITI, Moscow, 1988, pp. 163–224; English transl., Encyclopaedia Math. Sci., Springer-Verlag, Berlin–Heidelberg–New York, 1994, pp. 157–234.
24. G. I. Ol′shanskiĭ, *Unitary representations of infinite-dimensional pairs (G, K) and the formalism of R. Howe*, Representations of Lie Groups and Related Topics (A. M. Vershik and D. P. Zhelobenko, eds.), Gordon and Breach, New York and London, 1990, pp. 269–464.
25. M. Vergne, *Groupe simplectique et seconde quantification*, C. R. Acad. Sci. Paris Sér. A **285** (1977), 191–194.
26. A. Pressley and G. B. Segal, *Loop groups*, Clarendon Press, Oxford, 1986.
27. M. Sato, T. Miwa, and M. Jimbo, *Holonomic quantum fields.* I., Publ. Res. Inst. Math. Sci. **14** (1978), 223–267.
28. V. P. Potapov, *Multiplicative structure of J-nonexpanding matrix functions*, Trudy Moskov. Mat. Obshch. **4** (1955), 125–136; English transl. in Amer. Math. Soc. Transl. (2), vol. 15, Amer. Math. Soc., Providence, RI, 1960, pp. 131–143.
29. A. Ya. Azizov and I. S. Iokhvidov, *Linear operators in spaces with an indefinite metric*, "Nauka", Moscow, 1986; English transl., Wiley, New York, 1989.
30. G. I. Pyatetskiĭ-Shapiro, *The geometry of classical regions and automorphic functions*, "Fizmatgiz", Moscow, 1961; English transl., *Automorphic functions and geometry of classical domains*, Gordon and Breach, New York and London, 1969.
31. M. Kashiwara and M. Vergne, *On Segal–Shale–Weil representation and harmonic polynomials*, Invent. Math. **44** (1978), no. 1, 1–47.
32. M. G. Krein and Yu. L. Shmul'yan, *On linear-fractional transformations with operator coefficients*, Math. Research (1967), no. 2, 3, Kishinev, 64–96; English transl. **103** (1974), Amer. Math. Soc., Providence, RI, 125–152.
33. Yu. L. Shmul'yan, *The theory of linear relations and spaces with an indefinite matrix*, Funktsional. Anal. i Prilozhen. **10** (1976), no. 1, 67–72; English transl. in Functional Anal. Appl. **10** (1977), no. 1.

34. ———, *Generalized linear-fractional transforms of operator balls*, Sibirsk. Mat. Zh. **9** (1978), no. 2, 418–425; English transl. in Siberian Math. J. **19** (1978).
35. H. Klingen, *Über die analytischen Abbildungen verallgemeinerter Einheitskreise auf sich*, Math. Ann. **132** (1956), 134–144.
36. Yu. A. Neretin, *On extension of representations of classical group representations to representations of categories*, Algebra i Analiz **3** (1991), no. 1, 176–202; English transl. in Leningrad Math. J. **3** (1992).
37. J. Dieudonné, *La géometrie des groupes classiques*, Springer-Verlag, Heidelberg, 1955.
38. M. Takeuchi, *Basic transformations of symmetric R-spaces*, Osaka J. Math. **25** (1988), no. 2, 259–297.
39. A. B. Goncharov, *Generalized conformal structures on manifolds*, Selecta Math. Soviet. **6** (1987), no. 4, 306–340.
40. G. E. Shilov and Fak Dyk Tyn, *Integral, measure, derivative in a linear space*, "Nauka", Moscow, 1967. (Russian)
41. G. I. Ol′shanskiĭ, *Invariant cones in Lie algebras, Lie semigroups, and holomorphic discrete series*, Funktsional. Anal. i Prilozhen. **15** (1981), no. 4, 53–66; English transl. in Functional Anal. Appl. **15** (1981), no. 4.
42. Yu. A. Neretin, *On the complex semigroup containing a group of circle diffeomorphisms*, Funktsional. Anal. i Prilozhen. **21** (1987), no. 1, 82–83; English transl. in Functional Anal. Appl. **21** (1987), no. 1.
43. ———, *Holomorphic extensions of the representations of a complex group of circle diffeomorphisms*, Mat. Sb. **180** (1989), no. 5, 635–657; English transl., Math. USSR-Sb. **67** (1990), 75–98.
44. ———, *Infinite-dimensional groups. Their mantles, trains, and representations*, Topics in Representation Theory (A. A. Kirillov, ed.), Advances in Soviet Math., vol. 2, Amer. Math. Soc., Providence, RI, 1991, pp. 103–171.
45. G. I. Ol′shanskiĭ, *Unitary representations of (G, K)-pairs related to the infinite symmetric group $S(\infty)$*, Algebra i Analiz **1** (1989), no. 4, 178–209; English transl., Leningrad Math. J. **1** (1990), 983–1014.
46. ———, *On semigroups related to infinite-dimensional groups*, Topics in Representation Theory (A. A. Kirillov, ed.), Advances in Soviet Math., vol. 2, Amer. Math. Soc., Providence, RI, 1991, pp. 67–101.
47. Yu. A. Neretin, *Categories of bistochastic measures and representations of some infinite-dimensional groups*, Mat. Sb. **183** (1992), no. 2; English transl., Math. USSR-Sb. **75** (1993), 197–219.
48. J. Hilgert, K. H. Hoffmann, and D. Lawson, *Lie groups, convex cones and semigroups*, Oxford Univ. Press, Oxford, 1989.
49. E. Nelson, *Free Markov fields*, J. Funct. Anal. **12**, 211–227.
50. J. B. Epperson, *The hypercontractive approach to exactly bounding an operator with complex Gauss kernel*, J. Funct. Anal. **87** (1989), 1–30.
51. D. Bakry, *L'hypercontractivity et son utilisation en theory des semi-groupes*, Lecture Notes in Math., vol. 1581, Springer-Verlag, Berlin–Heidelberg, 1994, pp. 1–114.
52. M. Nazarov, *Oscillator semigroup over non archimedean field*, J. Funct. Anal. **128** (1995), no. 2, 384–438.
53. D. Mumford, M. Nori, and P. Norman, *Tata lectures on Theta* III, Birkhäuser, Basel, 1991.
54. I. Segal, *Tensor algebras over Hilbert spaces.* I, Trans. Amer. Math. Soc **18** (1956), 106–134.
55. ———, *Tensor algebras over Hilbert spaces.* II, Ann. Math. **63** (1956), 297–303.
56. Yu. A. Neretin and G. I. Olshanskii, *Boundary values of holomorphic functions, singular unitary representations of groups $O(p, q)$ and their limits $q \to \infty$*, Zap. Nauchn. Semin. POMI **223** (1995), 1–80; English transl. in J. Sov. Math (to appear).
57. E. F. Beckebah and R. Bellman, *Inequalities*, Fourth edition, Springer-Verlag, Berlin–Heidelberg, 1983.
58. J. D. Adams, *Discrete spectrum of dual reductive pair $(O(p, q), Sp(2m))$*, Inv. Math. **74** (1983), 449–475.

Translated by N. K. KULMAN

Amer. Math. Soc. Transl.
(2) Vol. **175**, 1996

Ergodic Unitarily Invariant Measures on the Space of Infinite Hermitian Matrices

Grigori Olshanski and Anatoli Vershik

In memory of F. A. Berezin

ABSTRACT. Let H be the space of all Hermitian matrices of infinite order and $U(\infty)$ be the inductive limit of the chain $U(1) \subset U(2) \subset \dots$ of compact unitary groups. The group $U(\infty)$ operates on the space H by conjugations, and our aim is to classify the ergodic $U(\infty)$-invariant probability measures on H by making use of a general asymptotic approach proposed in Vershik's note [**V**]. The problem is reduced to studying the limit behavior of orbital integrals of the form

$$\int_{B\in\Omega_n} e^{i\,\mathrm{tr}(AB)} M_n(dB),$$

where A is a fixed $\infty \times \infty$ Hermitian matrix with finitely many nonzero entries, Ω_n is a $U(n)$-orbit in the space of $n \times n$ Hermitian matrices, M_n is the normalized $U(n)$-invariant measure on the orbit Ω_n, and $n \to \infty$.

We also present a detailed proof of an ergodic theorem for inductive limits of compact groups that has been announced in [**V**].

There is a remarkable link between our subject and Schoenberg's [**S2**] theory of totally positive functions, and our approach leads to a new proof of Schoenberg's [**S2**] main theorem, originally proved by function-theoretic methods.

On the other hand, our results have a representation-theoretic interpretation, because the ergodic $U(\infty)$-invariant measures on H determine irreducible unitary spherical representations of an infinite–dimensional Cartan motion group.

The present paper is closely connected with a series of articles by S. V. Kerov and the authors on the asymptotic representation theory of "big" groups, but it can be read independently.

§0. Introduction

1. Ergodic measures. The description of ergodic invariant measures for group actions is a traditional problem of ergodic theory. It is well known that it is

1991 *Mathematics Subject Classification.* Primary 22E65, 28C10.

Both authors were partially supported by the International Science Foundation under Grant MQV000. The first author was also supported by the Russian Foundation for Basic Research under Grant 95-01-00814.

not always possible that this problem can be solved in a satisfactory way. It is a surprising fact that for certain infinite-dimensional (or "big"[1]) groups, there exist nice actions whose ergodic measures can be completely described. The simplest examples are given by the classical theorems due to B. de Finetti and I. J. Schoenberg.

We recall that, by de Finetti's theorem, the ergodic measures on a product space X^∞ that are invariant under the group of permutations of the copies of X (the infinite symmetric group), are exactly the product measures with identical factors. Schoenberg's theorem [**S1**] states that the ergodic $O(\infty)$-invariant measures on the space $\mathbb{R}^\infty$ are exactly the Gaussian product measures (by $O(\infty)$ we denote the inductive limit group $\varinjlim O(n)$).[2]

These classical examples and a number of more complicated ones were discussed in the note [**V**] by one of the authors. In that note, a general "ergodic method" for inductive limits of compact groups was proposed. This method was further developed in the papers [**VK1, VK2, KV**] by Vershik and Kerov. The aim of the present paper is to give a detailed exposition of the method on a model example considered in [**V**].

Let $H(n)$ denote the space of $n \times n$ complex Hermitian matrices, and let $H = \varprojlim H(n)$ be the space of all infinite Hermitian matrices. Let $U(\infty) = \varinjlim U(n)$ be the group of infinite unitary matrices $u = [u_{ij}]$ such that $u_{ij} = \delta_{ij}$ when $i + j$ is large enough. The group $U(\infty)$ operates on the space H by conjugations, and we are interested in the class $\mathcal{M}$ of all ergodic $U(\infty)$-invariant Borel probability measures on H.

Further, let $H(\infty) = \varinjlim H(n)$ be the space of $\infty \times \infty$ Hermitian matrices with finitely many nonzero entries. The spaces $H(\infty)$ and H are in a natural duality, and any measure on H is uniquely determined by its characteristic function (Fourier transform), which is a function on $H(\infty)$.

CLASSIFICATION THEOREM. *The characteristic functions of the measures $M \in \mathcal{M}$ are exactly those of the form*

$$f(A) = e^{i\gamma_1 \operatorname{tr} A - \gamma_2 \operatorname{tr}(A^2)/2} \det \left(\prod_{k=1}^{\infty} \frac{e^{-ix_k A}}{1 - ix_k A} \right), \qquad A \in H(\infty), \tag{0.1}$$

where $\gamma_1, \gamma_2, x_1, x_2, \dots$, the parameters of the measure M, are real numbers such that $\gamma_2 \geqslant 0$ and $\sum x_k^2 < \infty$.

The Classification Theorem implies that any measure $M \in \mathcal{M}$ can be written as the convolution product of a Gaussian measure and a (finite or countable) family of non-Gaussian "elementary" ergodic measures. The latter are essentially supported by rank one matrices and are related to Wishart distributions, well-known in multivariate statistical analysis. Let us emphasize that in this theorem, the structure of the answer turns out to be much more complicated than in de Finetti's and

[1]The term "big groups" was suggested by one of the authors in [**V**]. This term has no rigorous definition: it can mean "infinite-dimensional" or "out of the class of locally compact groups" or something similar. For instance, the group $S(\infty) := \varinjlim S(n)$, the inductive limit of finite symmetric groups, is a discrete, hence a locally compact group, but its properties are very similar to that of the infinite-dimensional group $U(\infty) := \varinjlim U(n)$, so we prefer to rank $S(\infty)$ among the "big" groups.

[2]About this theorem, see also Berg–Christensen–Ressel [**BCR**].

Schoenberg's [**S1**] theorems: up to trivial exceptions, the ergodic measures $M \in \mathcal{M}$ are neither product nor Gaussian measures.

Note that formula (0.1) can be rewritten in the following form:

$$f(A) = \prod_{a \in \operatorname{Spec}(A)} F(a), \qquad A \in H(\infty), \tag{0.2}$$

where Spec(A) is the collection of eigenvalues of A (taken with multiplicities) and

$$F(a) = e^{i\gamma_1 a - \gamma_2 a^2/2} \prod_{k=1}^{\infty} \frac{e^{-ix_k a}}{1 - ix_k a}, \qquad a \in \mathbb{R}. \tag{0.3}$$

The function $F(a)$ has a simple meaning. Let us regard the matrices $B \in H$ as random matrix variables defined on the probability space (H, M). Then $F(a)$ is the characteristic function for the distribution of any diagonal entry B_{ii}, $i = 1, 2, \dots$.[3]

As shown by Pickrell [**Pi2**], the Classification Theorem can be derived from a deep function-theoretic result due to Schoenberg [**S2**]. In the present paper, we prove the Classification Theorem in an entirely different way: our method consists in studying how an ergodic measure $M \in \mathcal{M}$ is approximated by finite-dimensional "orbital measures". (By an *orbital measure* we mean the $U(n)$-invariant probability measure supported by a $U(n)$-orbit $\Omega_n \subset H(n)$, where $n = 1, 2, \dots$.)

We would like to emphasize that such an approach not only leads to the description (0.1) of ergodic measures, but also provides us with additional information about them. Namely, we find a characterization of those sequences $\{M_n\}$ of orbital measures that weakly converge, as $n \to \infty$, to an ergodic measure $M \in \mathcal{M}$, so we can understand how M "grows up" from orbital measures. In particular, we can relate the parameters $\gamma_1, \gamma_2, x_1, x_2, \dots$ to the asymptotic behavior of the eigenvalues of large Hermitian matrices.

A characterization of the convergent sequences M_n is the main result of the paper, Theorem 4.1. The statement is too long to be reproduced here in detail, but roughly speaking, the picture is as follows. Given $n = 1, 2, \dots$, pick any matrix from the orbit Ω_n supporting M_n and represent the collection of the eigenvalues of that matrix as an n-point configuration on the real line. Then this configuration, being contracted with scaling factor $1/n$, must converge, as $n \to \infty$, to a (countable) point configuration $\{x_1, x_2, \dots\} \subset \mathbb{R}$. (Here the x_k's coincide with the parameters appearing in (0.1); in fact, there are also additional restrictions on the growth of the orbits related to the parameters γ_1 and γ_2.)

2. Relationships with representations and totally positive functions. We are interested in the measures of class $\mathcal{M}$ for several reasons:

1) (The initial motivation of [**V**].) The action of the group $U(\infty)$ on the space H is a natural example of a "big" group action. As compared with the action of the group $O(\infty)$ on the space $\mathbb{R}^\infty$ (Schoenberg's case [**S1**]) or the action of $U(\infty)$ on $\mathbb{C}^\infty$, this example is of the next level of complexity. So, from the viewpoint of ergodic theory, it is interesting to find ergodic measures and to compare the result with the one in Schoenberg's case.

[3]This implies that the whole distribution of the matrix entries is uniquely determined by the distribution of any diagonal entry! Note that this property holds for ergodic measures M only: it follows from the so-called Multiplicativity Theorem, see Theorem 2.1 below.

2) The ergodic measures of class $\mathcal{M}$ determine some irreducible unitary representations of a certain infinite-dimensional group $G(\infty)$.

3) These measures are closely related to Schoenberg's totally positive functions.

4) They are also related to harmonic analysis on the group $U(\infty)$ (cf. [**KOV**]).

We shall briefly comment on items 2) and 3).

The group $G(\infty)$ is a model example of an infinite-dimensional Cartan motion group. It can be defined as the inductive limit $G(\infty) = \varinjlim G(n)$, where $G(n)$ stands for the semidirect product $U(n) \ltimes H(n)$. There is a natural bijective correspondence between ergodic $U(\infty)$-invariant probability measures on H and irreducible unitary representations of the group $G(\infty)$ that are spherical with respect to the subgroup $U(\infty) \subset G(\infty)$. Under this correspondence, the spherical functions of $G(\infty)$, when restricted to $H(\infty) := \varinjlim H(n)$, coincide with the characteristic functions of ergodic measures.

It should be noted that all the results and constructions concerning the ergodic measures of class $\mathcal{M}$ can be translated into the language of representation theory. In particular, weak convergence of measures turns into uniform compact convergence of spherical functions, and the problem that is solved in the present paper is a particular case of the following general problem in representation theory of "big" (inductive limit) groups: given a (irreducible) unitary representation of an inductive limit group $\mathcal{G} = \varinjlim \mathcal{G}(n)$, study its approximation by (irreducible) unitary representations of the growing subgroups $\mathcal{G}(n)$ as $n \to \infty$.

(About various asymptotic results in representation theory of "big" groups, see the series of papers by the authors and S. V. Kerov, [**VK1–VK3, Ke, KV, O1–O7**].)

A remarkable feature of representation theory of "big" groups is its connection with analytic problems of total positivity theory (see Thoma [**T**], Boyer [**Bo**], Vershik and Kerov [**VK1, VK2**], Voiculescu [**Vo1**]; for a systematic exposition of total positivity theory, see Karlin's fundamental monograph [**K**]).

To describe this connection, we need the definition of totally positive (TP) functions: these are real-valued functions $\varphi(t)$ on $\mathbb{R}$ such that for any n and any choice of real numbers $t_1 < \cdots < t_n$ and $s_1 < \cdots < s_n$, the determinant of the $n \times n$ matrix $[\varphi(t_i - s_j)]$ is nonnegative. Similarly, one also defines two-sided and one-sided TP sequences: these are TP functions defined on the 1-dimensional lattice $\mathbb{Z} \subset \mathbb{R}$ or its nonnegative part $\mathbb{Z}_+$, respectively.

It turns out that the Fourier transform of TP functions φ on $\mathbb{R}$ leads exactly to functions F of the form (0.3); this claim is Schoenberg's main theorem in [**S2**]. Thus, there is a correspondence $M \leftrightarrow \varphi$ between ergodic measures $M \in \mathcal{M}$ and TP functions φ (normalized by the condition $\int \varphi(t)\,dt = 1$) that can be stated as follows: $\varphi(t)\,dt$ coincides with the distribution of (any) diagonal entry B_{ii}, where B stands for a random Hermitian matrix distributed according to M.

Further, there exists a similar correspondence between characters of the group $U(\infty)$ or of the infinite symmetric group $S(\infty)$ and two-sided or one-sided TP sequences, respectively.

This correspondence can be used in two directions:

On the one hand, old theorems in total positivity theory, obtained by analytic tools in [**ASW, E1, E2, S2, T, K**], can be applied to representation-theoretic problems (classification of spherical functions or characters). Such an approach is adopted by Thoma [**T**], Boyer [**Bo**], and Pickrell [**Pi2**].

But on the other hand, if we can classify spherical functions or characters independently, then we can prove some theorems on TP functions or TP sequences in a new way.

It is the second approach that is adopted in Vershik and Kerov's papers [**VK1, VK2**], and also in the present paper: as a corollary of our main result, we obtain a new derivation of Schoenberg's classification for TP functions on the real line.

Also note a recent paper by Okounkov [**Ok**], where a different (direct representation-theoretic) method is used; the result of [**Ok**] gives yet another way to classify one-sided TP sequences.

3. The method. Now let us describe our techniques in more detail.

The starting point of our approach is a general approximation theorem for ergodic measures (see Theorem 3.2 below[4]). When applied to our concrete situation, it implies that any ergodic measure $M \in \mathcal{M}$ can be approximated by a sequence $\{M_n\}$ of orbital measures (see Theorem 3.3 below). Then the main problem is to understand what sequences $\{M_n\}$ are weakly convergent, as $n \to \infty$, and what are their limits.

An attempt was made to do this in [**V**], but the solution proposed there was incomplete because of a gap in the calculations. Nevertheless, the method itself was correct, and a refinement of the calculations leads to the right result.

To study the weak convergence of orbital measures M_n, we deal with their characteristic functions f_n,

$$f_n(A) = \int_{B \in \Omega_n} e^{i\,\mathrm{tr}(AB)} M_n(dB), \qquad A \in H(n), \tag{0.4}$$

where $\Omega_n \subset H(n)$ is the $U(n)$-orbit carrying M_n. We calculate the Taylor decomposition at the origin of a characteristic function f_n and then analyze the asymptotics of its Taylor coefficients as $n \to \infty$. Note that these Taylor coefficients are nothing but the moments of the measure M_n. (In fact, due to the symmetry of f_n, it is convenient to rewrite its Taylor decomposition as a series of Schur polynomials.) A nonevident fact is that weak convergence of orbital measures can always be controlled by the moments. Such a phenomenon was first discovered in [**VK2**]; there it was applied to an allied problem: classifying characters of the group $U(\infty)$.

Note that we are able to generalize our results to the spaces of real symmetric and quaternionic Hermitian matrices. Instead of Schur polynomials, we must then use Jack symmetric polynomials. As a further generalization, one could consider all matrix spaces of the form $H = \varprojlim H(n)$, where $H(n)$ ranges over one of the 10 series of classical symmetric spaces of Euclidean type, the parameter n being the rank of the space (see [**O5, Pi1**] and especially [**Pi2**] for a discussion of these spaces H). Characteristic functions of orbital measures on the spaces $H(n)$ are sometimes called generalized Bessel functions; for 3 of the 10 spaces H, they can be expressed in terms of elementary functions, and for the 7 other spaces, these are certain multidimensional special functions. The problem consists in studying their limiting behavior as $n \to \infty$. We conjecture that our approach can be transferred to all the spaces H.

4. Contents. The present paper is organized as follows.

[4]There is also another version of this theorem that deals with spherical functions. See Theorem 3.5 below.

In §1, we introduce ergodic measures and explain their relationship with spherical unitary representations.

In §2, we formulate the so-called Multiplicativity Theorem for characteristic functions of ergodic measures (Theorem 2.1). This important result states that a $U(\infty)$-invariant probability measure M on H is ergodic if and only if its characteristic function is multiplicative in a certain sense; it follows that the set of ergodic measures is closed under convolution. Then we state the classification result (Theorem 2.9), which shows that any ergodic measure is a convolution product of certain "elementary" measures. We also discuss a number of corollaries of these theorems.

Section 3 is devoted to a proof of a general approximation theorem that has been announced in [**V**] (Theorem 3.2 below). This result may be viewed as an ergodic theorem for actions of general inductive limits of compact groups.

In §4, we state the main result of the paper, Theorem 4.1, and outline its proof.

Section 5 contains a preliminary result, needed for the proof of Theorem 4.1. There we decompose the characteristic function of an orbital measure into a series of Schur symmetric functions.[5]

In §6, we prove Theorem 4.1.

In §7, following Pickrell's arguments in [**Pi2**], we establish the equivalence of two classification problems: that of ergodic invariant measures $M \in \mathcal{M}$ and that of totally positive functions φ on $\mathbb{R}$. Note that our own contribution to the results of this section is very modest. Here we only aimed to clarify some technical details of the correspondence $M \leftrightarrow \varphi$ and to explain how our main result implies Schoenberg's classification.[6]

In the final §8, we explain a connection between our main theorem and the main result of the remarkable work [**CS**] by Curry and Schoenberg.

A preliminary version of the present paper appeared as the preprint [**OV**].

5. Acknowledgements. Already in the sixties, the second named author (A. V.) had discussions with F. A. Berezin about the subject of the present work and its links with mathematical physics.

The preparation of the preliminary version [**OV**] of the paper was completed while the first named author (G. O.) was a guest of the Department of Mathematical Sciences, University of Tokyo. It is a pleasure to thank Masatoshi Noumi and the Department of Mathematical Sciences for their kind invitation to Tokyo and hospitality.

Special thanks are due to Andreĭ Okounkov who brought to our attention the important relation with the work [**CS**] by Curry and Schoenberg.

Both authors also acknowledge the financial support by the International Science Foundation during the final phase of the work.

§1. Ergodic measures and spherical representations

Let $H(n)$ denote the real vector space formed by complex Hermitian $n \times n$ matrices, $n = 1, 2, \dots$. There is a natural embedding

$$H(n) \to H(n+1), \qquad A \mapsto \begin{bmatrix} A & 0 \\ 0 & 0 \end{bmatrix},$$

[5]It would be interesting to find analogues of this result for all generalized Bessel functions.

[6]Note, however, that we are not completely satisfied by the way we obtain Schoenberg's theorem. We think there should exist a more conceptual derivation of this result by approximation methods. A similar remark can be made *à propos* TP sequences as well.

and we denote by $H(\infty)$ the corresponding inductive limit space $\varinjlim H(n)$. Then $H(\infty)$ is identified with the space of infinite Hermitian matrices with a finite number of nonzero entries. We equip $H(\infty)$ with the inductive limit topology. In particular, a function $f\colon H(\infty) \to \mathbb{C}$ is continuous if its restriction to $H(n)$ is continuous for any n.

Let H stand for the space of all infinite Hermitian matrices. For $A \in H$ and $n = 1, 2, \ldots$, we denote by $\theta_n(A) \in H(n)$ the upper left $n \times n$ corner of A. Using the projections $\theta_n\colon H \to H(n)$, $n = 1, 2, \ldots$, we may identify H with the projective limit space $\varprojlim H(n)$. We equip H with the corresponding projective limit topology. In particular, H is a Borel space.

There is a natural pairing

$$H(\infty) \times H \to \mathbb{R}, \qquad (A, B) \mapsto \operatorname{tr}(AB).$$

Using it, we may regard H as the algebraic dual space of $H(\infty)$.

Note that H can be identified, in an obvious manner, with the space $\mathbb{R}^\infty = \mathbb{R} \times \mathbb{R} \times \cdots$. Under this identification, $H(\infty) \subset H$ turns into $\mathbb{R}_0^\infty := \bigcup_{n \geqslant 1} \mathbb{R}^n$, and the pairing defined above becomes the standard pairing between $\mathbb{R}_0^\infty$ and $\mathbb{R}^\infty$.

Given a Borel probability measure M on H, we define its Fourier transform, or characteristic function, as the following function on $H(\infty)$:

$$f(A) = f_M(A) = \int_H e^{i \operatorname{tr}(AB)} M(dB). \tag{1.1}$$

We need the following statement:

PROPOSITION 1.1. *The Fourier transform* (1.1) *establishes a bijective correspondence between Borel probability measures M on H and continuous positive definite normalized functions f on $H(\infty)$.*

Here "normalized" means that $f(0) = 1$.

PROOF. This is an immediate corollary of Kolmogorov's consistency theorem and Bochner's theorem. Indeed, let us identify $(H(\infty), H)$ with $(\mathbb{R}_0^\infty, \mathbb{R}^\infty)$. Then Kolmogorov's theorem implies that Borel probability measures on $\mathbb{R}^\infty$ are just the projective limits of probability measures on the $\mathbb{R}^n$'s, and Bochner's theorem allows us to restate this fact in terms of characteristic functions. □

Let $U(n)$ be the group of unitary $n \times n$ matrices, $n = 1, 2, \ldots$. For any n, we embed $U(n)$ into $U(n+1)$ using the mapping $u \mapsto \begin{bmatrix} u & 0 \\ 0 & 1 \end{bmatrix}$. Let $U(\infty) = \varinjlim U(n)$ denote the corresponding inductive limit group. We regard $U(\infty)$ as the group of infinite unitary matrices $u = [u_{ij}]_{i,j=1}^\infty$ with a finite number of entries $u_{ij} \neq \delta_{ij}$. The group $U(\infty)$ acts by conjugations both on $H(\infty)$ and H, and the pairing between these two spaces is clearly $U(\infty)$-invariant.

Let us recall some basic facts concerning ergodic measures.

DEFINITION 1.2. Let $\mathcal{X}$ be a Borel space, let $\mathcal{G}$ be a group of Borel transformations of $\mathcal{X}$, and let M be a $\mathcal{G}$-invariant probability Borel measure on $\mathcal{X}$.

(i) A Borel subset $\mathcal{Y} \subset \mathcal{X}$ is said to be *$\mathcal{G}$-invariant* mod 0 if $M((g\mathcal{Y}) \triangle \mathcal{Y}) = 0$ for any $g \in \mathcal{G}$.

(ii) The measure M is said to be *ergodic* with respect to $\mathcal{G}$ if the M-volume of any $\mathcal{G}$-invariant mod 0 Borel subset $\mathcal{Y} \subset \mathcal{X}$ is equal either to 0 or 1.

PROPOSITION 1.3. *The following conditions on a Borel probability measure M on $\mathcal{X}$ are equivalent:*

(i) *M is ergodic;*

(ii) *M is an extreme point of the convex set formed by all Borel probability measures on $\mathcal{X}$;*

(iii) *the subspace of $\mathcal{G}$-invariant vectors in the Hilbert space $L^2(\mathcal{X}, M)$ is exhausted by constant functions.*

PROOF. See, e.g., Phelps [**Ph,** Proposition 10.4]. □

For compact group actions on locally compact spaces, ergodic measures coincide with invariant measures supported by the orbits, so that for such actions, classifying orbits and classifying ergodic measures are equivalent problems. However, for "big" groups like $U(\infty)$, these two problems are quite different: the first one seems to be out of reach, while the second one has, in certain cases, a nice solution. According to a general principle of ergodic theory, ergodic measures may be viewed as true substitutes of orbits.

Let $\mathcal{M}$ denote the set of all ergodic $U(\infty)$-invariant Borel probability measures M on H, and let $\mathcal{F}$ denote the set of the corresponding characteristic functions $f = f_M$.

The following result, which follows from Propositions 1.1 and 1.3, gives a useful characterization of the class $\mathcal{F}$:

PROPOSITION 1.4. *The functions $f \in \mathcal{F}$ are exactly the extreme points of the convex set formed by all continuous $U(\infty)$-invariant positive definite normalized functions on $H(\infty)$.* □

Let $G(n) = U(n) \ltimes H(n)$ be the semidirect product of $U(n)$ and the additive group of the vector space $H(n)$. The elements $g \in G(n)$ are the pairs $(u, A) \in U(n) \times H(n)$ with the multiplication rule

$$(u, A) \cdot (v, B) = (uv, v^{-1}Av + B).$$

In the same way, we define the group $G(\infty) = U(\infty) \ltimes H(\infty)$. The group $G(\infty)$ can be also viewed as the inductive limit group $\varinjlim G(n)$, and we equip it with the inductive limit topology. Using the embeddings

$$u \mapsto (u, 0), \quad A \mapsto (1, A), \qquad u \in U(\infty), \quad A \in H(\infty),$$

we may identify $U(\infty)$ and $H(\infty)$ with the corresponding subgroups in $G(\infty)$.

Let T be a unitary representation of $G(\infty)$ in a Hilbert space $\mathcal{H}(T)$ (we tacitly assume T is continuous with respect to the inductive limit topology of the group $G(\infty)$). Then T is said to be *spherical* if it is irreducible and the subspace $\mathcal{H}(T)^{U(\infty)}$ of $U(\infty)$-invariants in $\mathcal{H}(T)$ is nonzero.

Suppose T is spherical. It is known that the space $\mathcal{H}(T)^{U(\infty)}$ is one-dimensional (see Olshanski [**O5,** Theorem 23.6]). A vector $h \in \mathcal{H}(T)^{U(\infty)}$ of norm 1 is called a *spherical vector* of T, and the corresponding matrix element

$$\varphi_T(g) = (T(g)h, h), \qquad g \in G(\infty),$$

is called the *spherical function* of T. Since φ_T does not change when h is multiplied by a complex number of absolute value 1, φ_T is an invariant of T; moreover, it uniquely determines T. Further, as φ_T is bi-invariant with respect to the subgroup

$U(\infty) \subset G(\infty)$, it is uniquely determined by its restriction $\varphi_T|H(\infty)$ to $H(\infty)$, which is a continuous $U(\infty)$-invariant function.

PROPOSITION 1.5. *There is a natural bijective correspondence $T \leftrightarrow M$ between (equivalence classes of) spherical representations T of the group $G(\infty)$ and ergodic measures $M \in \mathcal{M}$. Under that correspondence, the functions $\varphi_T|H(\infty)$ coincide with the characteristic functions f_M. Given $M \in \mathcal{M}$, the corresponding representation T can be realized in the Hilbert space $L^2(H, M)$ as follows:*

$$(T(u)\,\Psi)(B) = \Psi(u^{-1}Bu), \qquad u \in U(\infty) \subset G(\infty),$$
$$(T(A)\,\Psi)(B) = e^{i\,\mathrm{tr}(AB)}\Psi(B), \qquad A \in H(\infty) \subset G(\infty),$$

where $\Psi \in L^2(H, M)$ and $B \in H$. Finally, in this realization, the spherical vector is the constant function $\Psi_0(B) \equiv 1$.

PROOF. Indeed, it is enough to notice that spherical functions, when restricted to $H(\infty)$, are just the extreme $U(\infty)$-invariant continuous positive definite functions, normalized at $0 \in H(\infty)$ and then apply Proposition 1.4. □

§2. The Classification Theorem

Denote by $D(\infty)$ the subspace of diagonal matrices in $H(\infty)$. An element of $D(\infty)$ will be written as $\mathrm{diag}(a_1, a_2, \dots)$, where $a_1, a_2, \dots \in \mathbb{R}$ and $a_k = 0$ for k large enough. Since any matrix in $H(\infty)$ can be diagonalized under the action of the group $U(\infty)$, a $U(\infty)$-invariant function f on $H(\infty)$ is uniquely determined by $f\,|\,D(\infty)$, the restriction of f to $D(\infty)$.

Assume f is an $U(\infty)$-invariant function on $H(\infty)$, $f(0) = 1$. Let us say that f is *multiplicative* if

$$f(\mathrm{diag}\,(a_1, a_2, \dots)) = F(a_1)F(a_2)\cdots, \tag{2.1}$$

where F is a function on $\mathbb{R}$ such that $F(0) = 1$. In other words,

$$f(A) = F(a_1)F(a_2)\cdots, \qquad A \in H(\infty), \tag{2.2}$$

where $a_1, a_2, \dots$ are the eigenvalues of A. Note that $F(a) = f(\mathrm{diag}\,(a, 0, 0, \dots))$, i.e., $F = f\,|\,H(1)$.

THEOREM 2.1 (Multiplicativity Theorem). *Let f be a continuous $U(\infty)$-invariant positive definite function on $H(\infty)$, $f(0) = 1$. Then f is extreme (i.e., $f \in \mathcal{F}$) if and only if f is multiplicative.*

PROOF. See, e.g., Olshanski [**O5,** Theorem 23.8]. There the Multiplicativity Theorem is stated for allied groups (like $GL(n, \mathbb{C})$) but the proof is exactly the same. □

Not that there are many theorems of this type and many different methods to prove them: see Thoma [**T**], Ismagilov [**I1**] and [**I2**] (Ismagilov's method is also explained in Olshanski [**O4,** Section 2.5]), Nessonov [**N**], Olshanski [**O1**], Pickrell [**Pi1**], Vershik–Kerov [**VK4**], Voiculescu [**Vo1**] and [**Vo2**], Stratila–Voiculescu [**SV**].

Theorem 2.1 implies that a function $f \in \mathcal{F}$ is uniquely determined by the function $F := f|H(1)$, which is a continuous function in one real variable. Note that

$$F(a) = F_M(a) = \int_H e^{iaB_{11}} M(dB), \qquad a \in \mathbb{R},\ B \in H, \tag{2.3}$$

where $M \in \mathcal{M}$ corresponds to f.

DEFINITION 2.2. Let $\mathcal{F}_1 = \{F\}$ denote the class of all functions on $\mathbb{R}$ of the form $F = f|H(1)$ where f ranges over $\mathcal{F}$. In other words, a function F on $\mathbb{R}$ belongs to $\mathcal{F}_1$ if and only if F is continuous, $F(0) = 1$, and the corresponding function (2.2) on $H(\infty)$ is positive definite.

Clearly, the classification of the measures $M \in \mathcal{M}$ is reduced to that of the functions $F \in \mathcal{F}_1$.

Theorem 2.1 has a number of corollaries.

COROLLARY 2.3. *The class $\mathcal{F}_1$ is stable under pointwise multiplication.* □

This implies that the class $\mathcal{F}$ is also stable under multiplication, and the class $\mathcal{M}$ is stable under convolution.

COROLLARY 2.4. *If a sequence $F_1, F_2, \dots \in \mathcal{F}_1$ pointwise converges to a continuous function F on $\mathbb{R}$, then $F \in \mathcal{F}_1$.* □

COROLLARY 2.5. *For a real γ, the Dirac measure concentrated at $\gamma \cdot 1 \in H$ belongs to $\mathcal{M}$. The corresponding function $F \in \mathcal{F}_1$ is*

$$F(a) = e^{i\gamma a}, \qquad a \in \mathbb{R}. \tag{2.4}$$

PROOF. Since scalar matrices are $U(\infty)$-invariant, our measure is invariant. It is clearly ergodic. It is evident that the corresponding function from $\mathcal{F}_1$ is given by (2.4). □

COROLLARY 2.6. *Given $\gamma \geqslant 0$, let M be the Gaussian distribution on H such that, for a matrix $B \in H$, the diagonal entries B_{ii} and the off-diagonal entries $\operatorname{Re} B_{ij}$, $\operatorname{Im} B_{ij}$, $i < j$, are independent Gaussian variables with mean 0 and variance γ. Then $M \in \mathcal{M}$, and the corresponding function $F \in \mathcal{F}_1$ is*

$$F(a) = e^{-\gamma a^2/2}, \qquad a \in \mathbb{R}. \tag{2.5}$$

PROOF. It is easily verified that M is invariant and its characteristic function satisfies (2.1), where F is given by (2.5). □

COROLLARY 2.7. *Let ω denote the Gaussian measure on $\mathbb{C}$ with density given by $\pi^{-1}\exp(-|z|^2)$, $z \in \mathbb{C}$. Given $y \in \mathbb{R}$, let M be the image of the Gaussian product measure $\omega^{\otimes\infty}$ on $\mathbb{C}^\infty$ under the following Borel mapping $\mathbb{C}^\infty \to H$:*

$$\mathbb{C}^\infty \ni \xi \mapsto y(-1 + \xi^*\xi) = B \in H. \tag{2.6}$$

(Here $\xi = (\xi_1, \xi_2, \dots) \in \mathbb{C}^\infty$ is regarded as a row vector, so that the (i,j)-entry of the matrix B is equal to $y(-1 + \bar{\xi}_i\xi_j)$, $i,j = 1,2,\dots$.) Then $M \in \mathcal{M}$ and the corresponding function $F \in \mathcal{F}_1$ is given by

$$F(a) = \frac{e^{-iya}}{1 - iya}, \qquad a \in \mathbb{R}. \tag{2.7}$$

PROOF. The invariance of M is obvious. A direct calculation shows that the characteristic function of M satisfies (2.1) with F given by (2.7). □

We shall call the measures defined in Corollaries 2.5–2.7 the *elementary ergodic measures.*

Note that we could omit the term -1 in (2.6) and then the numerator in (2.7) would disappear. However, due to this term, the mean of M is equal to zero and the function (2.7) has the property

$$F(a) = \frac{e^{-iya}}{1 - iya} = 1 - \frac{3}{2}\, a^2 y^2 + O(y^3) \quad \text{as } y \to 0, \tag{2.8}$$

i.e., the term of degree 1 in y vanishes. This is important for the following construction.

PROPOSITION 2.8. *The class $\mathcal{F}_1$ contains all functions of the form*

$$F_{\gamma_1,\gamma_2,x}(a) = e^{i\gamma_1 a - \gamma_2 a^2/2} \prod_k \frac{e^{-ix_k a}}{1 - ix_k a}, \tag{2.9}$$

where $\gamma_1 \in \mathbb{R}$ and $\gamma_2 \geqslant 0$ are arbitrary constants, and $x = (x_1, x_2, \dots)$ is a sequence of real numbers such that $\sum x_k^2 < \infty$.

PROOF. Suppose first the sequence x is finite. Then (2.9) is a finite product of functions that belong to the class $\mathcal{F}_1$ due to Corollaries 2.5, 2.6, and 2.7. By Definition 2.2, their product also belongs to $\mathcal{F}_1$. Finally, if the sequence x is infinite, then the assumption $\sum x_k^2 < \infty$ and the estimate (2.8) imply that the product in (2.9) is convergent for any $a \in \mathbb{R}$; moreover, the result is a continuous function. Then Corollary 2.4 implies that this function belongs to $\mathcal{F}_1$. □

COMMENTS. 1) The order of the x_k's is unessential, so that $x = (x_k)$ is a point configuration (or multiset) rather than a sequence.

2) If $\sum |x_k| < \infty$, then (2.9) may be rewritten as

$$F_{\gamma_1,\gamma_2,x}(a) = e^{i\overline{\gamma}_1 a - \gamma_2 a^2/2} \prod_k \frac{1}{1 - ix_k a}, \qquad \overline{\gamma}_1 := \gamma_1 - \sum_k x_k. \tag{2.10}$$

3) The function (2.9) admits a holomorphic continuation to the horizontal strip $\{z \in \mathbb{C} \mid |\operatorname{Im} z| < \varepsilon\}$, where $\varepsilon^{-1} = \sup |x_k|$.

4) The parameters γ_1, γ_2, x are uniquely determined by the function $F_{\gamma_1,\gamma_2,x}$.

5) The characteristic function (2.2) corresponding to the function (2.9) can be written as

$$f(A) = e^{i\gamma_1 \operatorname{tr} A - \gamma_2 \operatorname{tr}(A^2)/2} \det\left(\prod_k \frac{e^{-ix_k A}}{1 - ix_k A}\right), \qquad A \in H(\infty). \tag{2.11}$$

THEOREM 2.9 (Classification Theorem). *The class $\mathcal{F}_1$ is exhausted by the functions of the form (2.9). Thus, the characteristic functions of the ergodic $U(\infty)$-invariant Borel probability measures on H are just the functions of the form (2.11), where $\gamma_1, \gamma_2, x_1, x_2, \dots$ are real parameters such that $\gamma_2 \geqslant 0$ and $\sum x_k^2 < \infty$.*

This result gives a description of ergodic measures $M \in \mathcal{M}$: any such M is a convolution of the elementary ergodic measures constructed in Corollaries 2.5–2.7.

The proof will be given in §4: we shall derive Theorem 2.9 from a more general result, Theorem 4.1.

Note that the elementary ergodic measures of Corollaries 2.5 and 2.6 are infinitely divisible with respect to convolution, whereas those of Corollary 2.7 are not.

REMARK 2.10. The construction of Corollary 2.7 can be generalized as follows. Fix $k = 1, 2, \dots$, consider the space $\mathbb{C}^{k\times\infty}$ of all complex matrices Ξ with k rows and infinitely many columns, and equip it with the Gaussian product measure

$$\omega^{k\times\infty} := \omega^{\otimes\infty} \otimes \cdots \otimes \omega^{\otimes\infty}. \tag{2.12}$$

Let $z, x_1, \dots, x_k$ be any real numbers, and let $X = \operatorname{diag}(x_1, \dots, x_k)$ denote the diagonal matrix of order k with diagonal entries $x_1, \dots, x_k$. Consider the mapping

$$\mathbb{C}^{k\times\infty} \ni \Xi \mapsto z \cdot 1 + \Xi^* X \Xi = B \in H. \tag{2.13}$$

Then the image of the Gaussian measure (2.12) under the mapping (2.13) is an ergodic measure $M_{z;x_1,\dots,x_k}$. Its parameters are

$$\gamma_1 = z + x_1 + \cdots + x_k, \qquad \gamma_2 = 0, \qquad x = (x_1, \dots, x_k, 0, 0, \dots). \tag{2.14}$$

From Theorem 2.9 one can deduce that the measures $M_{z;x_1,\dots,x_k}$, $k = 1, 2, \dots$, form a weakly dense subset of $\mathcal{M}$. Further, let $H_{\leqslant k}$ denote the closed subspace of H formed by matrices of rank $\leqslant k$. One can prove that the measures $M_{0;x_1,\dots,x_k}$ are just those measures $M \in \mathcal{M}$ that are supported by $H_{\leqslant k}$. Finally, note that the measures $M_{0;x_1,\dots,x_k}$ with $x_1 = \cdots = x_k$ are infinite-dimensional analogs of the well-known Wishart distributions (see e.g., Muirhead [**Mu**]).

REMARK 2.11. Let $H_+(n) \subset H(n)$ and $H_+ \subset H$ denote the subsets of nonnegative definite matrices. Then H_+ coincides with $\varprojlim H_+(n)$ and is a closed cone in H. One can prove that a measure $M \in \mathcal{M}$ is supported by H_+ if and only if its parameters satisfy the conditions

$$\gamma_2 = 0, \qquad x_1 \geqslant 0, \quad x_2 \geqslant 0, \quad \dots, \quad \sum x_k \leqslant \gamma_1 < \infty. \tag{2.15}$$

REMARK 2.12. Let $M \in \mathcal{M}$ and let $F \in \mathcal{F}_1$ be the corresponding function (2.9). We may regard (H, M) as a probability space and the matrix elements B_{ij} of a matrix $B \in H$ as random variables. Let μ denote the distribution of the real random variable B_{11}; then F is the characteristic function of μ. We know that M is completely determined by μ. Let us describe the structure of μ. It follows from (2.9) that μ is the convolution of a (not necessarily centered) normal distribution with a family of distributions possessing characteristic functions of the form (2.7). It is easily verified that if $y > 0$, then the distribution with characteristic function (2.7) is supported by the half-line $t \geqslant -y$ and has the density

$$t \mapsto y^{-1} e^{-y^{-1}(t+y)}, \qquad t \geqslant -y. \tag{2.16}$$

This is the shifted exponential distribution with variance y^2 and mean 0. If the parameter y is negative, it suffices to replace t by $-t$. Thus, μ is the convolution of a normal distribution with a family of modified exponential distributions.

REMARK 2.13. Let $M \in \mathcal{M}$ be an ergodic measure for which not all the parameters x_k vanish. Then the characteristic function $f(A)$, see (2.11), cannot be factorized into a product of factors each of which depends on a single matrix element A_{ij} only. This means that M is not a product measure, so that the matrix elements B_{pq}, considered as random variables defined on (H, M), are not independent on the whole (in the particular case of the measures $M_{z;x_1,\dots,x_k}$, this is seen from their construction via the mapping (2.13)). However, certain matrix elements are independent. For instance, it is evident that the diagonal elements are independent. More generally, for any $(p_1, q_1), \dots, (p_m, q_m)$ such that $i \neq j$ implies $p_i \neq p_j$, $q_i \neq q_j$, $p_i \neq q_j$, the matrix elements $B_{p_1 q_1}, \dots, B_{p_m q_m}$ are independent. This claim can be deduced from (2.2).

According to Proposition 1.5, Theorem 2.9 also gives a complete description of spherical representations of the group $G(\infty)$. We shall now discuss the possibility of extending spherical representations to some topological completions of the inductive group $G(\infty)$.

Let us regard $U(\infty)$ as a group of unitary operators in the complex coordinate Hilbert space ℓ_2, and let $\overline{U}(\infty) \supset U(\infty)$ stand for the group of all unitary operators in ℓ_2. Note that $\overline{U}(\infty)$ is a topological group with respect to the weak operator topology (which coincides on unitary operators with the strong operator topology). Further, let $H(\infty)_1 \supset H(\infty)$ (respectively, $H(\infty)_2 \supset H(\infty)$) denote the space of the trace class (respectively, Hilbert–Schmidt) Hermitian operators in ℓ_2, equipped with the topology defined by the trace norm $\|\cdot\|_1$ (respectively, by the Hilbert–Schmidt norm $\|\cdot\|_2$).

One can check that the actions

$$\overline{U}(\infty) \times H(\infty)_1 \to H(\infty)_1, \qquad \overline{U}(\infty) \times H(\infty)_2 \to H(\infty)_2,$$

where $(u, A) \mapsto uAu^{-1}$, are continuous (cf. Shale [**Sha**]). Thus we may form the semidirect products

$$G(\infty)_1 = \overline{U}(\infty) \ltimes H(\infty)_1, \qquad G(\infty)_2 = \overline{U}(\infty) \ltimes H(\infty)_2,$$

which are topological groups with respect to the corresponding product topologies. Note that the inductive limit group $G(\infty) = U(\infty) \ltimes H(\infty)$ is contained as a dense subgroup both in $G(\infty)_1$ and $G(\infty)_2$.

COROLLARY 2.14. (i) *Any spherical representation of the group $G(\infty)$ admits an extension to a continuous representation of the topological group $G(\infty)_1$.*

(ii) *It can be continued to $G(\infty)_2$ if and only if the parameter γ_1 in* (2.11) *vanishes.*

PROOF. (i) Indeed, for any function F of the form (2.9), the corresponding function f on $H(\infty)$, which is defined by (2.11), can be extended to a continuous function on $H(\infty)_1$. Then the latter function can be extended to a continuous $\overline{U}(\infty)$-bi-invariant function on the group $G(\infty)_1$, and our claim follows.

(ii) We argue as in (i), using the fact that f is continuous with respect to the Hilbert–Schmidt norm if and only if $\gamma_1 = 0$. □

§3. Approximation of ergodic measures by orbital measures

Let $\mathcal{X}$ be a separable metric space and $C(\mathcal{X})$ be the Banach space of bounded continuous functions on $\mathcal{X}$. By a *measure* on $\mathcal{X}$ we shall always mean a Borel probability measure. Recall that a sequence $\{\nu_n\}$ of measures on $\mathcal{X}$ is said to be *weakly convergent to a measure* ν (notation: $\nu_n \Rightarrow \nu$) if $\langle\psi, \nu_n\rangle \to \langle\psi, \nu\rangle$ as $n \to \infty$ for any $\psi \in C(\mathcal{X})$, see, e.g., [**Bi, Pa1**].

PROPOSITION 3.1. *There exists a countable set $\Psi \subset C(\mathcal{X})$ of functions with the following property: a sequence of measures $\nu_1, \nu_2, \dots$ weakly converges to a measure ν provided convergence occurs for any function $\psi \in \Psi$.*

PROOF. See, e.g., [**Pa1,** Chapter II, Theorem 6.6]. □

Let $\mathcal{K}(1) \subset \mathcal{K}(2) \subset \dots$ be an ascending chain of compact groups, and let $\mathcal{K} = \varinjlim \mathcal{K}(n)$ be the corresponding inductive limit topological group. We assume there is a jointly continuous action $(u, x) \mapsto u \cdot x$ of the group $\mathcal{K}$ on the space $\mathcal{X}$. Let m_n denote the normalized Haar measure on $\mathcal{K}(n)$. Given a point $x \in \mathcal{X}$, let $m_n(x)$ denote the image of m_n under the mapping $u \mapsto u \cdot x$, where u ranges over $\mathcal{K}(n)$, i.e., $m_n(x)$ is the unique $\mathcal{K}(n)$-invariant probability measure supported by the orbit $\mathcal{K}(n) \cdot x$.

THEOREM 3.2 (Vershik [**V,** Theorem 1]). *Let ν be an ergodic $\mathcal{K}$-invariant Borel probability measure on $\mathcal{X}$. Then there exists a point $x \in \mathcal{X}$ such that $m_n(x) \Rightarrow \nu$ as $n \to \infty$. Moreover, the set of all points $x \in \mathcal{X}$ with this property is of full measure with respect to ν.*

PROOF. Given $\psi \in C(\mathcal{X})$, put

$$\psi_n(x) = \langle\psi, m_n(x)\rangle = \int_{\mathcal{K}(n)} \psi(u \cdot x)\, m_n(du), \qquad x \in \mathcal{X}, \tag{3.1}$$

and let $\overline{\psi}$ denote the constant function

$$\overline{\psi}(x) \equiv \langle\psi, \nu\rangle \cdot 1 = \left(\int_{\mathcal{X}} \psi(x)\,\nu(dx)\right) \cdot 1. \tag{3.2}$$

To prove the theorem, it is enough to verify that $\psi_n \to \overline{\psi}$ almost everywhere for any $\psi \in C(\mathcal{X})$. Indeed, it will follow that, given an arbitrary countable family of functions $\Psi \subset C(\mathcal{X})$, there exists a subset $\mathcal{X}'$ of full measure such that

$$\langle\psi, m_n(x)\rangle \to \langle\psi, \nu\rangle \qquad \text{for any } \psi \in \Psi \text{ and any } x \in \mathcal{X}'. \tag{3.3}$$

Taking as Ψ a subset from Proposition 3.1, we shall obtain $m_n(x) \Rightarrow \nu$ for all $x \in \mathcal{X}'$.

Now we proceed with the proof that, given $\psi \in C(\mathcal{X})$, we have $\psi_n \to \overline{\psi}$ almost everywhere.

Remark that the ψ_n are bounded Borel functions. Indeed, given $x \in \mathcal{X}$ and n, the function $u \mapsto \psi(u \cdot x)$ is continuous on $\mathcal{K}(n)$, so that the integral in (3.1) can be approximated by its Riemannian integral sums. It follows that ψ_n is a pointwise limit of continuous functions, so it is a Borel function. Its boundedness is immediate.

Now it is enough to verify the following two claims:

Claim 1. $\psi_n \to \overline{\psi}$ in the metric of $L^2(\mathcal{X}, \nu)$.

Claim 2. For almost all points $x \in \mathcal{X}$, the following limit exists

$$\psi_\infty(x) := \lim_{n\to\infty} \psi_n(x).$$

Indeed, since the ψ_n's are uniformly bounded, Claim 2 will imply that $\psi_n \to \psi_\infty$ in $L^2(\mathcal{X},\nu)$, whence $\psi_\infty = \overline{\psi}$ almost everywhere, so that $\psi_n \to \overline{\psi}$ almost everywhere.

To verify Claim 1, consider the natural unitary representation of the group $\mathcal{K}$ in the Hilbert space $L^2(\mathcal{X},\nu)$. Let P_n denote the orthogonal projection in this space onto the subspace of $\mathcal{K}(n)$-invariant vectors, $n = 1, 2, \dots$. Then P_n strongly converges, as $n \to \infty$, to P, the orthoprojection onto the subspace of $\mathcal{K}$-invariant vectors. In particular, $P_n\psi \to P\psi$ in the metric of $L^2(\mathcal{X},\nu)$. Since the measure ν is ergodic, the only $\mathcal{K}$-invariant vectors are the constants (this is the only place where we use the assumption that ν is ergodic). Therefore, $P\psi$ is a constant function, which clearly equals $\overline{\psi}$. Finally, $P_n\psi = \psi_n$, so that $\psi_n \to \overline{\psi}$ in $L^2(\mathcal{X},\nu)$.

Now we shall prove Claim 2 by a method similar to the one used in the proof of Birkhoff–Khinchine's individual ergodic theorem (see, e.g., Parthasarathy [**Pa2**, §49]).

Without loss of generality one may assume that ψ is real-valued. For $N = 1, 2, \dots$, put

$$E_N = E_N(\psi) = \Big\{x \in \mathcal{X} \mid \sup_{1\leqslant n\leqslant N} \psi_n(x) > 0\Big\}, \tag{3.4}$$

$$E_\infty = E_\infty(\psi) = \Big\{x \in \mathcal{X} \mid \sup_{1\leqslant n\leqslant \infty} \psi_n(x) > 0\Big\} = \bigcup_{N=1}^{\infty} E_N(\psi). \tag{3.5}$$

Note that each E_N is a Borel subset, whence E_∞ also is a Borel subset.

The following claim is an analog of the maximal ergodic theorem.

Claim 3. We have

$$\int_{E_\infty} \psi(x)\nu(dx) \geqslant 0. \tag{3.6}$$

Indeed, since (E_N) is a monotone family of sets, it is enough to check that

$$\int_{E_N} \psi(x)\nu(dx) \geqslant 0, \qquad N = 1, 2, \dots. \tag{3.7}$$

We can write each E_N as a disjoint union of Borel subsets,

$$E_N = E_{1N} \cup \dots \cup E_{NN}, \tag{3.8}$$

where

$$E_{mN} = \{x \in \mathcal{X} \mid \psi_m(x) > 0,\ \psi_i(x) \leqslant 0 \text{ for } m+1 \leqslant i \leqslant N\}. \tag{3.9}$$

Then it is enough to show that

$$\int_{E_{mN}} \psi(x)\nu(dx) \geqslant 0, \quad 1 \leqslant m \leqslant N. \tag{3.10}$$

To do this, we remark that E_{mN} is invariant relative to the action of $\mathcal{K}(m)$. Since ν is an invariant measure, it follows that

$$\int_{E_{mN}} \psi(x)\nu(dx) = \int_{E_{mN}} \psi(u\cdot x)\nu(dx), \qquad u \in \mathcal{K}(m). \tag{3.11}$$

Since the function $(u, x) \to \psi(u \cdot x)$ is continuous on $\mathcal{K}(m) \times E_{mN}$, we may integrate the right-hand side of (3.11) over $\mathcal{K}(m)$ (with respect to the Haar measure) and then interchange the integrals over $\mathcal{K}(m)$ and over E_{mN}. This yields

$$\int_{E_{mN}} \psi(x)\,\nu(dx) = \int_{E_{mN}} \psi_m(x)\,\nu(dx). \tag{3.12}$$

By definition (3.9), ψ_m is positive on E_{mN}, so that (3.12) is nonnegative.

Thus, we have checked Claim 3.

Further, for arbitrary real $a < b$, put

$$\mathcal{X}_{ab} = \{x \in \mathcal{X} \mid \underline{\lim}\, \psi_n(x) < a < b < \overline{\lim}\, \psi_n(x)\}. \tag{3.13}$$

This is a Borel subset. Let us establish the double inequality

$$a\,\nu(\mathcal{X}_{ab}) \geqslant \int_{\mathcal{X}_{ab}} \psi(x)\,\nu(dx) \geqslant b\,\nu(\mathcal{X}_{ab}), \tag{3.14}$$

which will imply $\nu(\mathcal{X}_{ab}) = 0$.

Indeed, $\mathcal{X}_{ab}$ is a $\mathcal{K}$-invariant Borel subset of $\mathcal{X}$. Let us replace $\mathcal{X}$ by $\mathcal{X}_{ab}$ and apply Claim 3 to the functions

$$\psi' := (\psi - b)|_{\mathcal{X}_{ab}} \quad \text{and} \quad \psi'' := (a - \psi)|_{\mathcal{X}_{ab}}. \tag{3.15}$$

By definition (3.13) of $\mathcal{X}_{ab}$,

$$E_\infty(\psi') = E_\infty(\psi'') = \mathcal{X}_{ab}, \tag{3.16}$$

whence

$$\int_{\mathcal{X}_{ab}} \psi'(x)\nu(dx) \geqslant 0, \qquad \int_{\mathcal{X}_{ab}} \psi''(x)\nu(dx) \geqslant 0, \tag{3.17}$$

which is equivalent to (3.14).

Finally, applying (3.14) to various couples of rational numbers $a < b$, we see that

$$\underline{\lim}\, \psi_n(x) = \overline{\lim}\, \psi_n(x) \tag{3.18}$$

almost everywhere. This completes the proof of Claim 2 and of the theorem. □

We apply Theorem 3.2 to $\mathcal{X} = H$ and $\mathcal{K} = U(\infty)$. Note that the assumptions of Theorem 3.2 are satisfied. Indeed, H is homeomorphic to a separable metric space, because it is essentially a copy of $\mathbb{R}^\infty$, and, further, the action $U(\infty) \times H \to H$ is jointly continuous.

For $n = 1, 2, \dots$, denote by $\mathcal{M}_n$ the set of $U(n)$-invariant probability measures that are supported by the $U(n)$-orbits in the space $H(n)$. These measures will be called *orbital measures*. Since $H(n)$ is contained in H, we may view orbital measures as measures on the space H.

THEOREM 3.3. *For any ergodic measure $M \in \mathcal{M}$, there exists a sequence $\{M_n \in \mathcal{M}_n\}$ of orbital measures such that $M_n \Rightarrow M$ as $n \to \infty$.*

PROOF. We shall write $m_n(B)$ instead of $m_n(x)$; here B is a matrix from H. By the first claim of Theorem 3.2, there exists $B \in H$ such that $m_n(B) \Rightarrow M$ as $n \to \infty$. Consider the projections $\theta_n \colon H \to H(n)$ defined at the beginning of §1 and remark that $\theta_n(m_n(B))$ coincides with the orbital measure in $H(n)$ corresponding to the matrix $\theta_n(B)$. Let us take this orbital measure as M_n. If $k \leqslant n$, then $m_n(B)$ and M_n have the same image under θ_k; fixing k and letting $n \to \infty$, we see that the measures $\theta_k(M_n)$ on $H(k)$ weakly converge to the measure $\theta_k(M)$. Now identify H with $\mathbb{R}^\infty$ and recall that on $\mathbb{R}^\infty$ weak convergence of probability measures is equivalent to weak convergence of their finite-dimensional projections (see, e.g., [**Bi,** Chapter 1, §3]). Applying this to our sequence $\{M_n\}$, we see that $M_n \Rightarrow M$. □

It is convenient to analyze weak convergence of measures in terms of characteristic functions. Assume $\nu, \nu_1, \nu_2, \dots$ are Borel probability measures on H and $f, f_1, f_2, \dots$ denote their characteristic functions. We need the following simple claim, which again is essentially a well-known fact about the space $\mathbb{R}^\infty$.

PROPOSITION 3.4. *Weak convergence $\nu_n \Rightarrow \nu$ on H is equivalent to uniform convergence $f_n \to f$ on compact subsets in $H(\infty)$.* □

(Note that any compact subset in $H(\infty)$ is always contained in $H(n)$ for sufficiently large n.)

PROOF. We again pass to finite-dimensional projections and then use the fact that weak convergence of probability measures on $\mathbb{R}^n$ is equivalent to uniform convergence, on compact sets, of their characteristic functions. □

To obtain Theorem 3.3, we could use, instead of Theorem 3.2, another general result, where we have to specialize $\mathcal{G} = G(\infty)$, $\mathcal{K} = U(\infty)$:

THEOREM 3.5 (Olshanski [**O3,** Theorem 2.5] [**O5,** Theorem 22.10]). *Let $\mathcal{G} = \varinjlim \mathcal{G}(n)$ be an inductive limit of separable locally compact groups and $\mathcal{K}$ be a subgroup of $\mathcal{G}$. Let $\mathcal{K}(n) = \mathcal{K} \cap \mathcal{G}(n)$, so that $\mathcal{K} = \varinjlim \mathcal{K}(n)$.*

Then any extreme $\mathcal{K}$-bi-invariant continuous positive definite function f on $\mathcal{G}$, normalized at unity, can be approached, uniformly on compact sets of the group $\mathcal{G}$, by a sequence $\{f_n\}$ of extreme $\mathcal{K}(n)$-bi-invariant continuous positive definite normalized functions on the subgroups $\mathcal{G}(n)$. □

Note, however, that Theorem 3.2, due to its second claim, provides us with more detailed information on the approximation process than Theorem 3.5.

REMARK 3.6. Note that Claim 2 in the proof of Theorem 3.2 can also be deduced from Doob's theorem on convergence of (reversed) martingales (see [**D,** Chapter VII, Theorem 4.2]). Indeed, denote by $\mathcal{B}_n$ the σ-algebra of all $\mathcal{K}(n)$-invariant Borel subsets of the space $\mathcal{X}$. We have $\mathcal{B}_1 \supseteq \mathcal{B}_2 \supseteq \dots$, so that, by Doob's theorem, as $n \to \infty$, the conditional expectation $E(\psi|\mathcal{B}_n)$ of the bounded Borel function ψ converges almost everywhere, with respect to ν, to a function, which is the conditional expectation $E(\psi|\mathcal{B}_\infty)$, where $\mathcal{B}_\infty = \bigcap \mathcal{B}_n$. To derive Claim 2, we only need to show that $E(\psi|\mathcal{B}_n) = \psi_n$ almost everywhere with respect to ν, $n = 1, 2, \dots$. By definition of conditional expectation, this means that ψ_n is $\mathcal{B}_n$-measurable and the following condition holds:

$$A \in \mathcal{B}_n \implies \int_A \psi(x)\nu(dx) = \int_A \psi_n(x)\nu(dx). \tag{3.19}$$

Since ψ_n is a $\mathcal{K}(n)$-invariant Borel function, it is $\mathcal{B}_n$-measurable. Further, since the action of $\mathcal{K}$ on $\mathcal{X}$ is jointly continuous, the function $(u,x) \mapsto \psi(u \cdot x)$ is continuous on $\mathcal{K} \times \mathcal{X}$, hence is a Borel function on $\mathcal{K}(n) \times A$. By Fubini's theorem,

$$\int_{\mathcal{K}(n)} \left(\int_A \psi(u \cdot x) \nu(dx) \right) m_n(du) = \int_A \left(\int_{\mathcal{K}(n)} \psi(u \cdot x) m_n(du) \right) \nu(dx). \tag{3.20}$$

The left-hand side of (3.20) is equal to $\int_A \psi(x) \nu(dx)$, because A and ν are $\mathcal{K}(n)$-invariant, whereas the right-hand side is equal to $\int_A \psi_n(x) \nu(dx)$ by the definition of ψ_n.

§4. Main Theorem

We shall deal with a sequence $\{M_n \in \mathcal{M}_n\}$, $n = 1, 2, \dots$, of orbital measures. By f_n we denote the characteristic function of M_n; recall that

$$f_n(A) = \int_{\Omega_n} e^{i \operatorname{tr}(AB)} M_n(dB), \qquad A \in H(n), \tag{4.1}$$

where Ω_n stands for the $U(n)$-orbit that carries M_n. Let $\Lambda(n) = (\lambda_1(n), \dots, \lambda_n(n))$ be the common spectrum of all the matrices $B \in \Omega_n$. Then Ω_n may be specified as the orbit containing the diagonal matrix $\operatorname{diag} \Lambda(n)$ with diagonal entries $(\lambda_1(n), \dots, \lambda_n(n))$. The eigenvalues $\lambda_1(n), \dots, \lambda_n(n)$ may be arranged in any order; it will be convenient for us to separate the positive and the negative eigenvalues and to regard $\Lambda(n)$ as a double sequence formed by positive and negative eigenvalues, respectively, written in decreasing order of their absolute values:

$$\Lambda(n) = (\Lambda'(n), \Lambda''(n)), \tag{4.2}$$

where

$$\begin{aligned} \Lambda'(n) &= (\lambda_1'(n) \geqslant \lambda_2'(n) \geqslant \dots \geqslant 0), \\ \Lambda''(n) &= (\lambda_1''(n) \leqslant \lambda_2''(n) \leqslant \dots \leqslant 0). \end{aligned} \tag{4.3}$$

The possible zero values may be included either in $\Lambda'(n)$ or in $\Lambda''(n)$; in fact, we prefer to view both $\Lambda'(n)$ and $\Lambda''(n)$ as infinite sequences with a finite number of nonzero terms.

THEOREM 4.1 (Main Theorem). *Let $\{M_n \in \mathcal{M}_n\}$ be an infinite sequence of orbital measures defined by a sequence $\{\Omega_n \subset H(n)\}$ of $U(n)$-orbits. For $n = 1, 2, \dots$, pick a matrix B_n from the orbit Ω_n and write the collection $\Lambda(n)$ of its eigenvalues as a double sequence* (4.2). *(Note that $\Lambda(n)$ does not depend on the choice of B_n.)*

(i) *Suppose that the following limits exist*:

$$x_k' = \lim_{n \to \infty} \frac{\lambda_k'(n)}{n} \geqslant 0, \quad x_k'' = \lim_{n \to \infty} \frac{\lambda_k''(n)}{n} \leqslant 0, \qquad k = 1, 2, \dots, \tag{4.4}$$

$$\gamma_1 = \lim_{n \to \infty} \frac{1}{n} \sum_k (\lambda_k'(n) + \lambda_k''(n)) = \lim_{n \to \infty} \frac{1}{n} \operatorname{tr} B_n, \tag{4.5}$$

$$\tilde{\gamma}_2 = \lim_{n \to \infty} \frac{1}{n^2} \sum_k ((\lambda_k'(n))^2 + (\lambda_k''(n))^2) = \lim_{n \to \infty} \frac{1}{n^2} \operatorname{tr}(B_n^2). \tag{4.6}$$

Then the measures M_n weakly converge to an ergodic measure $M \in \mathcal{M}$ with the multiplicative characteristic function f defined by

$$f(A) = \prod_{a \in \operatorname{Spec}(A)} F(a), \qquad A \in H(\infty), \tag{4.7}$$

where $\operatorname{Spec}(A)$ *stands for the collection* $\{a_1, a_2, \dots, 0, 0, \dots\}$ *of the eigenvalues of* A *and*

$$F(a) = e^{i\gamma_1 a - \gamma_2 a^2/2} \prod_k \frac{e^{-ix'_k a}}{1 - ix'_k a} \prod_k \frac{e^{-ix''_k a}}{1 - ix''_k a}, \qquad a \in \mathbb{R}; \tag{4.8}$$

here γ_1 *is given by* (4.5), *the parameters* $x'_k \geqslant 0$ *and* $x''_k \leqslant 0$ *are given by* (4.4), *and, finally,*

$$\gamma_2 = \tilde{\gamma}_2 - \sum_k ((x'_k)^2 + (x''_k)^2), \tag{4.9}$$

where $\tilde{\gamma}_2$ *is given by* (4.6).

(ii) *Conversely, if the measures* M_n *weakly converge to a probability measure on* H, *then the limits* (4.4)–(4.6) *exist.*

COMMENT. It follows from the definition of the parameters x'_k, x''_k, and $\tilde{\gamma}_2$ that $\sum((x'_k)^2 + (x''_k)^2) \leqslant \tilde{\gamma}_2$, so that $\gamma_2 \geqslant 0$.

COROLLARY 4.2. *Let* $\{M_n \in \mathcal{M}_n\}$ *be a sequence of orbital measures that weakly converges to a Borel probability measure* M *on* H. *Then* $M \in \mathcal{M}$. □

DERIVATION OF THEOREM 2.9 FROM THEOREM 4.1. Let $F \in \mathcal{F}_1$ and let $M \in \mathcal{M}$ be the corresponding ergodic measure. We must show that F is of the form (2.9). By Theorem 3.3, there exists a sequence $\{M_n \in \mathcal{M}_n\}$ that weakly converges to M. Next, by claim (ii) of Theorem 4.1, the limits (4.4)–(4.6) exist. Finally, by claim (i) of Theorem 4.1, F is of the form (4.8) that coincides with (2.9) up to a reordering of the points in $x = (x_1, x_2, \dots)$ only; we recall (see Comment 1 to Proposition 2.8) that we may order the parameters $x_1, x_2, \dots$ in (2.9) in any way. □

OUTLINE OF THE PROOF OF THEOREM 4.1. In §5, we establish a preliminary result—we expand the orbital integral (4.1) into a series of Schur polynomials. The proof of the theorem is given in §6; it is divided into three steps.

In Step 1, we check that under assumptions (4.4)–(4.6),

$$f_n(\operatorname{diag}(a, 0, 0, \dots)) \to F_{\gamma_1, \gamma_2, x}(a), \qquad a \in \mathbb{R}, \tag{4.10}$$

where $\operatorname{diag}(\cdots)$ stands for a diagonal matrix and $F_{\gamma_1,\gamma_2,x}$ is given by (4.8) or, equivalently, by (2.9).

In Step 2, we generalize this to arbitrary diagonal matrices:

$$f_n(\operatorname{diag}(a_1, \dots, a_k, 0, 0, \dots)) \to \prod_{i=1}^{k} F_{\gamma_1,\gamma_2,x}(a_i), \qquad (a_1, \dots, a_k) \in \mathbb{R}^k. \tag{4.11}$$

This proves claim (i).

Finally, in Step 3, using a simple trick, we show that conditions (4.4)–(4.6) are indeed necessary (claim (ii)).

Let us emphasize that to prove Theorem 2.9 only, one could avoid Step 2. However, we need this step to characterize the convergent sequences $\{M_n\}$.

REMARK 4.3. To check (4.10) or, more generally, (4.11), we consider the Taylor series decomposition at zero for the left-hand side and show that its coefficients tend, as $n \to \infty$, to the corresponding Taylor coefficients for the right-hand side. Note that these coefficients are nothing but the moments of the measures.

Moreover, it follows from claim (ii) and the proof of (i) that whenever a sequence $\{M_n \in \mathcal{M}_n\}$ weakly converges to a probability measure M, the moments of M_n must tend to the moments of M. Thus, in our situation, weak convergence $M_n \to M$ is always controlled by moments—a fact that is not at all evident a priori.

Such a "moment method" also works in allied classification problems, related to characters of $U(\infty)$ (see Vershik–Kerov [**VK2**]) and spherical functions of $GL(\infty, \mathbb{C})$ (see Nessonov [**N**]). We conjecture that it can be used for all families of classical symmetric spaces. (About spherical functions on infinite-dimensional symmetric spaces, see Olshanski [**O5**] and Pickrell [**Pi1**].)

REMARK 4.4. Assume that in the spectrum $\Lambda(n) = (\lambda_1(n), \ldots, \lambda_n(n))$ there are at most k nonzero eigenvalues, where k does not depend on n. Then we can prove claim (i) of Theorem 4.1 directly, i.e., without using moments. Indeed, let

$$\lim_{n\to\infty} \frac{\lambda_i(n)}{n} = x_i, \quad 1 \leqslant i \leqslant n, \qquad \lambda_i(n) = 0, \quad i > k. \tag{4.12}$$

Then the measure M_n can be viewed as the image of the normalized Haar measure of the group $U(n)$ under the mapping

$$U(n) \ni u \mapsto u^* \operatorname{diag}(\lambda_1(n), \ldots, \lambda_k(n), 0, \ldots, 0)\, u = B \in H(n) \subset H. \tag{4.13}$$

The matrix B can be rewritten as follows:

$$B = (\Xi(n))^* X(n)\, \Xi(n), \tag{4.14}$$

where $\Xi(n)$ denotes the $k \times n$ matrix formed by the first k rows of the matrix $u \in U(n)$ multiplied by the scalar $\sqrt{n}$, and

$$X(n) = \operatorname{diag}\left(\frac{\lambda_1(n)}{n}, \ldots, \frac{\lambda_k(n)}{n}\right). \tag{4.15}$$

Now let $n \to \infty$. Then $X(n) \to X := \operatorname{diag}(x_1, \ldots, x_k)$, because of (4.12). Further, fix $m = 1, 2, \ldots$ and regard the $k \times m$ matrix formed by the first m columns of $\Xi(n)$ as a random matrix variable (with respect to the Haar measure of $U(n)$). It is well-known that the limit distribution of this matrix is given by the Gaussian product measure $\omega^{k\times m}$ (product of km copies of ω, cf. (2.12)), where ω stands for the Gaussian measure on $\mathbb{C}$ specified in Corollary 2.7. This fact is proved, e.g., in Olshanski [**O5**, Lemma 5.3]. It follows that the measures M_n weakly converge, as $n \to \infty$, to the measure $M_{0;x_1,\ldots,x_k} \in \mathcal{M}$ defined in Remark 2.10.

REMARK 4.5. In general, for a convergent sequence of orbital measures, the eigenvalues in the spectrum $\Lambda(n)$ must grow linearly in n as $n \to \infty$. But if the limiting ergodic measure is Gaussian (that is, $x \equiv 0$), then the order of growth of the eigenvalues becomes equal to $\sqrt{n}$. Example: let

$$\Lambda(n) = (\underbrace{\sqrt{\gamma n}, \ldots, \sqrt{\gamma n}}_{[n/2]}, \underbrace{-\sqrt{\gamma n}, \ldots, -\sqrt{\gamma n}}_{[(n+1)/2]}),$$

then the limiting measure is the Gaussian measure $M_{0,\gamma,0}$.

§5. Expanding spherical functions into series of Schur polynomials

In this section, we fix $n = 1, 2, \dots$. Let $(a_1, \dots, a_n) \in \mathbb{C}^n$, $(\lambda_1, \dots, \lambda_n) \in \mathbb{R}^n$, and suppose

$$A = \operatorname{diag}(a_1, \dots, a_n), \qquad \Lambda = \operatorname{diag}(\lambda_1, \dots, \lambda_n) \tag{5.1}$$

are the corresponding diagonal matrices. We fix Λ and deal with the $U(n)$-orbit $\Omega \subset H(n)$ passing through Λ. Let f_Λ stand for the characteristic function of the invariant probability measure supported by Ω. Then

$$f_\Lambda(A) = \int_{U(n)} e^{i\,\mathrm{tr}(Au\Lambda u^{-1})}\, du, \tag{5.2}$$

where du is the normalized Haar measure on $U(n)$.

Since f_Λ is an entire function of $(a_1, \dots, a_n) \in \mathbb{C}^n$, it admits an everywhere convergent Taylor series expansion. But since $f_\Lambda(a_1, \dots, a_n)$ is also symmetric in $a_1, \dots, a_n$, it is more convenient to rewrite this Taylor series as a series of Schur polynomials s_μ,

$$f_\Lambda(A) = f_\Lambda(a_1, \dots, a_n) = \sum_\mu c_\mu s_\mu(a_1, \dots, a_n), \tag{5.3}$$

where μ ranges over the set of all Young diagrams with at most n rows. (About Schur polynomials, see, e.g., Macdonald [**M**].)

We shall use some standard notation concerning Young diagrams: $\mu \vdash m$ means that μ is a partition of m (i.e., m equals $|\mu|$, the number of boxes in μ), $\ell(\mu)$ is the number of (nonzero) rows in μ, $(p,q) \in \mu$ denotes the box of μ lying on the intersection of pth row and qth column, $\dim \mu$ is the dimension of the irreducible representation of the symmetric group $S(m)$, $m = |\mu|$, that corresponds to the diagram μ.

THEOREM 5.1. *The coefficients in* (5.3) *are given by the following formula*:

$$c_\mu = \prod_{(p,q)\in\mu} \frac{1}{n+q-p} \cdot s_\mu(i\lambda_1, \dots, i\lambda_n) = i^{|\mu|} \prod_{(p,q)\in\mu} \frac{1}{n+q-p} \cdot s_\mu(\lambda_1, \dots, \lambda_n). \tag{5.4}$$

PROOF. *Step* 1. Let π be an irreducible representation of $U(n)$, $\operatorname{Dim}\pi$ be its dimension, and χ be its normalized character:

$$\chi(g) = \frac{\mathrm{tr}\,\pi(g)}{\operatorname{Dim}\pi}, \qquad g \in U(n).$$

It is well known that χ satisfies the following functional equation:

$$\int_{U(n)} \chi(guhu^{-1})\, du = \chi(g)\chi(h), \qquad g, h \in U(n). \tag{5.5}$$

Now let us take as π the irreducible polynomial representation with highest weight $\mu = (\mu_1, \dots, \mu_n)$. Then

$$\chi(g) = \frac{s_\mu(\mathrm{Spec}\,(g))}{\operatorname{Dim}_n \mu}, \qquad g \in U(n), \tag{5.6}$$

where, given a $n \times n$ matrix g, Spec(g) stands for the collection $(z_1, \dots, z_n)$ of its eigenvalues, and $\mathrm{Dim}_n \mu$ denotes the dimension of the representation π of the group $U(n)$.

Substituting (5.6) into (5.5), we obtain

$$\int_{U(n)} s_\mu(\mathrm{Spec}\,(guhu^{-1}))\,du = \frac{1}{\mathrm{Dim}_n \mu}\, s_\mu(\mathrm{Spec}\,(g))\, s_\mu(\mathrm{Spec}\,(h)). \tag{5.7}$$

By analytic continuation, this formula holds for arbitrary $n \times n$ complex matrices g and h.

Step 2. Let us come back to the orbital integral (5.2), where A and Λ are given by (5.1). We may write

$$e^{i\,\mathrm{tr}(Au\Lambda u^{-1})} = \sum_{m \geqslant 0} \frac{1}{m!}\, p_1^m(\mathrm{Spec}\,(iAu\Lambda u^{-1})), \tag{5.8}$$

where $p_1(z_1, \dots, z_n) = z_1 + \dots + z_n$ is the first power sum.

Recall the well-known identity (see, e.g., Macdonald [**M,** Chapter I, (7.8)]):

$$p_1^m(z_1, \dots, z_n) = \sum_{\substack{\mu \vdash m \\ \ell(\mu) \leqslant n}} \dim \mu \cdot s_\mu(z_1, \dots, z_n). \tag{5.9}$$

Using it, we may rewrite (5.8) as follows:

$$e^{i\,\mathrm{tr}(Au\Lambda u^{-1})} = \sum_{m \geqslant 0} \sum_{\substack{\mu \vdash m \\ \ell(\mu) \leqslant n}} \frac{\dim \mu}{m!}\, s_\mu(\mathrm{Spec}\,(iAu\Lambda u^{-1})). \tag{5.10}$$

Integrating both sides of (5.10) over $u \in U(n)$ and applying the functional equation (5.7) with $g = A$, $h = i\Lambda$, we obtain

(5.11)

$$\int_{U(n)} e^{i\,\mathrm{tr}(Au\Lambda u^{-1})}\,du = \sum_{m \geqslant 0} \sum_{\substack{\mu \vdash m \\ \ell(\mu) \leqslant n}} \frac{\dim \mu}{m!\,\mathrm{Dim}_n \mu}\, s_\mu(a_1, \dots, a_n)\, s_\mu(i\lambda_1, \dots, i\lambda_n).$$

Step 3. For a box $(p, q) \in \mu$, let $h(p, q)$ denote the corresponding hook length. Recall the well-known formulas

$$\dim \mu = m! \prod_{(p,q) \in \mu} \frac{1}{h(p,q)}, \qquad \mathrm{Dim}_n \mu = \prod_{(p,q) \in \mu} \frac{n + q - p}{h(p,q)}$$

(see, e.g., [**M,** Chapter I, §3, Example 4]). It follows that

$$\frac{\dim \mu}{m!\,\mathrm{Dim}_n \mu} = \prod_{(p,q) \in \mu} \frac{1}{n + q - p}.$$

Substituting this into (5.11), we obtain (5.4). □

(After work on this paper was completed, we learned that the argument presented above was used much earlier by James, see his survey paper [**J,** (60)].)

COROLLARY 5.2. *The orbital integral* (5.2) *is given by the following explicit formula:*

$$f_\Lambda(A)=\int_{U(n)} e^{i\,\mathrm{tr}(Au\Lambda u^{-1})}\,du=\frac{(n-1)!\cdots 0!\,\det\left[e^{ia_j\lambda_k}\right]_{j,k=1}^n}{V(a_1,\dots,a_n)\,V(i\lambda_1,\dots,i\lambda_n)}. \tag{5.12}$$

Here $V(\cdots)$ *is the Vandermonde determinant,*

$$V(z_1,\dots,z_n)=\prod_{1\leqslant j<k\leqslant n}(z_j-z_k),$$

and the coordinates $a_1,\dots,a_n$*, as well as* $\lambda_1,\dots,\lambda_n$*, are assumed to be pairwise distinct.*

PROOF. The statement of Theorem 5.1 may be rewritten as follows:

$$f_\Lambda(A)=(n-1)!\cdots 0!\sum_{\substack{\mu\\ \ell(\mu)\leqslant n}}\frac{s_\mu(a_1,\dots,a_n)\,s_\mu(i\lambda_1,\dots,i\lambda_n)}{(\mu_1+n-1)!\,(\mu_2+n-2)!\cdots\mu_n!}.$$

Using the determinant formula for Schur polynomials, we obtain

$$\begin{aligned}&\frac{V(a_1,\dots,a_n)\,V(i\lambda_1,\dots,i\lambda_n)}{(n-1)!\cdots 0!}\,f_\Lambda(A)\\ &\quad=\sum_{\substack{\mu\\ \ell(\mu)\leqslant n}}\frac{\det\left[a_j^{\mu_k+n-k}\right]_{j,k=1}^n\cdot\det\left[(i\lambda_j)^{\mu_k+n-k}\right]_{j,k=1}^n}{(\mu_1+n-1)!\,(\mu_2+n-2)!\cdots\mu_n!}\\ &\quad=\sum_{m_1>\cdots>m_n\geqslant 0}\frac{\det\left[a_j^{m_k}\right]_{j,k=1}^n\cdot\det\left[(i\lambda_j)^{m_k}\right]_{j,k=1}^n}{m_1!\,m_2!\cdots m_n!}.\end{aligned}$$

Since the numerator in the latter expression is symmetric in $m_1,\dots,m_n$ and vanishes if some of these numbers are equal, we may drop the assumption $m_1>\cdots>m_n$. Then we obtain

$$\begin{aligned}&\frac{V(a_1,\dots,a_n)\,V(i\lambda_1,\dots,i\lambda_n)}{(n-1)!\cdots 0!}\,f_\Lambda(A)\\ &\quad=\frac{1}{n!}\sum_{m_1,\dots,m_n\geqslant 0}\frac{\det\left[a_j^{m_k}\right]_{j,k=1}^n\cdot\det\left[(i\lambda_j)^{m_k}\right]_{j,k=1}^n}{m_1!\,m_2!\cdots m_n!}\\ &\quad=\frac{1}{n!}\sum_{m_1,\dots,m_n\geqslant 0}\frac{1}{m_1!\cdots m_n!}\sum_{\sigma,\tau}\mathrm{sgn}(\sigma)\,\mathrm{sgn}(\tau)(a_{\sigma_1}\cdot i\lambda_{\tau_1})^{m_1}\cdots(a_{\sigma_n}\cdot i\lambda_{\tau_n})^{m_n},\end{aligned}$$

where $\sigma=(\sigma_1,\dots,\sigma_n)$ and $\tau=(\tau_1,\dots,\tau_n)$ range over all the permutations of $1,\dots,n$.

Changing the order of the summation, we obtain

$$\frac{1}{n!}\sum_{\sigma,\tau}\mathrm{sgn}(\sigma)\,\mathrm{sgn}(\tau)\,e^{ia_{\sigma_1}\lambda_{\tau_1}}\cdots e^{ia_{\sigma_n}\lambda_{\tau_n}}=\det\left[e^{ia_j\lambda_k}\right]_{j,k=1}^n,$$

which completes the proof. □

REMARK 5.3. Formula (5.12) is well known and can be proved in a number of different ways. For instance, it can be obtained from the Gelfand–Naimark calculation of spherical functions on $GL(n,\mathbb{C})$ (see Gelfand–Naimark [**GN**]) by a passage to the limit. Another way consists in using the radial parts of invariant differential operators. Yet another method can be found in [**BGV,** Section 7.5] (we are grateful to Michel Duflo for the latter reference). Note also that writing the above calculations in reverse order, we can deduce Theorem 5.1 from formula (5.12).

COROLLARY 5.4. *Let us substitute* $A = \operatorname{diag}(a,0,\dots,0)$ *into* (5.2). *Then*

$$f_\Lambda(\operatorname{diag}(a,0,\dots,0)) = \int_{U(n)} e^{ia(u\Lambda u^{-1})_{11}}\,du = \sum_{m\geqslant 0} \frac{h_m(i\lambda_1,\dots,i\lambda_n)}{n(n+1)\cdots(n+m-1)}\,a^m, \tag{5.13}$$

where h_m *is the* m*th complete symmetric function.*

PROOF. Indeed, note that $s_\mu(a,0,\dots,0)$ vanishes unless $\mu=(m)$, where $m=0,1,\dots$. Thus the summation is really taken over the diagrams $\mu=(m)$ only. For these diagrams, s_μ reduces to h_m and the product $\prod_{(p,q)\in\mu}(n+q-p)$ turns into $n(n+1)\cdots(n+m-1)$, so that we obtain (5.13). □

REMARK 5.5. The expansion (5.3) of Theorem 5.1 may be written in the following form, which emphasizes the symmetry between A and Λ:

$$f_\Lambda(A) = \sum_{\substack{\mu\\ \ell(\mu)\leqslant n}} i^{|\mu|}\Bigg(\prod_{(p,q)\in\mu}\frac{1}{n+q-p}\Bigg) s_\mu(\lambda_1,\dots,\lambda_n)\,s_\mu(a_1,\dots,a_n). \tag{5.14}$$

The symmetry can already be seen from (5.2).

§6. Proof of Theorem 4.1

We start with the following simple claim.[7]

PROPOSITION 6.1. *Let* $f, f_1, f_2, \dots$ *be analytic functions on* $\mathbb{R}^k$ *satisfying the following conditions*:

(i) $f_1, f_2, \dots$ *are positive definite and normalized at the origin*;

(ii) *the Taylor coefficients of* f_n *at the origin tend to the corresponding Taylor coefficients of* f *as* $n\to\infty$;

(iii) *the Taylor decomposition of* f_n *converges absolutely and uniformly on* n *in a neighborhood of the origin that does not depend on* n.

Then $f_n\to f$ *uniformly on compact subsets of* $\mathbb{R}^k$.

PROOF. This is a standard exercise. It is clear that $f_n\to f$ in a neighborhood of the origin. Consider the probability measures $\nu_1,\nu_2,\dots$ on $\mathbb{R}^k$ that correspond to the functions $f_1, f_2,\dots$ by Bochner's theorem. By Paul Lévy's classical continuity theorem (see, e.g., Shiryaev's textbook [**Shi,** Chapter III, §3]), the measures ν_n weakly converge to a probability measure ν. Let $\tilde f$ stand for the characteristic function of ν. Then $f_n\to\tilde f$ uniformly on compact sets of $\mathbb{R}^k$. Further, (ii) and (iii) imply that $\tilde f$ is analytic in a neighborhood of the origin, and this implies that $\tilde f$ is analytic on the whole space $\mathbb{R}^k$. Therefore, $\tilde f = f$ and our claim follows. □

[7] See also Remark 6.2.

We proceed with the proof of the theorem. The notation of §4 is maintained.

Step 1. Let us fix a sequence $\{\Lambda(n)\}$ such that the limits (4.4), (4.5), and (4.6) exist. Let us abbreviate

$$f_n(a) = f_n(\operatorname{diag}(a, 0, 0, \dots)), \qquad a \in \mathbb{R}. \tag{6.1}$$

Note that $f_n(a)$ is the characteristic function of the probability measure $\nu_n := M_n^{(1)}$, the image of the measure M_n under the projection $\theta_1 \colon H(n) \to H(1) = \mathbb{R}$.

The purpose of this step is to prove that

$$\lim_{n\to\infty} f_n(a) = F_{\gamma_1,\gamma_2,x}(a), \qquad a \in \mathbb{R}, \tag{6.2}$$

uniformly on bounded sets in $\mathbb{R}$. (Recall that γ_1 is given by (4.5), γ_2 is given by (4.6) and (4.9), and $x = (x_1', x_2', \dots; x_1'', x_2'', \dots)$, where the latter parameters are defined by (4.4). The function $F_{\gamma_1,\gamma_2,x}$ is given by (4.8) or, equivalently, by (2.9).)

To do this, let us expand both sides of (6.2) into Taylor series:

$$f_n(a) = \sum_{m\geqslant 0} c_m^{(n)} a^m, \tag{6.3}$$

$$F_{\gamma_1,\gamma_2,x}(a) = \sum_{m\geqslant 0} c_m^{(\infty)} a^m. \tag{6.4}$$

By Proposition 6.1, it is enough to verify the following two claims. First,

$$\lim_{n\to\infty} c_m^{(n)} = c_m^{(\infty)}, \qquad m = 0, 1, 2, \dots. \tag{6.5}$$

Second, the series (6.3) converges absolutely and uniformly on n in a sufficiently small neighborhood of the origin, i.e.,

$$|c_m^{(n)}| \leqslant C_1 C_2^m, \qquad m = 0, 1, 2, \dots, \tag{6.6}$$

where the constants $C_1 > 0$, $C_2 > 0$ do not depend on n.

By (5.13), we have

$$\begin{aligned} c_m^{(n)} &= \frac{h_m(i\lambda_1(n), \dots, i\lambda_n(n))}{n(n+1)\cdots(n+m-1)} \\ &= \frac{n^m}{n(n+1)\cdots(n+m-1)}\, h_m\!\left(\frac{i\lambda_1(n)}{n}, \dots, \frac{i\lambda_n(n)}{n}\right). \end{aligned} \tag{6.7}$$

It is clear that in both claims (6.5) and (6.6) we may replace $c_m^{(n)}$ by

$$\tilde{c}_m^{(n)} := h_m\!\left(\frac{i\lambda_1(n)}{n}, \dots, \frac{i\lambda_n(n)}{n}\right). \tag{6.8}$$

Further, instead of the series $\sum \tilde{c}_m^{(n)} a^m$ and $\sum c_m^{(\infty)} a^m$, it is more convenient to deal with their logarithms $\ln(\sum \tilde{c}_m^{(n)} a^m)$ and $\ln(\sum c_m^{(\infty)} a^m)$, respectively.

Now recall a well-known identity from the theory of symmetric functions,

$$\sum_{m\geqslant 0} h_m(\,\cdot\,) a^m = \exp\!\left(\sum_{m\geqslant 1} p_m(\,\cdot\,)\frac{a^m}{m}\right), \tag{6.9}$$

where $p_m(\,\cdot\,)$ are the power sum symmetric functions (see, e.g., [**M**]). It follows from (6.8) and (6.9) that

$$\ln\Big(\sum_{m\geqslant 0}\tilde c_m^{(n)}a^m\Big)=\sum_{m\geqslant 1}p_m\left(\frac{i\lambda_1(n)}{n},\dots,\frac{i\lambda_n(n)}{n}\right)\frac{a^m}{m}. \tag{6.10}$$

On the other hand, by definition (2.9) of the function $F_{\gamma_1,\gamma_2,x}$,

$$\ln\Big(\sum_{m\geqslant 0}c_m^{(\infty)}a^m\Big)=i\gamma_1 a-\frac{1}{2}\gamma_2 a^2+\sum_{m\geqslant 2}p_m(ix)\frac{a^m}{m}, \tag{6.11}$$

where

$$p_m(ix)=i^m p_m(x)=i^m\sum_{k=1}^{\infty}((x_k')^m+(x_k'')^m). \tag{6.12}$$

Note that the sum in the right-hand side of (6.12) is convergent, because, due to assumption (4.6), we have $p_2(x)<\infty$ (see the Comment to Theorem 4.1).

Thus, our first claim reduces to the existence of the limits

$$\lim_{n\to\infty}p_1\left(\frac{i\lambda_1(n)}{n},\dots,\frac{i\lambda_n(n)}{n}\right)=i\gamma_1, \tag{6.13}$$

$$\lim_{n\to\infty}p_2\left(\frac{i\lambda_1(n)}{n},\dots,\frac{i\lambda_n(n)}{n}\right)=-\gamma_2+p_2(ix), \tag{6.14}$$

$$\lim_{n\to\infty}p_m\left(\frac{i\lambda_1(n)}{n},\dots,\frac{i\lambda_n(n)}{n}\right)=p_m(ix),\qquad m\geqslant 3, \tag{6.15}$$

and our second claim reduces to an estimate of the form

$$\left|p_m\left(\frac{i\lambda_1(n)}{n},\dots,\frac{i\lambda_n(n)}{n}\right)\right|\leqslant C_1'(C_2')^m,\qquad m\geqslant 3, \tag{6.16}$$

where $C_1'>0$, $C_2'>0$ are some constants not depending on n.

Clearly, (6.13) is just the assumption (4.5). Further, (6.14) immediately follows from (4.6) and (4.9). Indeed, by (4.6),

$$\lim_{n\to\infty}p_2\left(\frac{i\lambda_1(n)}{n},\dots,\frac{i\lambda_n(n)}{n}\right)=-\lim_{n\to\infty}p_2\left(\frac{\lambda_1(n)}{n},\dots,\frac{\lambda_n(n)}{n}\right)=-\tilde\gamma_2.$$

Now, by (4.9),

$$-\tilde\gamma_2=-\gamma_2-\sum_k((x_k')^2+(x_k'')^2)=-\gamma_2+p_2(ix).$$

Let us verify (6.15). Using the notation (4.2), (4.3), we have

$$p_m\left(\frac{i\lambda_1(n)}{n},\dots,\frac{i\lambda_n(n)}{n}\right)=\sum_{r\geqslant 1}\left(\frac{i\lambda_r'(n)}{n}\right)^m+\sum_{r\geqslant 1}\left(\frac{i\lambda_r''(n)}{n}\right)^m, \tag{6.17}$$

$$p_m(ix)=\sum_{r\geqslant 1}(ix_r')^m+\sum_{r\geqslant 1}(ix_r'')^m. \tag{6.18}$$

To deduce (6.15) from (4.4), it suffices to show that both sums in the right-hand side of (6.17) converge absolutely and uniformly on n. Let us examine the first sum (for the second one the arguments are just the same). Since $m \geqslant 3$ and $\lambda'_1(n) \geqslant \lambda'_2(n) \geqslant \cdots$, we have for $N = 1, 2, \ldots$

$$\begin{aligned}\left|\sum_{r\geqslant N}\left(\frac{i\lambda'_r(n)}{n}\right)^m\right| = \sum_{r\geqslant N}\left(\frac{\lambda'_r(n)}{n}\right)^m \leqslant \left(\frac{\lambda'_N(n)}{n}\right)^{m-2}\sum_{r\geqslant N}\left(\frac{\lambda'_r(n)}{n}\right)^2 \\ \leqslant \frac{\lambda'_N(n)}{n}p_2\left(\frac{\lambda_1(n)}{n},\ldots,\frac{\lambda_n(n)}{n}\right).\end{aligned} \tag{6.19}$$

By (4.6), $p_2(\lambda_1(n)/n, \ldots, \lambda_n(n)/n)$ remains bounded as $n \to \infty$. Finally, since $x'_N \to 0$ as $N \to \infty$ and since $\lambda'_N(n)/n \to x'_N$ as $n \to \infty$, the value of $\lambda'_N(n)/n$ may be made arbitrarily small provided first N and then n are chosen large enough.

Let us verify (6.16). For $m \geqslant 3$

$$\begin{aligned}&\left|p_m\left(\frac{i\lambda_1(n)}{n},\ldots,\frac{i\lambda_n(n)}{n}\right)\right| \\ &\leqslant p_2\left(\frac{\lambda_1(n)}{n},\ldots,\frac{\lambda_n(n)}{n}\right)\cdot\sup_{1\leqslant r\leqslant n}\left(\frac{\lambda_r(n)}{n}\right)^{m-2} \\ &\leqslant p_2\left(\frac{\lambda_1(n)}{n},\ldots,\frac{\lambda_n(n)}{n}\right)p_2\left(\frac{\lambda_1(n)}{n},\ldots,\frac{\lambda_n(n)}{n}\right)^{(m-2)/2}.\end{aligned} \tag{6.20}$$

Since $p_2(\lambda_1(n)/n, \ldots, \lambda_n(n)/n)$ remains bounded as $n \to \infty$, we obtain (6.16).

This completes Step 1.

Step 2. The purpose of this step is to prove that under the same assumptions as in Step 1, we have a more general result: for any fixed $k = 1, 2, \ldots,$

$$\lim_{n\to\infty} f_n(\operatorname{diag}(a_1, \ldots, a_k, 0, \ldots, 0)) = \prod_{p=1}^{k} F_{\gamma_1,\gamma_2,x}(a_p) \tag{6.21}$$

uniformly on bounded subsets in $\mathbb{R}^k$.

Let us abbreviate

$$f_n(a_1, \ldots, a_k) = f_n(\operatorname{diag}(a_1, \ldots, a_k, 0, \ldots, 0)) \tag{6.22}$$

and note that $f_n(a_1, \ldots, a_k)$ is again the characteristic function of some probability measure ν_n on $\mathbb{R}^k$. Namely, ν_n is the radial part of the measure $M_n^{(k)} = \theta_k(M_n)$ on $H(k)$ with respect to the projection

$$H(k) \ni A \mapsto \operatorname{Spec}(A) \in \mathbb{R}^k. \tag{6.23}$$

Our arguments are similar to those of Step 1. We expand both sides of (6.21) into multidimensional Taylor series. However, since these are symmetric functions of $(a_1, \ldots, a_k)$, we prefer to rewrite the Taylor series as series of Schur polynomials $s_\mu(a_1, \ldots, a_k)$, where μ ranges over the set of all Young diagrams with $\ell(\,\cdot\,) \leqslant k$:

$$f_n(a_1, \ldots, a_k) = \sum c_\mu^{(n)} s_\mu(a_1, \ldots, a_k), \tag{6.24}$$

$$\prod_{p=1}^{k} F_{\gamma_1,\gamma_2,x}(a_p) = \sum c_\mu^{(\infty)} s_\mu(a_1, \ldots, a_k). \tag{6.25}$$

As in Step 1, applying Proposition 6.1, we reduce our problem to verifying the following two claims:

First,

$$\lim_{n\to\infty} c_\mu^{(n)} = c_\mu^{(\infty)} \quad \text{for any } \mu \text{ with } \ell(\mu) \leqslant k, \tag{6.26}$$

and, second,

$$|c_\mu^{(n)}| \leqslant C_1 C_2^{|\mu|}, \tag{6.27}$$

where $C_1 > 0$, $C_2 > 0$ are some constants not depending on n.

By (5.4),

$$\begin{aligned} c_\mu^{(n)} &= \prod_{(p,q)\in\mu} \frac{1}{n+q-p} \cdot s_\mu(i\lambda_1(n), \dots, i\lambda_n(n)) \\ &= \prod_{(p,q)\in\mu} \frac{n}{n+q-p} \cdot s_\mu\left(\frac{i\lambda_1(n)}{n}, \dots, \frac{i\lambda_n(n)}{n}\right). \end{aligned} \tag{6.28}$$

Thus, in both claims, (6.26) and (6.27), we may replace $c_\mu^{(n)}$ by

$$\tilde{c}_\mu^{(n)} := s_\mu\left(\frac{i\lambda_1(n)}{n}, \dots, \frac{i\lambda_n(n)}{n}\right). \tag{6.29}$$

Let us prove (6.26) with $c_\mu^{(n)}$ replaced by $\tilde{c}_\mu^{(n)}$.

Recall the Jacobi–Trudi identity expressing the Schur functions in terms of the complete symmetric functions (see, e.g., [**M,** Chapter I, (3.4)]):

$$s_\mu = \det[h_{\mu_i-i+j}]_{i,j=1}^k, \qquad \ell(\mu) \leqslant k. \tag{6.30}$$

It follows from (6.8), (6.29), and (6.30) that

$$\tilde{c}_\mu^{(n)} = \det[\tilde{c}_{\mu_i-i+j}^{(n)}]_{i,j=1}^k. \tag{6.31}$$

By Step 1, $\tilde{c}_m^{(n)} \to c_m^{(\infty)}$ as $n \to \infty$, so that

$$\lim_{n\to\infty} \tilde{c}_\mu^{(n)} = \det[c_{\mu_i-i+j}^{(\infty)}]_{i,j=1}^k. \tag{6.32}$$

Further, it is well known that for an arbitrary formal series $\sum_{m\geqslant 0} c_m a^m$ with $c_0 = 1$, we have

$$\prod_{p=1}^k \left(\sum_{m=0}^\infty c_m a_p^m\right) = \sum_{\substack{\mu \\ \ell(\mu)\leqslant k}} \det[c_{\mu_i-i+j}]_{i,j=1}^k s_\mu(a_1, \dots, a_k). \tag{6.33}$$

Applying (6.33) to the left-hand side of (6.25), we conclude that

$$c_\mu^{(\infty)} = \det[c_{\mu_i-i+j}^{(\infty)}]_{i,j=1}^k = \lim_{n\to\infty} \tilde{c}_\mu^{(n)}. \tag{6.34}$$

Thus, we have verified the first claim. As for the second claim, an estimate of type (6.27) for $\tilde{c}_\mu^{(n)}$ follows at once from (6.31) and from the estimate (6.6) for $\tilde{c}_m^{(n)}$ proved in Step 1.

This completes Step 2.

It follows that $f_n \to f$ (where f is given by (4.7)) uniformly on compact subsets of $H(\infty)$. Clearly, f is a continuous positive definite normalized function on $H(\infty)$. By Proposition 1.1, it is the characteristic function of a $U(\infty)$-invariant Borel probability measure M on the space H, invariant under the action of $U(\infty)$.

Then, by Proposition 3.4, M_n weakly converges to M. Since f is multiplicative, M is ergodic. Thus, we have verified claim (i) of Theorem 4.1.

Step 3. Let us fix a sequence $\{\Lambda(n)\}$, where $\Lambda(n) = (\lambda_1(n), \dots, \lambda_n(n))$, and let $f_n(a)$, $a \in \mathbb{R}$, be defined as in Step 1. Let us assume that

$$\lim_{n\to\infty} f_n(a) = f(a), \qquad a \in \mathbb{R}, \tag{6.35}$$

uniformly on bounded subsets in $\mathbb{R}$, where $f(a)$ is a function on $\mathbb{R}$ (of course, f is automatically continuous). We shall prove that then $\{\Lambda(n)\}$ must satisfy the assumptions (4.4)–(4.6) of Theorem 4.1(i).

Suppose first that

$$\sup_n \left\{ p_2\left(\frac{\lambda_1(n)}{n}, \dots, \frac{\lambda_n(n)}{n}\right) + \left(p_1\left(\frac{\lambda_1(n)}{n}, \dots, \frac{\lambda_n(n)}{n}\right)\right)^2 \right\} < \infty. \tag{6.36}$$

Then, given an infinite subset $N \subseteq \{1, 2, \dots\}$, there exists a possibly smaller infinite subset $N' \subseteq N$ such that the limits (4.4), (4.5), and (4.6) exist provided n goes to infinity inside N'. Then, by Step 1, $f_n \to F_{\gamma_1,\gamma_2,x}$ as $n \to \infty$ inside N', so that $F_{\gamma_1,\gamma_2,x} = f$. For any other N and N', the parameters γ_1, γ_2, and x will be the same, because they are uniquely determined by the function itself, see Comment 4 after Proposition 2.8. It follows that the limits (4.4)–(4.6) exist as n ranges over the set of all natural numbers.

Suppose now that (6.36) does not hold. We shall show that this leads to a contradiction with the initial assumption (6.35).

Indeed, since the expression $\{\cdots\}$ in (6.36) is a homogeneous function of $\Lambda(n)$, we can choose an infinite subset $N \subseteq \{1, 2, \dots\}$ and a sequence of positive numbers $\{\varepsilon_n \mid n \in N\}$ such that $\lim_{n\in N} \varepsilon_n = 0$ and

$$\lim_{n\subset N} \left\{ p_2\left(\varepsilon_n\frac{\lambda_1(n)}{n}, \dots, \varepsilon_n\frac{\lambda_n(n)}{n}\right) + \left(p_1\left(\varepsilon_n\frac{\lambda_1(n)}{n}, \dots, \varepsilon_n\frac{\lambda_n(n)}{n}\right)\right)^2 \right\} = 1. \tag{6.37}$$

Then, replacing N by a smaller infinite subset N', we can arrange so that for the sequence $\{\varepsilon_n\Lambda(n)\}$, the limits (4.4)–(4.6) will exist provided n goes to infinity inside N'. Moreover, at least one of the corresponding parameters γ_1, $\tilde{\gamma}_2$ will be nonzero.

Note that the effect of multiplying $\Lambda(n)$ by ε_n is the same as that of multiplying a by ε_n. Thus, by Step 1,

$$\lim_{n\in N'} f_n(\varepsilon_n a) = F_{\gamma_1,\gamma_2,x}, \qquad a \in \mathbb{R}. \tag{6.38}$$

Since at least one of the parameters γ_1, $\tilde{\gamma}_2$ of the function $F_{\gamma_1,\gamma_2,x}$ is nonzero, it follows from the definition of this function that it is not equal identically to 1. But since $F_{\gamma_1,\gamma_2,x}$ is analytic, the same is true in an arbitrarily small neighborhood of the point $a = 0$. Then, comparing (6.38) with (6.35), we arrive at a contradiction.

Thus, we have verified claim (ii) of Theorem 4.1. □

REMARK 6.2. Note that the estimates (6.6) and (6.16) are not necessary to assert the convergence of the functions f_n. Indeed, Proposition 6.1 may be replaced by the following stronger claim:

Let $f_1, f_2, \dots$ and f be smooth positive definite functions on $\mathbb{R}^k$, normalized at the origin. Expand them into Taylor series at the origin and assume that each Taylor coefficient of f_n tends to the corresponding coefficient of f as $n \to \infty$.

Finally, assume that the moment problem defined by the coefficients of f has a unique solution (the latter condition is satisfied, e.g., if f is analytic).

Then the sequence (f_n) converges to f uniformly on compact subsets of $\mathbb{R}^n$.

By virtue of this claim, the verification of the uniform convergence of the Taylor expansions may be omitted.

§7. Total positivity

DEFINITION 7.1. Let $\varphi(t)$ be a real nonnegative measurable function on $\mathbb{R}$. Then φ is said to be a *totally positive* function if for $n = 1, 2, \dots$

$$\det[\varphi(t_i - s_j)]_{i,j=1}^n \geqslant 0 \quad \text{for any } t_1 < \dots < t_n \text{ and } s_1 < \dots < s_n. \tag{7.1}$$

It will be convenient for us to include in the definition the following additional assumption: φ is summable and $\int \varphi(t)\,dt = 1$, i.e., $\varphi(t)\,dt$ is a probability measure on $\mathbb{R}$. (In [**S2**], functions satisfying both conditions are called *Pólya frequency functions.* The second condition is in fact not restrictive, see [**S2,** Lemma 4].)

PROPOSITION 7.2 (Schoenberg [**S2,** p. 341, Lemma 5]). *The set of totally positive functions is stable under convolution.*

PROOF. For two summable functions φ and ψ, the convolution $\varphi*\psi$ is correctly defined and the following formula is readily verified:

$$\begin{aligned} &\det[(\varphi * \psi)(t_i - s_j)]_{i,j=1}^n \\ &\quad = \frac{1}{n!}\int_{\mathbb{R}^n} \det[\varphi(t_i - u_k)]_{i,k=1}^n \cdot \det[\psi(u_k - s_j)]_{k,j=1}^n \, du_1 \cdots du_n. \end{aligned} \tag{7.2}$$

Now suppose φ and ψ are totally positive and let $t_1 < \dots < t_n$, $s_1 < \dots < s_n$. Then the integrand in (7.2) is nonnegative for all (pairwise distinct) $u_1, \dots, u_n$, because both determinants have the same sign, equal to that of $\prod_{k>l}(u_k - u_l)$. □

PROPOSITION 7.3 (Schoenberg [**S2,** pp. 335 and 343]). *The densities of the normal and exponential distributions,*

$$\psi_\gamma(t) = \frac{1}{\sqrt{2\pi\gamma}}\, e^{-t^2/(2\gamma)}, \qquad \gamma > 0, \tag{7.3}$$

and

$$\varphi_y(t) = \begin{cases} y^{-1} e^{-y^{-1}t}, & t \geqslant 0, \\ 0, & t < 0, \end{cases} \qquad y > 0, \tag{7.4}$$

are totally positive. □

Note that the result remains true after the shift $t \mapsto t + \text{const}$ of the argument or the change of sign $t \mapsto -t$.

Note also that $\{\psi_\gamma\}$ and $\{\varphi_y\}$ are one-parametric semigroups with respect to the convolution product.

THEOREM 7.4 (Schoenberg's theorem on totally positive functions, see Schoenberg [**S2**], Karlin [**K**]). *The Fourier transforms $\widehat{\varphi}$ of totally positive functions φ are* *just the functions $F_{\gamma_1,\gamma_2,x}$, defined in* (2.9), *where at least one of the parameters $\gamma_2, x_1, x_2, \dots$ is nonzero.* □

This fundamental result shows that a totally positive function φ is a convolution product $\varphi_0 * \varphi_1 * \varphi_2 * \cdots$, where $\varphi_0(t)\,dt$ is a normal distribution and $\varphi_k(t)\,dt$, $k = 1, 2, \dots$, are, up to transformations $t \mapsto \pm t + \text{const}$, exponential distributions. (We have to exclude $\gamma_2 = x_1 = x_2 = \cdots = 0$, because the inverse Fourier transform of the corresponding function F is the Dirac measure.)

Comparing Theorem 2.9 and Theorem 7.4, we obtain the following correspondence $M \leftrightarrow \varphi$ between the ergodic measures $M \in \mathcal{M}$ (except the Dirac measures on scalar matrices) and the totally positive functions φ on the real line:

$$\theta_1(M)(dt) = \varphi(t)\,dt$$

(recall that the mapping θ_1 assigns to a matrix $B \in H$ its matrix element $B_{11} \in \mathbb{R}$). In other words, this means that the distribution of the random variable B_{11} with respect to the probability distribution M on the matrices $B \in H$ is given by the density φ.

The easy part of Theorem 7.4 consists in verifying the fact that $F_{\gamma_1,\gamma_2,x} = \widehat{\varphi}$ with a totally positive φ. This is done by making use of Propositions 7.2 and 7.3 and an evident passage to the limit (cf. Proposition 2.8).

The hard part of Theorem 7.4 is to prove that the Fourier transform $\widehat{\varphi}$ of any totally positive function φ is of the form (2.9). Our purpose is to show that this claim is equivalent to Theorem 2.9.

DEFINITION 7.5 (Karlin [**K,** p. 49]). A real smooth nonnegative function $\varphi(t)$ on $\mathbb{R}$, $\int \varphi(t)\,dt = 1$, is called *extended totally positive* if

$$\det\,[\varphi^{(i-1)}(v_j)]_{i,j=1}^n \geqslant 0, \qquad n = 1, 2, \dots, \; v_1 > \cdots > v_n. \tag{7.5}$$

PROPOSITION 7.6. (i) *Any smooth totally positive function φ is extended totally positive.*

(ii) *Conversely, if φ is an extended totally positive function, then the function $\varphi * \psi_\gamma$, where ψ_γ was defined by* (7.3), *is totally positive for any $\gamma > 0$.*

PROOF. (i) By definition of total positivity, for any pairwise distinct real $t_1, \dots, t_n$ and any $s_1 < \cdots < s_n$,

$$\prod_{p>q} (t_p - t_q)^{-1} \cdot \det\,[\varphi(t_i - s_j)]_{i,j=1}^n \geqslant 0. \tag{7.6}$$

Putting $s_1 = -v_1, \dots, s_n = -v_n$ and letting $t_1, \dots, t_n \to 0$ in (7.6), we obtain (7.5).

(ii) By Theorem 2.1 in Karlin [**K,** p. 50], if φ verifies strict inequalities in (7.5), then it is totally positive. So it suffices to prove that if we replace φ by $\varphi * \psi_\gamma$, then the inequalities in (7.5) become strict.

For $n = 1$ this is evident, because φ is nonnegative and not identically equal to zero whereas ψ_γ is strictly positive. For $n > 1$ this argument is generalized as follows.

First, remark that the functions $\varphi, \varphi', \varphi'', \dots$ are linearly independent. Indeed, if this is not true, then φ satisfies a linear differential equation with constant coefficients, whence $|\varphi(t)| \to \infty$ as $t \to \infty$ or $t \to -\infty$. But this contradicts the assumption $\varphi \in L^1(\mathbb{R})$.

Next, substitute $\psi = \psi_\gamma$ into formula (7.2) and repeat the argument used in the proof of (i). Then we obtain, for any $s_1 < \cdots < s_n$,

$$\begin{aligned}(7.7)\qquad &\det[(\varphi * \psi_\gamma)^{(i-1)}(-s_j)]_{i,j=1}^n \\ &= \frac{1}{n!}\int_{\mathbb{R}^n} \det[\varphi^{(i-1)}(-u_k)]_{i,k=1}^n \det[\psi_\gamma(u_k - s_j)]_{k,j=1}^n \, du_1 \cdots du_n \,.\end{aligned}$$

Arguing just as in the proof of Proposition 7.2, we see that the integrand in (7.7) is nonnegative.

Finally, as noted in Schoenberg [**S2,** p. 336], the second determinant in the integrand is nonzero provided $u_1, \dots, u_n$ are pairwise distinct. On the other hand, since $\varphi, \varphi', \dots$ are linearly independent, the first determinant in the integrand does not vanish for certain $(u_1, \dots, u_n) \in \mathbb{R}^n$, hence on an open subset of $\mathbb{R}^n$. We conclude that the integrand is everywhere nonnegative and strictly positive on an open subset, so that the integral (7.7) is strictly positive. □

Note that Proposition 7.6 corresponds to a part of Pickrell's proof in [**Pi2,** pp. 154–155]. There it is claimed that an analytic extended totally positive function is totally positive; however, the arguments are too sketchy and seem to be incomplete. To avoid this difficulty, we modified the claim somewhat and used a trick suggested by Boyer's paper [**Bo,** p. 218].

Note also that not all totally positive functions are smooth; for instance, the function (7.4) is not smooth at 0. Thus, the class of extended totally positive functions, as defined above, does not coincide with the class of totally positive functions (although the two classes are very close, as is seen from Proposition 7.6). For this reason, to use property (7.5) we must first smooth totally positive functions.

The next theorem is Pickrell's main calculation in [**Pi2,** pp. 154–155]. It is simple but instructive. For completeness and for reader's convenience, we give the proof (which is presented here in slightly more detail than in [**Pi2**]).

THEOREM 7.7 (Pickrell [**Pi2,** pp. 154–155]). *Let φ be a smooth nonnegative function on $\mathbb{R}$, $\int \varphi(t)\,dt = 1$, and let $F = \widehat{\varphi}$ be its Fourier transform. Then F belongs to the class $\mathcal{F}_1$ (see definition (2.2)) if and only if φ is extended totally positive.*

PROOF. Given $n = 1, 2, \dots$ and $A \in H(n)$, denote by $a_1, \dots, a_n$ the eigenvalues of A and put

$$f_n(A) = F(a_1) \cdots F(a_n);$$

this is a continuous $U(n)$-invariant function on $H(n)$. Let us fix n and show that positive definiteness of f_n is equivalent to condition (7.5).

By Bochner's theorem, f_n is positive definite if and only if its inverse Fourier transform is a measure. This condition is equivalent to the following one: for any function $\Psi \geqslant 0$ from the Schwartz space $\mathcal{S}(H(n))$,

$$(7.8)\qquad \langle f_n, \widehat{\Psi}\rangle := \int_{H(n)} f_n(A)\overline{\widehat{\Psi}(A)}\, dA \geqslant 0 \,.$$

Since f_n is $U(n)$-invariant, one may assume Ψ is $U(n)$-invariant too.

Let $D(n)$ denote the subspace of diagonal matrices in $H(n)$. We identify $D(n)$ with $\mathbb{R}^n$ and write elements of $D(n)$ as $\operatorname{diag}(a_1, \dots, a_n)$ where $(a_1, \dots, a_n) \in \mathbb{R}^n$.

It is well known and easily verified that the radial part of the Lebesgue measure on $H(n)$ with respect to the action of $U(n)$ is the measure

$$\text{const}\, V^2(a_1,\dots,a_n)\,da_1\cdots da_n,\qquad \text{const}>0 \tag{7.9}$$

on $D(n)$, where

$$V(a_1,\dots,a_n)=\prod_{p<q}(a_p-a_q).$$

It follows that

$$\langle f_n,\widehat{\Psi}\rangle = \text{const}\int_{D(n)} V(a_1,\dots,a_n)\,\widehat{\varphi}(a_1)\cdots\widehat{\varphi}(a_n) \times \overline{V(a_1,\dots,a_n)\,\widehat{\Psi}(\operatorname{diag}(a_1,\dots,a_n))}\,da_1\cdots da_n. \tag{7.10}$$

Put

$$\theta(t_1,\dots,t_n)=V(t_1,\dots,t_n)\,\Psi(\operatorname{diag}(t_1,\dots,t_n)). \tag{7.11}$$

We shall show that the integral (7.10) is equal, up to a positive factor, to

$$\int_{\mathbb{R}^n}\det[\varphi^{(j-1)}(t_k)]_{j,k=1}^n\theta(t_1,\dots,t_n)\,dt_1\cdots dt_n. \tag{7.12}$$

Indeed, the function $V(a_1,\dots,a_n)\widehat{\varphi}(a_1)\cdots\widehat{\varphi}(a_n)$ is the Fourier transform of

$$V\left(i\frac{\partial}{\partial t_1},\dots,i\frac{\partial}{\partial t_n}\right)\cdot\varphi(t_1)\cdots\varphi(t_n)=i^{n(n-1)/2}\det[\varphi^{(n-j)}(t_k)]_{j,k=1}^n. \tag{7.13}$$

On the other hand, using the $U(n)$-invariance of Ψ, we have

$$\begin{aligned} &V(a_1,\dots,a_n)\,\widehat{\Psi}(\operatorname{diag}(a_1,\dots,a_n))\\ &\quad= V(a_1,\dots,a_n)\int_{T\in H(n)} e^{i\operatorname{tr}(\operatorname{diag}(a_1,\dots,a_n)T)}\Psi(T)\,dT\\ &\quad= V(a_1,\dots,a_n)\int_{H(n)}\left(\int_{U(n)} e^{i\operatorname{tr}(\operatorname{diag}(a_1,\dots,a_n)uTu^{-1})}du\right)\Psi(T)\,dT. \end{aligned} \tag{7.14}$$

Using formula (5.12) for the interior integral and again applying formula (7.8) for the radial part of the Lebesgue measure, we see that (7.14) is equal, up to a positive factor, to

$$(-i)^{n(n-1)/2}\int_{\mathbb{R}^n}\det[e^{ia_jt_k}]_{j,k=1}^n\cdot\theta(t_1,\dots,t_n)\,dt_1\cdots dt_n, \tag{7.15}$$

where θ was defined in (7.11). Expanding the determinant and using the fact that $\theta(t_1,\dots,t_n)$ is antisymmetric with respect to permutations of $t_1,\dots,t_n$, we conclude that (7.15) is equal to

$$n!\,(-i)^{n(n-1)/2}\hat{\theta}(a_1,\dots,a_n). \tag{7.16}$$

Now (7.13) and (7.16) imply that the integral (7.10) is equal, up to a positive factor, to

$$\begin{aligned} &(-1)^{n(n-1)/2}\int_{\mathbb{R}^n}\det[\varphi^{(n-j)}(t_k)]_{j,k=1}^n\cdot\theta(t_1,\dots,t_n)\,dt_1\cdots dt_n\\ &\qquad=\int_{\mathbb{R}^n}\det[\varphi^{(j-1)}(t_k)]_{j,k=1}^n\cdot\theta(t_1,\dots,t_n)\,dt_1\dots dt_n. \end{aligned} \tag{7.17}$$

Thus, we have verified (7.12).

Since the integrand in (7.12) is symmetric with respect to permutations of $t_1, \dots, t_n$, we see that (7.12) is equal, up to a positive factor, to

$$\int_{t_1 > \cdots > t_n} \det [\varphi^{(j-1)}(t_k)]_{j,k=1}^n \cdot \theta(t_1, \dots, t_n)\, dt_1 \dots dt_n . \tag{7.18}$$

Recall that θ is given by (7.11), where Ψ is a $U(n)$-invariant nonnegative function from the Schwartz space, and remark that $V(t_1, \dots, t_n) > 0$ in the domain $t_1 > \cdots > t_n$. It follows that (7.18) is nonnegative for any such Ψ if and only if φ is extended totally positive. □

COROLLARY 7.8. *Theorem 2.9 (classification of ergodic measures $M \in \mathcal{M}$ or of functions $F \in \mathcal{F}_1$) and Schoenberg's Classification Theorem 7.4 can be derived one from another.*

PROOF. Let us show that Theorem 2.9 implies Theorem 7.4. Let φ be a totally positive function. We must prove that the Fourier transform $F = \widehat{\varphi}$ is of the form (2.9), where at least one of the parameters $\gamma_2, x_1, x_2, \dots$ is nonzero.

First suppose φ is smooth. Then, by Proposition 7.6 (i), φ is extended totally positive. Next, by Theorem 7.7, $F \in \mathcal{F}_1$, and, finally, by Theorem 2.9, F is of the form (2.9).

The general case can be reduced to that of a smooth φ as follows. We again use the Gaussian totally positive function ψ_γ (see (7.3)) to smooth φ. Then, for any $\gamma > 0$, we have a smooth (even analytic) totally positive function $\varphi * \psi_\gamma$. This implies that $F(a)e^{-\gamma a^2/2}$ is of the form (2.9) for any $\gamma > 0$, whence F itself is of this form.

It should be added that F cannot be equal to $F_{\gamma_1,0,0}$, because we know that the inverse Fourier transform of F is a function and not a Dirac measure. So at least one of the parameters $\gamma_2, x_1, x_2, \dots$ is nonzero.

The inverse implication is verified similarly, by making use of Proposition 7.6 (ii). □

Thus, our proof of Theorem 2.9 leads to a new proof of Schoenberg's Theorem 7.4.

§8. Totally positive functions as limits of splines

After reading the preliminary version [**OV**] of the present paper, Andreĭ Okounkov remarked that the one-dimensional projections of orbital measures coincide with the so-called fundamental splines (= B-splines) whose limits were studied in an important paper by Curry and Schoenberg [**CS**]. The purpose of this section is to briefly discuss the relationship between the results of [**CS**] and our results.

We start by stating some classical facts used in [**CS**].

Fix real numbers $t_1 < \cdots < t_n$, called the *knots* ($n \geqslant 3$). There exists a (unique) function $M_{n-1}(t) = M_{n-1}(t; t_1, \dots, t_n)$ on $\mathbb{R}$ such that:

(i) On each open interval determined by adjacent knots, $M_{n-1}(t)$ is a polynomial of degree $n-2$.
(ii) $M_{n-1}(t)$ vanishes when $t < t_1$ or $t > t_n$.
(iii) $M_{n-1}(t)$ has $n-3$ continuous derivatives at each knot.
(iv) $\int M_{n-1}(t)\, dt = 1$.

In [**CS**], the function $M_{n-1}(t)$ is called the *fundamental spline* (with knots $t_1, \dots, t_n$). Another term, used in the modern literature, is *B-spline.*

The fundamental spline $M_{n-1}(t)$ is given by the following explicit formula:

$$M_{n-1}(t; t_1, \dots, t_n) = (n-1) \sum_{k=1}^{n} \frac{(\max(t_k - t, 0))^{n-2}}{\prod_{i \neq k}(t_k - t_i)}. \tag{8.1}$$

Let σ_{n-1} denote the standard $(n-1)$-dimensional simplex,

$$\sigma_{n-1} = \{(p_1, \dots, p_n) \mid 0 \leqslant p_1, \dots, p_n \leqslant 1,\ p_1 + \dots + p_n = 1\} \subset \mathbb{R}^n, \tag{8.2}$$

and let ξ denote the affine functional on the simplex taking values $t_1, \dots, t_n$ at its vertices. Then $M_{n-1}(t)$ coincides with the density of the image under ξ of the Lebesgue measure on σ_{n-1}, so normalized that the volume of σ_{n-1} is equal to 1. This implies, in particular, that $M_{n-1}(t)$ is nonnegative, so that $M_{n-1}(t)\,dt$ is a probability measure on $\mathbb{R}$.

Using a passage to the limit, one easily extends the definition of $M_{n-1}(t)$ to the case when some of the knots coincide.

For further properties of the functions $M_{n-1}(t)$ and for proofs of the facts mentioned above, see [**CS**] or, e.g., Babenko's textbook [**Ba**].

Now we are in a position to state the main result of Curry and Schoenberg [**CS**, Theorem 6]:

THEOREM 8.1. *Consider the class of probability measures on $\mathbb{R}$ which can be obtained as weak limits of measures of the form $M_{n-1}(t; t_1, \dots, t_n)\,dt$, where $n \to \infty$ and the knots $t_1, \dots, t_n$ depend on n. Then the characteristic functions of the measures of this class are exactly those given by formula* (4.8).

We shall briefly describe the method of proof used in [**CS**]. Given $t_1, \dots, t_n$, put

$$F_n(a) = \int_{-\infty}^{\infty} \left(1 - \frac{iat}{n}\right)^{-n} M_{n-1}(t; t_1, \dots, t_n)\,dt. \tag{8.3}$$

Since

$$\left(1 - \frac{iat}{n}\right)^{-n} = e^{iat}\left(1 + O\left(\frac{1}{n}\right)\right), \tag{8.4}$$

$F_n(a)$ may be viewed as the "approximate Fourier transform" of $M_{n-1}(t)$. Its advantage with respect to the ordinary Fourier image of $M_{n-1}(t)$ is that it is given by a very simple expression, namely

$$F_n(a) = \prod_{k=1}^{n} \left(1 - \frac{iat_k}{n}\right)^{-1}. \tag{8.5}$$

Now note that $F_n(ia)^{-1}$ is a polynomial in a with only real zeros, and use a well-known theorem, due to Laguerre and Pólya, which describes the class of entire functions that can be approximated by polynomials with real zeros: up to change of a variable $a \mapsto ia$, these are exactly the reciprocals to functions of type (4.8), see Hirschman–Widder [**HW**].

Note that this result on entire functions also plays an important role in Schoenberg's classification of totally positive functions (Theorem 7.4 above).

Comparing Theorem 8.1 with Theorem 7.4, we conclude that a probability measure on $\mathbb{R}$ (distinct from a Dirac measure) can be approximated by a sequence of fundamental splines with growing number n of knots if and only if it is given by a totally positive density.

The following fact, remarked by Andreĭ Okounkov, is crucial for our discussion.

PROPOSITION 8.2 (A. Yu. Okounkov). *Let $\lambda_1 \leqslant \dots \leqslant \lambda_n$ be real numbers and $n \geqslant 3$. Consider the $U(n)$-orbit in $H(n)$ passing through the diagonal matrix $\Lambda = \operatorname{diag}(\lambda_1, \dots, \lambda_n)$ and denote by $\mu(dB)$ the corresponding orbital measure.*

Then the image of μ under the projection $H(n) \ni B \mapsto B_{11} \in \mathbb{R}$ coincides with the fundamental spline $M_{n-1}(t; \lambda_1, \dots, \lambda_n)\, dt$.

PROOF. For $u \in U(n)$, we have

$$(u\Lambda u^{-1})_{11} = \sum_{k=1}^{n} u_{1k}\lambda_k (u^{-1})_{k1} = \sum_{k=1}^{n} |u_{1k}|^2 \lambda_k . \tag{8.6}$$

When the matrix u ranges over $U(n)$, its first row

$$z = (z_1, \dots, z_n) := (u_{11}, \dots, u_{1n}) \tag{8.7}$$

ranges over the unit sphere $S^{2n-1} \subset \mathbb{C}^n$, and under the mapping $u \mapsto z$, the orbital measure μ projects onto the normalized invariant measure on S^{2n-1}, which may be written as

$$\text{const}\ \frac{d(\operatorname{Re} z_1)\, d(\operatorname{Im} z_1) \cdots d(\operatorname{Re} z_n)\, d(\operatorname{Im} z_n)}{d(|z_1|^2 + \dots + |z_n|^2 - 1)} . \tag{8.8}$$

Further, under the mapping

$$(z_1, \dots, z_n) \mapsto (|z_1|^2, \dots, |z_n|^2) = (p_1, \dots, p_n) \tag{8.9}$$

of the sphere onto the simplex σ_{n-1}, the measure (8.8) projects onto the measure

$$\text{const}\ \frac{dp_1 \cdots dp_n}{d(p_1 + \dots + p_n - 1)}, \tag{8.10}$$

which coincides with the normalized Lebesgue measure on the simplex.

Finally, the right-hand side of (8.6), which may be rewritten as $\sum_{k=1}^{n} p_k \lambda_k$, is just the value at the point $(p_1, \dots, p_n) \in \sigma_{n-1}$ of the linear functional $\xi \colon \mathbb{R}^n \to \mathbb{R}$ taking values $\lambda_1, \dots, \lambda_n$ at the vertices of the simplex. By a property of the spline function $M_{n-1}(t)$ mentioned above, we conclude that the image of the orbital measure μ under the mapping $B \mapsto B_{11}$ is equal to $M_{n-1}(t; \lambda_1, \dots, \lambda_n)\, dt$. □

By virtue of Proposition 8.2, the one-dimensional projections of orbital measures admit a very nice analytic interpretation in terms of splines, and Curry–Schoenberg's result described in Theorem 8.1 turns out to be almost equivalent to the "one-dimensional part" of our Theorem 4.1, that is, to the results of Steps 1 and 3 in §6. The difference in the statements is that Curry and Schoenberg do not obtain necessary and sufficient conditions on the knots under which a sequence of fundamental splines would be weakly convergent; they are only interested in describing the limiting functions.

Theorems 7.4 and 8.1 together imply that totally positive functions are exactly those functions which may be approximated by fundamental splines with a growing number of knots. This fact seems to be highly nontrivial, because the fundamental splines themselves are not totally positive. We hope that the chain of relations traced in the present paper furnishes a certain explanation of this phenomenon.

REMARK 8.3. Recall that the *Dirichlet distribution* $\mathcal{D}(\theta_1, \dots, \theta_n)$ with parameters $\theta_1 > 0, \dots, \theta_n > 0$ is defined as the probability measure on the $(n-1)$-dimensional simplex (8.2) whose density with respect to the Lebesgue measure is given by

$$\text{const}\, p_1^{\theta_1 - 1} \cdots p_n^{\theta_n - 1}, \tag{8.11}$$

see Kingman [**Ki,** Section 9.1]. Now in Proposition 8.2 let us replace the space $H(n)$ of $n \times n$ Hermitian matrices by the space of $n \times n$ real symmetric (respectively, quaternion Hermitian) matrices. Then the one-dimensional projections of orbital measures coincide with various one-dimensional projections of the Dirichlet distribution $\mathcal{D}(1/2, \dots, 1/2)$ (respectively, of the Dirichlet distribution $\mathcal{D}(2, \dots, 2)$).

More generally, one can consider one-dimensional projections of the Dirichlet distribution $\mathcal{D}(\theta, \dots, \theta)$ with arbitrary parameter $\theta > 0$. If $\theta = 1, 2, 3, \dots$ then one-dimensional projections of this distribution are the fundamental splines with multiple knots: the multiplicity of each knot is equal to θ. For general $\theta > 0$ there is no such interpretation. However, for any θ, using the "moment method", one can still obtain an analog of Theorem 8.1. The limiting measures will have a characteristic function of the following form (cf. (4.8)):

$$F(a) = e^{i\gamma_1 a - \gamma_2 a^2/2} \prod_k \frac{e^{-i\theta x'_k a}}{(1 - i x'_k a)^\theta} \prod_k \frac{e^{-i\theta x''_k a}}{(1 - i x''_k a)^\theta}, \qquad a \in \mathbb{R}, \tag{8.12}$$

where the parameters are the same as in (4.8).

Finally, note that the parameter $\alpha = \theta^{-1}$ exactly corresponds to the parameter that appears in the theory of Jack's symmetric functions (see Stanley [**Sta**] or the 2nd edition (1995) of Macdonald's book [**M**]). In particular, the mth moment of a one-dimensional projection of the Dirichlet distribition $\mathcal{D}(\theta, \dots, \theta)$ is equal, up to a scalar factor, to $P_m(t_1, \dots, t_n; \theta^{-1})$, where $t_1, \dots, t_n$ stands for the parameters of the projection (i.e., the values of the corresponding affine functional ξ at the vertices of the simplex) and $P_m(\,\cdot\,; \alpha)$, $m = 1, 2, \dots$, are one-row Jack's symmetric functions with parameter α.

References

[ASW] M. Aissen, I. J. Schoenberg, and A. M. Whitney, *On the generating functions of totally positive sequences,* I, J. Analyse Math. **2** (1952), 93–103.

[Ba] K. I. Babenko, *Basic numerical analysis*, "Nauka", Moscow, 1986. (Russian)

[BCR] C. Berg, J. P. R. Christensen, and P. Ressel, *Harmonic analysis on semigroups. Theory of positive definite and related functions*, Springer-Verlag, Berlin, 1984.

[BGV] N. Berline, E. Getzler, and M. Vergne, *Heat kernels and Dirac operators*, Grundlehren Math. Wiss., vol. 298, Springer-Verlag, Berlin and New York, 1992.

[Bi] P. Billingsley, *Convergence of probability measures*, Wiley, New York, 1968.

[Bo] R. P. Boyer, *Infinite traces of AF-algebras and characters of* $U(\infty)$, J. Operator Theory **9** (1983), 205–236.

[CS] H. B. Curry and I. J. Schoenberg, *On Pólya frequency functions* IV. *The fundamental spline functions and their limits*, J. Analyse Math. **17** (1966), 71–107.

[D] J. L. Doob, *Stochastic processes*, Wiley, New York, 1953.

[E1] A. Edrei, *On the generating functions of totally positive sequences* II, J. Analyse Math. **2** (1952), 104–109.

[E2] ———, *On the generating function of a doubly-infinite, totally positive sequence*, Trans. Amer. Math. Soc. **74** (1953), 367–383.

[GN] I. M. Gelfand and M. A. Naimark, *Unitary representations of classical groups*, Moscow–Leningrad, 1950 (Russian); German transl., Akademie Verlag, Berlin, 1957.

[HW] I. I. Hirschman and D. V. Widder, *The convolution transform*, Princeton Univ. Press, Princeton, NJ, 1955.

[I1] R. S. Ismagilov, *Linear representations of groups of matrices with elements from a normed field*, Izv. Akad. Nauk SSSR Ser. Mat. **33** (1969), 1296–1323; English transl., Math. USSR-Izv. **3** (1969), 1219–1244.

[I2] ———, *Spherical functions over a normed field whose residue field is infinite*, Funktsional. Anal. i Prilozhen. **4** (1970), no. 1, 42–51; English transl., Functional Anal. Appl. **4** (1970), no. 1, 37–45.

[J] A. T. James, *Distributions of matrix variates and latent roots derived from normal samples*, Ann. Math. Stat. **35** (1964), 475–501.

[K] S. Karlin, *Total positivity*, vol. I, Stanford Univ. Press, Stanford, CA, 1968.

[Ke] S. Kerov, *Gaussian limit for the Plancherel measure of the symmetric group*, C. R. Acad. Sci. Paris, Sér. I **316** (1993), 303–308.

[KV] S. V. Kerov and A. M. Vershik, *The characters of the infinite symmetric group and probability properties of the Robinson–Schensted–Knuth algorithm*, SIAM J. Alg. Discr. Meth. **7** (1986), 116–124.

[KOV] S. Kerov, G. Olshanski, and A. Vershik, *Harmonic analysis on the infinite symmetric group. Deformation of the regular representation*, C. R. Acad. Sci. Paris, Sér. I **316** (1993), 773–778.

[Ki] J. F. C. Kingman, *Poisson processes*, Clarendon Press, Oxford, 1993.

[M] I. G. Macdonald, *Symmetric functions and Hall polynomials*, Clarendon Press, Oxford, 1979.

[Mu] R. J. Muirhead, *Aspects of multivariate statistical theory*, Wiley, New York, 1982.

[N] N. I. Nessonov, *The complete classification of representations of the group $GL(\infty)$ containing the identity representation of the unitary subgroup*, Mat. Sb. **130** (1986), 131–150; English transl., Math. USSR-Sb. **5** (1987), 122–147.

[Ok] A. Yu. Okounkov, *Thoma's theorem and representations of the infinite bisymmetric group*, Funktsional. Anal. i Prilozhen. **28** (1994), no. 2, 31–40; English transl., Functional Anal. Appl. **28** (1994), no. 2, 100–107.

[O1] G. I. Ol′shanskiĭ, *Unitary representations of the infinite-dimensional classical groups $U(p,\infty)$, $SO(p,\infty)$, $Sp(p,\infty)$ and the corresponding motion groups*, Funktsional. Anal. i Prilozhen. **12** (1978), no. 3, 32–44; English transl., Functional Anal. Appl. **12** (1978), no. 3, 185–195.

[O2] ———, *Infinite-dimensional classical groups of finite $\mathbb{R}$-rank: Description of representations and asymptotic theory*, Funktsional. Anal. i Prilozhen. **18** (1984), no. 1, 28–42; English transl., Functional Anal. Appl. **18** (1984), no. 1, 22–34.

[O3] ———, *Unitary representations of the group $SO(\infty,\infty)$ as limits of unitary representations of the groups $SO(n,\infty)$ as $n \to \infty$*, Funktsional. Anal. i Prilozhen. **20** (1986), no. 4, 46–57; English transl., Functional Anal. Appl. **20** (1986), no. 4, 292–301.

[O4] ———, *Unitary representations of (G,K)-pairs connected with the infinite symmetric group $S(\infty)$*, Algebra i Analiz **1** (1989), no. 4, 178–209; English transl., Leningrad Math. J. **1** (1989), 983–1014.

[O5] ———, *Unitary representations of infinite-dimensional pairs (G,K) and the formalism of R. Howe*, Representation of Lie Groups and Related Topics (A. M. Vershik and D. P. Zhelobenko, eds.), Advanced Studies in Contemporary Mathematics, vol. 7, Gordon and Breach Science Publishers, New York etc., 1990, pp. 269–463.

[O6] ———, *Irreducible unitary representations of the groups $U(p,q)$ sustaining passage to the limit as $q \to \infty$*, J. Soviet Math. **59** (1992), no. 5, 1102–1107.

[O7] ———, *Representations of infinite-dimensional classical groups, limits of enveloping algebras and Yangians*, Topics in Representation Theory (A. A. Kirillov, ed.), Advances in Soviet Math., vol. 2, Amer. Math. Soc., Providence, RI, 1991, pp. 1–66.

[OV] G. Olshanski and A. Vershik, *Ergodic unitarily invariant measures on the space of infinite Hermitian matrices*, Preprint UTMS 94–61 October 19, University of Tokyo, Department of Mathematical Sciences (1994).

[Pa1] K. R. Parthasarathy, *Probability measures on metric spaces*, Academic Press, New York–London, 1967.

[Pa2] ———, *Introduction to probability and measure*, 1980.

[Ph] R. R. Phelps, *Lectures on Choquet's theorem*, Van Nostrand, 1966.

[Pi1] D. Pickrell, *Separable representations for automorphism groups of infinite symmetric spaces*, J. Funct. Anal. **90** (1990), 1–26.

[Pi2] ———, *Mackey analysis of infinite classical motion groups*, Pacific J. Math. **150** (1991), 139–166.

[S1] I. J. Schoenberg, *Metric spaces and completely monotone functions*, Ann. of Math. **39** (1938), 811–841.

[S2] I. J. Schoenberg, *On Pólya frequency functions* I. *The totally positive functions and their Laplace transforms*, J. Analyse Math. **1** (1951), 331–374.

[Sha] D. Shale, *Linear symmetries of free boson fields*, Trans. Amer. Math. Soc. **103** (1962), 149–167.

[Shi] A. N. Shiryaev, *Probability*, "Nauka", Moscow, 1980; English transl., Springer-Verlag, Berlin, New York and Heidelberg, 1984.

[Sta2] R. P. Stanley, *Some combinatorial properties of Jack symmetric functions*, Advances in Math. **77** (1989), 76–115.

[SV] S. Stratila and D. Voiculescu, *A survey on representations of unitary group* $U(\infty)$, Spectral Theory, Banach Center Publications, vol. 8, Polish Science Publ., Warsaw, 1982, pp. 416–434.

[T] E. Thoma, *Die unzerlegbaren, positiv-definiten Klassenfunktionen der abzählbar unendlichen, symmetrischen Gruppe*, Math. Z. **85** (1964), 40–61.

[V] A. M. Vershik, *Description of invariant measures for the actions of some infinite-dimensional groups*, Dokl. Akad. Nauk SSSR **218** (1974), 749–752; English transl., Soviet Math. Dokl. **15** (1974), 1396–1400.

[VK1] A. M. Vershik and S. V. Kerov, *Asymptotic theory of characters of the symmetric group*, Funktsional. Anal. i Prilozhen. **15** (1981), no. 4, 17–27; English transl., Functional Anal. Appl. **15** (1981), no. 4, 246–255.

[VK2] ———, *Characters and factor representations of the infinite unitary group*, Dokl. Akad. Nauk SSSR **267** (1982), 272–276; English transl., Soviet Math. Dokl. **26** (1982), 570–574.

[VK3] ———, *Asymptotics of the largest and the typical dimensions of the irreducible representations of a symmetric group*, Funktsional. Anal. i Prilozhen. **19** (1985), no. 1, 25–36; English transl., Functional Anal. Appl. **19** (1985), no. 1, 21–31.

[VK4] ———, *The Grothendieck group of the infinite symmetric group and symmetric functions (with elements of the theory of the K_0-functor of AF-algebras)*, Representation of Lie Groups and Related Topics. Advanced Studies in Contemporary Mathematics (A. M. Vershik and D. P. Zhelobenko, eds.), vol. 7, Gordon and Breach Science Publishers, New York etc., 1990, pp. 39–117.

[Vo1] D. Voiculescu, *Représentations factorielles de type* II_1 *de* $U(\infty)$, J. Math. Pures Appl. **55** (1976), 1–20.

[Vo2] ———, *On extremal invariant functions of positive type on certain groups*, INCREST Preprint Series Math., 1978.

INSTITUTE FOR PROBLEMS OF INFORMATION TRANSMISSION, BOLSHOI KARETNYI PER. 19, 101447 MOSCOW GSP-4, RUSSIA

E-mail address: olsh@ippi.ac.msk.su

ST. PETERSBURG BRANCH OF THE STEKLOV MATHEMATICAL INSTITUTE, FONTANKA 27, 191011 ST. PETERSBURG, RUSSIA

E-mail address: vershik@pdmi.ras.ru

Translated by THE AUTHORS

Amer. Math. Soc. Transl.
(2) Vol. **175**, 1996

Cogitations over Berezin's Integral

V. P. Palamodov

§0. Introduction

A cornerstone of analysis on real supermanifolds (s-manifolds) is the construction of the Berezin integral. Recall that for an open subspace $U \subset \mathbb{R}^{n,\nu}$ the integral of a (super)density ρ is defined as follows:

$$\int_U \rho := \int_{U^e} \frac{\partial}{\partial \xi_\nu} \cdots \frac{\partial}{\partial \xi_1}(a)\, dx_1 \cdots dx_n, \tag{0.1}$$

where U^e denotes the underlying open set in $\mathbb{R}^n$ and $x = x_1, \dots x_n$, $\xi = \xi_1, \dots, \xi_\nu$ are even and odd coordinates on $\mathbb{R}^{n,\nu}$. Here we write $\rho = aD$, where a is a function on U with compact support and the symbol $D = D(x, \xi)$ denotes the generator of the line bundle Vol of volume forms (superdensities) related to this coordinate system. In particular, for the case of the space of one odd variable, this formula is equivalent to the following:

$$\int D(\xi) = 0, \qquad \int \xi D(\xi) = 1. \tag{0.2}$$

These key formulas were discovered by Berezin in the sixties [**B1**].

For any adjoining local coordinate systems (x, ξ), (y, η) the corresponding local sections $D(x, \xi)$, $D(y, \eta)$ of the bundle Vol are related by the famous Berezin formula

$$\frac{D(y,\eta)}{D(x,\xi)} = \det \left[\frac{\partial y}{\partial x} - \frac{\partial y}{\partial \xi} \left(\frac{\partial \eta}{\partial \xi} \right)^{-1} \frac{\partial \eta}{\partial x} \right] \det{}^{-1} \left(\frac{\partial \eta}{\partial \xi} \right). \tag{0.3}$$

The right-hand side is called the *Berezinian* of the coordinate change and is denoted $\mathrm{Ber}\,(y, \eta \,|\, x, \xi)$. The integral (0.1) calculated for both coordinate systems takes equal values [**B2, Pak, Le**]. Therefore this construction can be globalized for any s-manifold X provided the underlying even manifold X^e is oriented.

The complex of integral forms on an s-manifold [**Bn-L**] is an amplification of Berezin's construction. It is a substitute of the de Rham complex of an even manifold in the context of the integration theory. Bernstein and Leites found a version of the usual Stokes' formula for integral forms on an s-domain with smooth

The author acknowledges to the Center for Advanced Study at the Norwegian Academy of Science and Letters for support of this research.

boundary of codimension $(1,0)$. We initiate here another version of Stokes' theorem that involves s-domains whose boundary is the union of smooth parts of codimensions $(1,0)$ and $(0,1)$ (the latter does not exist for even manifolds). We prove that Berezin's integral over an s-domain of this kind is well defined, and we state a Stokes type theorem. This theorem is extended to the class of s-domains whose boundary has normal crossing singularities.

Our analysis of the Stokes theorem involves a whimsical odd part topology on a supermanifold. The $(0,1)$-part of boundary does not look like the $(1,0)$-part. In particular, the origin in an odd line $\mathbb{R}^{0,1}$ is an s-domain whose $(0,1)$-boundary is the line.

§1. Implications of odd topology on an odd space

We discuss the concept of integral starting with model cases.

Even case. Take the even line $\mathbb{R}$. The Riemann integral over the negative halfline $\mathbb{R}_- \subset \mathbb{R}$ of a 1-form w can be calculated by means of the Newton–Leibniz formula

$$\int_{\mathbb{R}_-} da = a(0) \equiv \int_{\partial \mathbb{R}_-} a, \tag{1.1}$$

where a is a primitive that vanishes at $-\infty$.

Odd case. Denote by Ξ the odd line. Its support is the one point space, denoted $\odot$, and the corresponding structure algebra is $\mathcal{O} = \mathbb{R}[\xi]/(\xi^2)$. The integration table (0.2) is different from that of the even case. Comparing it with (1.1), we can rearrange this formula as follows

$$\int_{\Xi} aD_{\Xi} = \tau(a)(\odot),$$

where $\tau = d/d\xi$. To interpret this equation like (1.1), we assume that the value of the function $\tau(a)$ at $\odot$ is the integral over this point and $\tau(a) = d(\tau \otimes a)$, where $\otimes$ means the tensor product over the field $\mathbb{R}$. Then we get the equation

$$\int_{\odot} d(\tau \otimes a) = \int_{\Xi} aD_{\Xi} \tag{1.2}$$

which is similar to (1.1) if we assume that

- *the odd line Ξ is the boundary of the point $\odot$;*
- *the restriction of the tensor $\tau \otimes a$ to this boundary is equal to aD_{Ξ}.*

Asserting that the boundary of point is an odd line, we get an unusual kind of topology. Nevertheless, this assumption agrees with the basic algebraic formula

$$\operatorname{str}(A) = \operatorname{tr}(A_{00}) - \operatorname{tr}(A_{11})$$

for the supertrace of the endomorphism A of a $\mathbb{Z}_2$-graded vector space V. In particular, $\operatorname{str}(E) = n - \nu$ for the identity endomorphism E, where $\dim V = (n, \nu)$. This means that the effective dimension of V is equal to $n - \nu$! Therefore the boundary operator we have introduced for an odd space has effective dimension -1 just like the standard boundary operator on an even manifold.

§2. Integral forms

By an *s-manifold* we mean any real supermanifold $(X, \mathcal{O}(X))$ of class C^∞. For basic definitions and notations see [**B-L, Le, B3**]. In particular, we denote by $\mathbb{R}^{n,\nu} = (\mathbb{R}^n, \mathcal{O}^{n,\nu})$ the coordinate s-space of dimension (n,ν) with even and odd coordinate functions x, ξ. A chart in an s-manifold X of dimension (n,ν) is an arbitrary embedding of s-manifolds: $U \to X$, where U is an open part of $\mathbb{R}^{n,\nu}$. The functions (x,ξ) are used as local coordinates on X. The sheaf Vol of volume forms is a locally free left $\mathcal{O}$-sheaf of rank 1. The mapping $\mathcal{O}\,|\,U \cong \mathrm{Vol}\,|\,U$, $a \mapsto aD(x,\xi)$ is an isomorphism for any chart (x,ξ). The sections $D(x,\xi)$ are connected by the transformations (0.3).

DEFINITION 2.1 [**Bn-L**]. The *sheaf of integral forms* of an s-manifold $(X,\mathcal{O})$ is defined as follows

$$\Upsilon = \Upsilon_* = \oplus \Upsilon_i, \qquad \Upsilon_{n-\nu-i} = (\Omega^i)' \otimes_{\mathcal{O}} \mathrm{Vol},$$

where Ω^* denotes the $\mathcal{O}$-sheaf of differential forms on X, $(\Omega^*)'$ means its left $\mathcal{O}$-dual and the tensor product taken for the structures of left $\mathcal{O}$-modules. (We write this product in the opposite order to that of [**Bn-L**].)

The *differential* of an integral form is defined in a chart (x,ξ) in the following way:

$$\delta \otimes aD(x,\xi) \mapsto \delta \cdot da\, D(x,\xi),$$

where $\delta \in (\Omega^*)'$ and the multiplication means the natural pairing $(\Omega^*)' \times \Omega^* \to \mathcal{O}$.

Let us present this formula in coordinate-free form. Consider the sheaf $\mathcal{T} \equiv \mathrm{Der}\,\mathcal{O}(X)$ of germs of tangent fields on X. Recall that a homogeneous element $v \in \mathcal{T}_p$, $p \in X$, of parity $k \in \mathbb{Z}_2$ is an $\mathbb{R}$-endomorphism of $\mathcal{O}(X)_p$ satisfying the Leibniz equation

$$v(ab) = v(a)b + (-1)^{kp(a)} av(b), \qquad a, b \in \mathcal{O}(X)_p,$$

where a is homogeneous of parity $p(a)$. This is a $\mathbb{Z}_2$-Lie algebra sheaf acting on $\mathcal{O}$. On the other hand, the sheaf $\mathcal{T}$ possesses a natural left $\mathcal{O}$-module structure that agrees with this representation. The latter means that

$$[u, av] = u(a)v + (-1)^{kp(a)} a[u,v], \qquad u, v \in \mathcal{T},\ a \in \mathcal{O}. \tag{2.1}$$

We say that $\mathcal{T}$ is a *$\mathcal{O}$-Lie algebra* when referring to this combination of structures.

The sheaf $\mathcal{T}$ is a locally free $\mathcal{O}$-module of rank $n+\nu$. For any chart $\theta\colon U \to X$, it is generated on $\theta(U)$ by the derivations $\partial/\partial x_i$, $i = 1,\dots,n$, $\partial/\partial \xi_j$, $j = 1,\dots\nu$; they are called *coordinate fields*.

By definition, we have $\Omega^1 \cong (\mathcal{T})' \equiv \mathcal{H}om_{\mathcal{O}}(\mathcal{T},\mathcal{O})$. Therefore $(\Omega^1)' \cong \mathcal{T}$ since $\mathcal{O}$ is a sheaf of regular $\mathbb{R}$-algebras. Whence $(\Omega^i)^* \cong \Lambda^i\mathcal{T}$, where the exterior product Λ^* is taken over the structure sheaf $\mathcal{O}$ and

$$\Upsilon_{n-\nu-i} \cong \Lambda^i \mathcal{T} \otimes_{\mathcal{O}} \mathrm{Vol}.$$

Since the $\mathcal{O}$-sheaves Vol, $\mathcal{T}$ are locally free and soft, any integral form $\sigma \in \Upsilon_{n-\nu-i}$ can be written as a finite sum

$$\sigma = \sum v_1 \wedge \cdots \wedge v_i \otimes \rho_v,$$

where ρ_v is the volume form, $v := (v_1, \ldots, v_i)$ are tangent fields on X and the usual commutation rule

$$u \wedge v = -(-1)^{p(u)p(v)} v \wedge u \tag{2.2}$$

is assumed.

PROPOSITION 2.1 ([**Le**]). *For any s-manifold X there exists an action $L(\cdot)$ of the $\mathbb{Z}_2$-Lie algebra sheaf $\mathcal{T}$ on Υ with the following properties:*

$$L(v)(a\rho) = v(a)\rho + (-1)^{p(a)p(v)} aL(v)\rho, \tag{2.3}$$

$$L(av)\rho = (-1)^{p(a)p(v)} L(v)(a\rho), \tag{2.4}$$

where $\rho \in \Upsilon$, $a \in \mathcal{O}$ and

$$L(v)D(x,\xi) = 0 \tag{2.5}$$

for any chart (x,ξ) on X and any coordinate field v on the chart.

The equation (2.5) means that $D(x,\xi)$ is always a flat section of the sheaf Vol with respect to this action of $\mathcal{T}$ (called the *Lie derivative*). This property together with (2.3) and (2.4) defines this action uniquely for any chart (x,ξ). To check that these actions coincide for adjoining charts Leites used an infinitesimal version of the transitivity equation for the Berezinian in (0.3).

Now we write a formula for the differential.

PROPOSITION 2.2. *For any volume form ρ and tangent fields $v_1, \ldots, v_i$ on X, we set*

$$\begin{aligned} d(v_1 \wedge \cdots \wedge v_i \otimes \rho) &= \sum_{k=1}^{i} (-1)^{k-1} \Bigg(\sum_{j=k+1}^{i} (-1)^{\varepsilon(k,j-1)} v_1 \wedge \cdots \wedge \hat{v}_k \wedge \cdots \\ &\qquad \wedge v_{j-1} \wedge [v_k, v_j] \wedge v_{j+1} \wedge \cdots \wedge v_i \otimes \rho \\ &\qquad + (-1)^{\varepsilon(k,i)} v_1 \wedge \cdots \wedge \hat{v}_k \wedge \cdots \wedge v_i \otimes L(v_k)\rho \Bigg), \end{aligned} \tag{2.6}$$

where $\varepsilon(k,j) = p(v_k)[p(v_{k+1}) + \cdots + p(v_j)]$ is the well-defined differential in the graded sheaf $\Upsilon(X)_*$. For any chart on X it agrees with the differential given by Definition 2.1.

The choice of numbers $\sigma(k,j)$ in this formula agrees with the standard sign convention for any homogeneous elements of $\mathbb{Z}_2$-graded modules. Note that the action of a field v_k on a density is consistent with the tensor product by virtue of (2.1), (2.3), (2.4). A proof of (2.6) can be done by a routine computation.

It follows that a sheaf complex of integral forms

$$\Upsilon_* \colon \cdots \xrightarrow{d} \Upsilon_i \xrightarrow{d} \Upsilon_{i+1} \to \cdots \to \Upsilon_{n-\nu} \to 0$$

is defined. Proposition 2.2 implies that this is a variant of Chevalley–Eilenberg chain complex [**C-E**] for the $\mathcal{O}$-Lie algebra $\mathcal{T}$ and for the $\mathcal{T}$-module structure in Vol defined in Proposition 2.1.

The natural pairing

$$\mathcal{T} \times \Omega^1 \to \mathcal{O}, \qquad (v \times w) \mapsto v \rhd w,$$

is a morphism of left $\mathcal{O}$-modules called an *interior product*. For any $i \geqslant j > 0$, it generates a morphism of $\mathcal{O}$-modules

$$\Lambda^i \mathcal{T} \times \mathcal{H}om_{\mathcal{O}}(\Lambda^j \mathcal{T}, \mathcal{O}) \to \Lambda^{i-j}\mathcal{T}$$

by the rule

$$(v_1 \wedge \cdots \wedge v_i) \times \omega \mapsto \sum_{\sigma} \varepsilon(\sigma)\,\varepsilon(v,\omega)(v_1 \wedge \cdots \wedge v_{i-j})\,\omega(v_{i-j+1}, \ldots, v_i),$$

where $\varepsilon(v, \omega) = (-1)^{p(\omega)[p(v_{i-j+1})+\cdots+p(v_i)]}$, the sum is taken over the set of all transpositions σ of elements $1, \ldots, i$, and the number $\varepsilon(\sigma) = \pm$ is defined according to (2.2). The interior product equals zero for $i < j$.

Any tangent field v on X generates *Lie derivatives* $L(v)$ acting on the sheaf Ω^*:

$$L(v)\omega = d(v \triangleright \omega) + v \triangleright d\omega.$$

They commute with the differential in Ω^*.

REMARK. Compare this construction with the de Rham complex $\Omega^*(X)$. The latter looks like the Chevalley–Eilenberg cochain complex for the $\mathcal{T}$-module $\mathcal{O}$:

$$\Omega^*(X)\colon \mathcal{O} \xrightarrow{\delta} \mathcal{H}om_{\mathcal{O}}(\mathcal{T}, \mathcal{O}) \xrightarrow{\delta} \cdots \to \mathcal{H}om_{\mathcal{O}}(\Lambda^i \mathcal{T}, \mathcal{O}) \xrightarrow{\delta} \mathcal{H}om_{\mathcal{O}}(\Lambda^{i+1}\mathcal{T}, \mathcal{O}) \to \ldots,$$

where

$$\begin{aligned}
&\delta\omega(v_0 \wedge \cdots \wedge v_i) \\
&\quad = \sum_{k=0}^{i} (-1)^k \Bigg(\sum_{j=0}^{k-1} (-1)^{\pi(k,j+1)} \\
&\qquad\qquad \times \omega(v_0 \wedge \cdots \wedge v_{j-1} \wedge [v_j, v_k] \wedge v_{j+1} \wedge \cdots \wedge \hat{v}_k \wedge \cdots \wedge v_i) \\
&\qquad\qquad + (-1)^{\pi(k)} L(v_k)\omega(v_0 \wedge \cdots \wedge \hat{v}_k \wedge \cdots \wedge v_i) \Bigg);
\end{aligned}$$

here $\pi(k, j) = p(v_k)[p(v_j) + \cdots + p(v_{k-1})]$, $\pi(k) = \pi(k, 0) + p(v_k)p(\omega)$. Therefore there is an $\mathcal{O}$-bilinear pairing

$$\Upsilon_* \times \Omega^* \to \Upsilon_*, \qquad (v_1 \wedge \cdots \wedge v_i \otimes \rho) \times \omega \mapsto (-1)^{p(\rho)p(\omega)}(v_1 \wedge \cdots \wedge v_i) \triangleright \omega \otimes \rho$$

that is consistent with the differentials. As far as I know, this complex is not as useful for integration theory as it is for even manifolds.

§3. The integral

Recall some properties of Berezin's integral. An s-manifold X is called *oriented* if the underlying manifold X^e is oriented.

PROPOSITION 3.1. *Let X be an oriented s-manifold. For any volume form α on X with compact support the integral*

$$\int_X \rho$$

that coincides with (0.1) *for an arbitrary chart θ if* $\operatorname{supp} \rho \subset \theta(U)$ *and U^e is endowed with the induced orientation is defined. This integral satisfies the equation*

$$\int_X d\sigma = 0 \tag{3.1}$$

for any integral form $\sigma \in \Gamma_c(X, \Upsilon_{n-\nu-1})$.

COROLLARY 3.2. *For any* $\rho \in \Gamma_c(X, \mathrm{Vol})$ *and any tangent field* v *on* X, *we have*

$$\int_X L(v)\,\rho = 0.$$

This equation follows from (3.1) if we set $\sigma = v \otimes \rho$.

PROOF OF PROPOSITION 3.1. Take an arbitrary volume form ρ on X with compact support and choose a finite set $\{\theta_i : U_i \to X,\ i \in I\}$ of charts such that the open sets $\theta_i(U_i)^e$ cover the set $\operatorname{supp}\rho$. Then choose a decomposition $\rho = \sum \rho_i$ such that $\operatorname{supp}\rho_i$ is a compact subset of $\theta_i(U_i)^e$ for any $i \in I$. Set

$$\int \rho := \sum_i \int_{U_i} \theta_i^*(\rho_i),$$

where each term in the right-hand side is defined by means of (0.1). The sum does not depend on the choice of charts and of the decomposition of the volume form by virtue of (0.3).

Let us check (3.1) for the form $\sigma = v \otimes \rho$, where ρ is a density with compact support. Take a covering $\{\theta_i\}$ of $\operatorname{supp}\rho$ and a decomposition $\rho = \sum \rho_i$ as above. Then $d\sigma = \sum L(v)\,\rho_i$, whence

$$\int d\sigma = \sum \int_{U_i} \theta_i^*(L(v)\,\rho_i) = \sum \int_{U_i} L(\theta_i^*(v))\,\theta_i^*(\rho) = 0.$$

The integrals in the right-hand side vanish by [**Le,** Lemma 2.4.8] and (3.1) follows.

§4. The "Stokes" theorem

Let us recall more basics ([**B3,** Chapter IV]). Let $(X, \mathcal{O}(X))$ be an s-manifold. For any point $p \in X$ the stalk $\mathcal{O}_p$ of the structure sheaf $\mathcal{O} = \mathcal{O}(X)$ is a $\mathbb{Z}_2$-commutative local $\mathbb{R}$-algebra with $\mathbb{Z}_2$-grading $\mathcal{O}_p = \mathcal{O}_p^0 \oplus \mathcal{O}_p^1$ equipped with the *residue morphism* $\operatorname{res}_p\colon \mathcal{O}_p \to \mathbb{R}$ that sends a function germ a to $a(p)$ (hence vanishes on $\mathcal{O}_p^1$).

Let $\mathcal{I}$ be a sheaf ideal in $\mathcal{O}(X)$, i.e., an $\mathcal{O}$-subsheaf of $\mathcal{O}$. The $\mathbb{R}$-subspace S of X determined by $\mathcal{I}$ is the subspace $S^e \subset X^e$ of common zeros of functions $a \in \mathcal{I}$ endowed with the algebra sheaf $\mathcal{O}(S) := \mathcal{O}(X)/\mathcal{I}$. This is a $\mathbb{Z}_2$-graded ringed space if the sheaf $\mathcal{I}$ is generated locally by homogeneous elements of $\mathcal{O}$. This subspace is called *closed* if S^e is closed. The space $(S, \mathcal{O}(S))$ is an s-submanifold if the sheaf ideal $\mathcal{I}$ is generated locally by homogeneous elements $g_1, \dots, g_m \in \mathcal{O}^0(X)$, $\gamma_1, \dots, \gamma_\mu \in \mathcal{O}^1(X)$ such that the differentials $dg_1, \dots, dg_m$, $d\gamma_1, \dots, d\gamma_\mu$ are linearly independent. The couple (m, μ) is said to be the *codimension* of S.

In particular, the underlying manifold X^e for X is the closed s-submanifold of X determined by the sheaf ideal $\mathcal{I}$ generated by $\mathcal{O}^1(X)$.

DEFINITION 4.1. A *morphism of* s-*manifolds* (or s-*morphism*) $g\colon (Y, \mathcal{O}(Y)) \to (X, \mathcal{O}(X))$ is a mapping $g^e\colon Y^e \to X^e$ of the underlying s-manifolds together with a morphism of $\mathbb{Z}_2$-graded sheaves of $\mathbb{R}$-algebras $G\colon g^{e*}(\mathcal{O}(X)) \to \mathcal{O}(Y)$ on Y that agrees with g^e. The latter implies that for any $q \in Y$ we have the relation $\operatorname{res}_p = \operatorname{res}_q G_q$, which holds where $p = g(q)$.

Domains. An s-domain in an s-manifold X is surrounded by fences like a domain in even geometry. The difference is that the boundary of the former consists of two parts: one of codimension $(1,0)$, another of codimension $(0,1)$. The construction of the first one is close to the usual one.

DEFINITION 4.2. We call *even fence* on an s-manifold X any even function f on X such that $df \neq 0$ on the subset $f^e = 0$ (where f^e denotes the restriction of f to X^e). We say that even fences f, f' are *equivalent* if $f' = af$ for some function a such that $a^e > 0$. We call *fence system* on X any system F of fences defined on open submanifolds $U_i \subset X$ such that

(i) $\bigcup U_i = X$;

(ii) for any adjoining sets U_i, U_j the corresponding fences are equivalent on $U_i \cap U_j$.

We say that an even fence system F on X *fences an s-domain* $Z = X_F$; any equivalent fence system fences the same s-domain. This s-domain is underlaid by the domain $Z^e := \{f^0(p) \leqslant 0,\, p \in X^e\}$. In a more formal way, we mean that the s-domain is the ringed $\mathbb{Z}_2$-space $(Z^e, \mathcal{O}(X) \mid Z^e)$ endowed with a class of equivalent fence systems.

Take an even fence system F and consider the system of equations $f = 0$, $f \in F$. The solution is a smooth closed s-submanifold of X of codimension $(1,0)$. For any equivalent fence system, we get the same s-submanifold underlaid by the manifold $\partial Z^e = \{f^0(p) = 0\}$. We call it the *even boundary* of Z and denote $\partial^0 Z$. There is a natural immersion of $\mathbb{Z}_2$-ringed spaces $b_0\colon \partial^0 Z \to Z$.

If X is oriented we use the induced orientation for Z^e. However, we endow the boundary $\partial^0 Z$ with a nonstandard orientation that is equal the standard one times the factor $(-1)^\nu$, i.e., the orientation of $\partial^0 Z$ is given by the frame $(-1)^\nu w_1 \wedge \cdots \wedge w_{n-1}$ if $df^e \wedge w_1 \wedge \cdots \wedge w_{n-1}$ is an orienting frame for X.

The odd counterpart of this construction is the following. We call *odd fence* on an s-manifold X any odd function φ on X that satisfies the same condition as above $d\varphi \neq 0$, $\varphi = 0$. Two odd fences φ, φ' are called *equivalent* if $\varphi' = \alpha\varphi$ for some function $\alpha \neq 0$. Here we need not assume that the function α^e is positive since the integral (1.2) does not require the orientation of the odd line. An odd fence system Φ on X is defined by the above conditions (i), (ii). In fact any odd fence system fences a submanifold $Y = X_\Phi$ in X of codimension $(0,1)$ and vice versa a $(0,1)$-submanifold is fenced by a class of equivalent odd fence systems. The manifold X is regarded as the *odd boundary* of Y and denoted by $\partial^1 Y$. To keep the symmetry with the even case, we assume that there exists a morphism $b_1\colon \partial^1 Y \to Y$ such that $b_1 \mid Y = \mathrm{id}(Y)$. Now we combine both constructions:

DEFINITION 4.3. Let even and odd fence systems F, Φ be given on an s-manifold X. We call an s-*domain fenced by these systems* the s-domain $Z = X_{F,\Phi}$ fenced by F in the $(0,1)$-submanifold X_Φ that is fenced by Φ. The boundary of Z consists of two parts

$$\partial Z = \partial^0 Z \cup \partial^1 Z,$$

where $\partial^0 Z := \{f = 0,\, f \in F,\, \varphi = 0,\, \varphi \in \Phi\}$ is the even boundary of Z in the s-manifold X_Φ and $\partial^1 Z = X_F$ is the s-domain in X fenced by the system F. There is a natural immersion $b_0\colon \partial^0 Z \to Z$ and we assume that there exists a morphism $b_1\colon \partial^1 Z \to Z$ such that $b_1 \mid Z = \mathrm{id}(Z)$. We call b_0, b_1 *boundary mappings*.

Note that the double boundaries $\partial^1\partial^0 Z$, $\partial^0\partial^1 Z$ coincide and we have the following commutative diagram

$$\begin{array}{ccccc} \partial^0\partial^1 Z & = & \partial^1\partial^0 Z & \xrightarrow{b_{01}} & \partial^1 Z \\ & & \big\downarrow b_{10} & & b_1 \big\downarrow \\ & & \partial^0 Z & \xrightarrow{b_0} & Z \end{array}$$

where b_{01}, b_{10} are the corresponding boundary morphisms. Note that $\partial^0 Z$ has no even boundary and $\partial^1 Z$ has no odd boundary.

Suppose that X is oriented. Then the domain $X_{F,\Phi}$ and parts of its boundary are equipped with induced orientations as above. Therefore the double boundaries $\partial^1\partial^0 Z$, $\partial^0\partial^1 Z$ are equipped with opposite orientations.

Integral forms on s-domains. Let $Z = X_{F,\Phi}$ be an s-domain in s-manifold X of dimension (n, ν) as above. We have the commutative diagram of immersions

$$\begin{array}{ccccc} Z & = & X_{F,\Phi} & \xrightarrow{(0,0)} & X_\Phi \\ & & {\scriptstyle (0,1)}\big\downarrow & & {\scriptstyle (0,1)}\big\downarrow \\ & & X_F & \xrightarrow{(0,0)} & X \end{array}$$

where the codimensions of the immersions are shown. Define

$$\Upsilon_{n-\nu-1-i}(Z) \cong \Lambda^i \mathcal{T}(X_F) \otimes_{\mathcal{O}(Z)} \mathrm{Vol}(X_\Phi), \tag{4.1}$$

where the sheaf $\Lambda^* \mathcal{T}(X_F) := \Lambda^* \mathcal{T}(X) \,|\, X_F$ is regarded as an $\mathcal{O}(Z)$-sheaf via the algebra morphism $b_1^*\colon \mathcal{O}(Z) \to \mathcal{O}(X_F)$. We call $\Upsilon(Z) = \bigoplus \Upsilon_j(Z)$ the *sheaf of volume forms* on the domain Z. It possesses a left $\mathcal{O}(X)$-module structure.

PROPOSITION 4.1. *There exists an action $L(\,\cdot\,)$ of the Lie algebra sheaf $\mathcal{T}(X)$ on the $\mathcal{O}(X)$-module* $\mathrm{Vol}(X_\Phi)$ *that satisfies conditions* (2.3), (2.4), *and* (2.5) *for any chart (x, ξ) on X_Φ and any coordinate field v on this chart.*

This follows from Proposition 2.1. We can define a differential d on that sheaf by formula (2.6), where L denotes the action of the sheaf $\mathcal{T}(X)$.

Pullbacks. We call a mapping $\theta\colon U \to X$ an *even (odd) fence* chart for an s-domain Z if it is a chart and the coordinate function $x_1(\xi_1)$ is a local fence for Z. For any integral form σ on Z we can construct its pullbacks with regard to the boundary morphisms. First take a volume form $\sigma = v_1 \wedge \cdots \wedge v_i \otimes D(X_\Phi)$ on Z and set

$$b_0^*(\sigma) = (-1)^{\nu-1}(v_1 \wedge \cdots \wedge v_i) \rhd dx_1 \otimes D(\partial^0 Z)$$

in any even fence chart (x, ξ). The result is well defined as an element of $\Upsilon(\partial^0 Z)$. We set similarly

$$b_1^*(\sigma) = (-1)^{\nu-1}(v_1 \wedge \cdots \wedge v_i) \rhd d\xi_1 \otimes D(X)$$

for any odd fence chart and the result is in $\Upsilon(X_F)$. Here By $D(X)$, $D(X_\Phi)$, and $D(\partial^0 Z)$ we denote the generator of the corresponding volume sheaf related to the corresponding chart θ, $\theta \,|\, \xi_1 = 0$ and $\theta \,|\, x_1 = 0$, $\xi_1 = 0$.

Integral on s-domain. Suppose that X is an oriented s-manifold and F is an even fence system on it. For an arbitrary form $\rho \in \Gamma_c(X, \mathrm{Vol}(X))$ define the integral

$$\int_{X_F} \rho = \sum \int_{U_{i-}} \theta_i^*(\rho_i), \tag{4.2}$$

where $\{U_i\}$ is a covering of $\operatorname{supp} \rho$ by even fence charts and $\rho = \sum \rho_i$ is the corresponding decomposition of the form (cf. Proposition 3.1).

PROPOSITION 4.2. *The integral* (4.2) *does not depend on the choice of fence charts and of the decomposition of the volume form.*

A proof follows from Corollary 6.2.

Let us summarize the above. Let Z be an s-domain in X given by the even and odd fence systems F, Φ. Taking a form $\sigma \in \Gamma_c(Z, \Upsilon_{n-\nu-1}(Z))$, we can define the integral of $b_1^*(\sigma)$ over $\partial^1 Z$, the integral of $b_0^*(\sigma)$ over the manifold $\partial^0 Z$, and the integral of $d\sigma$ over Z. Now we formulate

STOKES THEOREM FOR SMOOTH BOUNDARY. *Let Z be an s-domain in an oriented s-manifold X of dimension (n, ν). Then for any $\sigma \in \Gamma_c(Z, \Upsilon_{n-\nu-1}(Z))$ the following equation holds*:

$$\int_Z d\sigma = \int_{\partial^0 Z} b_0^*(\sigma) + \int_{\partial^1 Z} b_1^*(\sigma). \tag{4.3}$$

We omit the proof, which is routine.

§5. Generalization

Ordinary analysis includes Stokes' theorem for domains with singular boundary. In fact, the class of domains with normal crossing singular points is large enough for many applications. We say that a domain Z in an even manifold X has *normal crossing points* (is an nc-*domain*) if it is locally diffeomorphic to the model singular domain $\mathbb{R}_-^k \times \mathbb{R}^n$. The boundary ∂Z is again a domain with normal crossing boundary points.

The odd counterpart of the above construction can be given in the following way. Instead of $\mathbb{R}_-$ we consider the couple $\oslash = (\odot, \Xi)$, where $\odot$ is the origin of the odd line Ξ. We call it *a point with boundary in the odd line*. Next we replace the direct product by a *bouquet* operation applied to $\oslash$. This is a closed singular $\mathbb{Z}_2$-subspace

$$B^\kappa(\oslash) \subset \mathbb{R}^{0,\kappa}, \qquad \mathcal{O}(B^\kappa(\oslash)) := \mathcal{O}(\mathbb{R}^{0,\kappa})/I,$$

where I is the ideal generated by the monomials $\xi_i \xi_j$, $i \leqslant j$. Taking a geometrical point of view, we can visualize the bouquet as the union of coordinate lines in an odd space $\mathbb{R}^\kappa$. Actually, these lines are sub-s-manifolds of this space. Combining even and odd singularities, we take the model singular s-space in the form

$$\mathbb{R}_-^k \times B^\kappa(\oslash) \times \mathbb{R}^{n,\nu} \tag{5.1}$$

with some integers k, κ, n, ν.

DEFINITION 5.1. We call Z an s-domain with *normal crossing boundary* (an ncs-*domain*) if it is a $\mathbb{Z}_2$-graded $\mathbb{R}$-ringed space $(S, \mathcal{O}(S))$ that is locally isomorphic to a model space M of the form (5.1).

An ncs-domain W is called *semiregular* if it is locally isomorphic to a model space (5.1) with $\kappa \leqslant 1$. Any ncs-domain Z is locally a bouquet of semiregular ncs-domains $Z(j)$ over an underlying nc-domain. Therefore, we can define the integral $\int_Z \sigma$ for an oriented ncs-domain Z as the sum of the integrals $\int_{Z(ij)} \sigma_i$ over its local semiregular parts $Z(ij)$. The latter is calculated by a reduction to an integral over the set (5.1) with $\kappa = 0$, which is an s-domain according to §4. To check that the result does not depend on charts and on the decomposition of the form $\sigma = \sum \sigma_i$, we can use the corresponding generalization of Theorem 6.1.

The boundary ∂Z of an oriented ncs-domain Z is the union of ncs-domains with the induced orientations. Note that it can be decomposed in the whole as the union two parts $\partial Z = \partial^0 Z \cup \partial^1 Z$ of codimension $(1,0)$ and $(0,1)$ respectively.

STOKES THEOREM FOR SINGULAR BOUNDARY. *Let Z be an oriented ncs-domain. Then for an arbitrary integral form $\alpha \in \Gamma_c(Z, \Upsilon(Z))$ equation* (4.3) *holds.*

We omit the details, which are routine.

§6. Change of variables in Berezin's Integral

Here we prove that Berezin's formula is also valid for coordinate changes in s-domains.

Fix an open subspace $U \subset \mathbb{R}^{n,\nu}$ with coordinate functions x, ξ and consider the s-domain U_- with the fence function x_1. This domain is underlaid by the set $U^e_- = \{x \in U^e,\, x_1 \leqslant 0\}$. Take an arbitrary differential form w on U with compact support and define an integral over this domain as follows

$$\int_{U_-} w := \int_{U^e_-} (\varepsilon)^*[L(\tau_\nu) \cdots L(\tau_1)(w)],$$

where $\tau_i := \partial/\partial\xi_i$, $i = 1, \dots, \nu$, and $\varepsilon\colon U^e \to U$ is the canonical embedding. In particular, for a form $w = a\Lambda(dx) := a dx_1 \wedge \cdots \wedge dx_n$, we get

$$\int_{U_-} a\Lambda(dx) := \int_{U^e_-} (\varepsilon)^*[L(\tau_\nu) \cdots L(\tau_1)(a)]\Lambda(dx) \equiv \int_{U_-} aD(x,\xi) \tag{6.1}$$

since the form $\Lambda(dx)$ vanishes under the action of the coordinate fields.

THEOREM 6.1. *Let $V_- \subset \mathbb{R}^{n,\nu}$ be an open subspace with coordinates (y,η), V_- be the s-domain with fence function y_1, and $\lambda\colon U \to V$ be another fence chart for U_-. Then for any function $a \in \Gamma_c(V, \mathcal{O}^{n,\nu})$, we have*

$$\int_{V_-} a\Lambda(dy) = \int_{U_-} a \det\left[\frac{\partial y}{\partial x} - \frac{\partial y}{\partial \xi}\left(\frac{\partial \eta}{\partial \xi}\right)^{-1}\frac{\partial \eta}{\partial x}\right] \det{}^{-1}\left(\frac{\partial \eta}{\partial \xi}\right)\Lambda(dx).$$

Here and later we write $a, dy, \dots$ instead of $\lambda^*(a), \lambda^*(dy), \dots$.

COROLLARY 6.2. *The following equation holds*

$$\int_{V_-} aD(y,\eta) = \int_{U_-} a \operatorname{Ber}(y,\eta \,|\, x,\xi) D(x,\xi)$$

for any couple of equivalent fence charts.

This follows from (6.1) and Theorem 6.1.

REMARK. A similar assertion is valid for any nc-domain of the form $U \cap (\mathbb{R}^k_- \times \mathbb{R}^{n,\nu})$ as well.

LEMMA 6.3. *The equation*

$$\int_{V^e_-} L(v)\,w = 0$$

holds for any form $w \in \Gamma_c(V, \Omega)$ *and any tangent field* v *on* V *that satisfies the equations*

$$v(\eta_1) = \cdots = v(\eta_\nu) = 0, \qquad v(y_1) = e y_1,\ e \in \Gamma(V, \mathcal{O}). \tag{6.2}$$

LEMMA 6.4. *The equation*

$$\int_{V_-} y_1 w + dy_1 \wedge w' = 0$$

holds for any $w, w' \in \Theta \wedge \Gamma_c(V, \Omega^*)$, *where* Θ *is the linear span of forms* $d\eta_1, \dots, d\eta_\nu$.

PROOF OF THEOREM 6.1. First we note that the equivalence of the fences implies the following equation

$$y_1 = c x_1, \qquad c \in \Gamma(V, \mathcal{O}),\ c \,|\, V^e > 0. \tag{6.3}$$

Changing variables, we get the equations

$$dy = \frac{\partial y}{\partial x}\,dx + \frac{\partial y}{\partial \xi}\,d\xi, \qquad d\eta = \frac{\partial \eta}{\partial x}\,dx + \frac{\partial \eta}{\partial \xi}\,d\xi,$$

where $dx, dy,$ are understood as columns and find

$$dy - \frac{\partial y}{\partial \xi}\left(\frac{\partial \eta}{\partial \xi}\right)^{-1} d\eta = B_0\,dx, \qquad B_0 := \frac{\partial y}{\partial x} - \frac{\partial y}{\partial \xi}\left(\frac{\partial \eta}{\partial \xi}\right)^{-1}\frac{\partial \eta}{\partial x}.$$

Consider the product

$$\Lambda\left(dy - \frac{\partial y}{\partial \xi}\left(\frac{\partial \eta}{\partial \xi}\right)^{-1} d\eta\right) = \det B_0\,\Lambda(dx); \tag{6.4}$$

let us show that it can be substituted for $\Lambda(dy)$ in the integral over V_-. For this we note that the first factor is equal to

$$dy_1 - \sum_{i,j} \frac{\partial y_1}{\partial \xi_i}\,\delta_i^j\,d\eta_j, \tag{6.5}$$

where δ_i^j are the entries of the matrix $(\partial\eta/\partial\xi)^{-1}$. We derive from (6.3) that

$$\frac{\partial y_1}{\partial \xi_i} = x_1 \frac{\partial c}{\partial \xi_i} = y_1 c_1, \qquad c_1 \in \Gamma_c(V, \mathcal{O}). \tag{6.6}$$

Therefore the second term in (6.5) belongs to $y_1 \Theta \wedge \Gamma_c(V, \Omega^*)$ and we conclude that

$$\Lambda\left(dy - \frac{\partial y}{\partial \xi}\left(\frac{\partial \eta}{\partial \xi}\right)^{-1} d\eta\right) - \Lambda(dy) \in y_1 \Theta \wedge \Gamma_c(V, \Omega^*) + dy_1 \wedge \Theta \wedge \Gamma_c(V, \Omega^*).$$

According to Lemma 6.4, the integral of this form over V_- vanishes. Thus

$$(6.7)\quad \int_{V_-} a\,\Lambda(dy) = \int_{V_-} a \det B_0\,\Lambda(dx) \equiv \int_{V_-^e} L\Big(\frac{\partial}{\partial \eta_\nu}\Big)\cdots L\Big(\frac{\partial}{\partial \eta_1}\Big) a B_0\,\Lambda(dx).$$

We can change the domain V_-^e to U_-^e in the right-hand side by virtue of (6.3). Then

$$\frac{\partial}{\partial \xi} = {}'\Big(\frac{\partial \eta}{\partial \xi}\Big)\frac{\partial}{\partial \eta} + {}'\Big(\frac{\partial y}{\partial \xi}\Big)\frac{\partial}{\partial y},$$

where $'$ denotes the matrix conjugation. Whence

$$\frac{\partial}{\partial \eta_j} = \sum_i \delta_j^i \Big[\frac{\partial}{\partial \xi_i} - \sum_k \frac{\partial y_k}{\partial \xi_i}\frac{\partial}{\partial y_k}\Big], \qquad j = 1, \dots, \nu.$$

We can use this equation in (6.7) and omit each term $v_k := (\partial y_k/\partial \xi_i)(\partial/\partial y_k)$ by virtue of Lemma 6.3. Actually, this field vanishes identically on y_1 for any $k > 1$. The field v_1 satisfies the condition (6.2) because of (6.6). Therefore

$$L\Big(\frac{\partial}{\partial \eta_\nu}\Big)\cdots L\Big(\frac{\partial}{\partial \eta_1}\Big) a\Lambda(dx) = \delta_\nu \frac{\partial}{\partial \xi}\cdots \delta_1 \frac{\partial}{\partial \xi(a)}\Lambda(dx)$$

since $L(\partial/\partial\xi)\,dx = 0$. Then we use the following identity:

$$\delta_1 \frac{\partial}{\partial \xi}\cdots \delta_\nu \frac{\partial}{\partial \xi} = \frac{\partial}{\partial \xi_1}\cdots \frac{\partial}{\partial \xi_\nu} B_1, \qquad B_1 := \det\Big(\frac{\partial \eta}{\partial \xi}\Big)^{-1}.$$

This identity was proved in an equivalent form in [**B2**]. Therefore, the form to be integrated in (6.7) is equal to

$$L\Big(\frac{\partial}{\partial \xi_1}\Big)\cdots L\Big(\frac{\partial}{\partial \xi_\nu}\Big) a B_1 B_0\,\Lambda(dx),$$

since B_1 is an even function. Using the formula $B_0 B_1 = \operatorname{Ber}(y, \eta \,|\, x, \xi)$, we complete the proof of Theorem 6.1.

Proof of Lemma 6.3. We have $L(v)w = d(v \triangleright w) + v \triangleright dw$ and no term of $v \triangleright dw$ contains the product $dx_1 \wedge \cdots \wedge dx_n$, hence $\varepsilon^*(v \triangleright dw) = 0$. The form $v \triangleright w$ vanishes on the boundary of $\partial V_- = \{y_1 = 0\}$ because of (6.2). The integral of $d(v \triangleright w)$ over V_-^e vanishes by the Gauss–Ostrogradskiĭ theorem.

Proof of Lemma 6.4. We start with the formula

$$\begin{aligned} L(s)(u \wedge u') = {} & L(s)\,u \wedge u' + (-1)^{kp(u)} u \wedge L(s)\,u' \\ & + (-1)^{\deg(u)} s \triangleright u \wedge du' + (-1)^{\deg(u)+kp(u)} du \wedge s \triangleright u', \end{aligned}$$

where u, u' are arbitrary differential forms and s is any tangent field of parity k. This implies the following equation for any odd coordinate tangent field τ and any element $d\zeta \in \Theta$:

$$L(\tau)(u \wedge d\zeta) = L(\tau)\,u \wedge d\zeta + (-1)^{\deg(u)+p(u)}\tau(\zeta)\,du,$$

where $\tau \triangleright d\zeta = \tau(\zeta)$ is a constant. Therefore the second term of the right-hand side vanishes. Applying this formula once for the odd coordinate field τ', we get

$$L(\tau')\,L(\tau)(u \wedge d\zeta) = L(\tau')\,L(\tau)\,u \wedge d\zeta \pm \tau'(\zeta)\,L(\tau)\,du \pm \tau(\zeta)\,L(\tau')\,du$$

and so on. Then we apply the restriction morphism ε^*. It vanishes on the factor $d\zeta$. Hence the integral over V_-^e of any form that contains this factor vanishes. Other terms are of the form

$$L(\tau)\,L(\tau')\cdots du \tag{6.8}$$

for some odd coordinate fields $\tau, \tau', \dots$. Now we assume that

$$u = y_1 w + dy_1 \wedge w', \qquad w, w' \in \Gamma_c(V, \Omega^*) \tag{6.9}$$

and show that the integral of the form (6.8) over V_-^e vanishes. This will imply Lemma 6.4. We have

$$du = d(y_1 w + dy_1 \wedge w'). \tag{6.10}$$

For any odd coordinate field τ the form $\tau \triangleright du$ admits the representation (6.9) with some w, w', hence $L(\tau)\,du \equiv d(\tau \triangleright du)$ is of the form (6.10) again, and so on. Arguing inductively, we conclude that any term (6.8) belongs to the class (6.10). The equality

$$\int_{V_-^e} d(y_1 w + dy_1 \wedge w) = 0$$

is once more a corollary of the Gauss–Ostrogradskiĭ theorem, since the form (6.9) vanishes on the boundary of the domain.

References

[B1] F. A. Berezin, *The method of second quantization*, Moscow, 1965; English transl., Academic Press, New York–London, 1966.

[B2] ——, *Automorphisms of a Grassmanian algebras*, Mat. Zametki **1** (1967), no. 3, 180–184; English transl. in Math. Notes **1** (1967).

[B3] ——, *Introduction to algebra and analysis with anticommuting variables*, Izdat. Moskov. Univer., Moscow, 1983; English transl., *Introduction to superanalysis*, Reidel, Dordrecht, 1987.

[B-L] F. A. Berezin and D. A. Leites, *Supermanifolds*, Dokl. Akad. Nauk SSSR **224** (1975), no. 3, 505–508; English transl. in Soviet Math. Dokl. **16** (1975), no. 5, 1218–1222.

[Bn-L] I. N. Bernstein and D. A. Leites, *Integral forms and the Stokes' formula on supermanifolds*, Funktsional. Anal. i Prilozhen. **11** (1977), no. 1, 45–47; English transl. in Functional Anal. Appl. **11** (1977), no. 1.

[C-E] C. Chevalley and S. Eilenberg, *Cohomology theory of Lie groups and Lie algebras*, Trans. Amer. Math. Soc. **63** (1948), 85–124.

[Le] D. A. Leites, *Introduction to the theory of supermanifolds*, Uspekhi Mat. Nauk **35** (1980), no. 1, 1–64; English transl. in Russian Math. Surveys **35** (1980), no. 1.

[Pak] V. F. Pakhomov, *Automorphisms of the tensor product of abelian and Grassmanian algebras*, Mat. Zametki **16** (1974), no. 1, 624–629; English transl. in Math. Notes **16** (1974).

Mathematical College of the Independent University of Moscow, 18, Fotievoy Str., Moscow, Russia

E-mail address: palamo@tomogr.msk.su

Translated by THE AUTHOR

Amer. Math. Soc. Transl.
(2) Vol. **175**, 1996

Remarks on the Topology of the Hilbert Grassmannian

M. A. Shubin

§0. Introduction

The topology of ordinary Grassmannians in finite-dimensional spaces plays an important role in analysis and topology. In particular, it gives rise to characteristic classes of vector bundles and is also connected with important analytic objects such as generalized hypergeometric functions.

In this paper we make some elementary remarks about some aspects of the topology of the Hilbert Grassmannian, i.e., the Grassmannian of the closed linear subspaces in a separable Hilbert space. It turns out that some interesting features can be found already on the level of general topology. We identify the points of this Grassmannian with the selfadjoint projections in the given space. Then we can consider norm topology, strong and weak topology on the set of projections. The last two topologies in fact coincide as the topologies on the Grassmannian and they define a separable metrizable topology there. However, if we take closures of the set of selfadjoint projections in the strong and weak topologies in the set of all bounded linear operators, then the results will be different: the set of projections is closed in the strong topology but it is not closed in the weak topology. The closure in the weak topology is the "operator segment" $[0,1]$: the set of all operators A such that $0 \leqslant A \leqslant I$, where I is the identity operator and the inequality is understood in the standard operator sense (as the inequality of the corresponding quadratic forms). This operator segment is a compact set in the weak operator topology.

We also consider the natural splitting of the Grassmannian by the dimensions of the image and the kernel of the projection (unlike the finite-dimensional situation, it is not sufficient to take the dimension of the image only: if the image is infinite-dimensional, then the kernel still can have any dimension). We describe the closures of the corresponding subsets in all topologies.

A possible motivation for the study of the topology of the Hilbert Grassmannian is that it might give an approach to the invariant subspace problem. Namely, any bounded linear operator in Hilbert space acts on the Grassmannian and the problem reduces to finding a fixed point of this action in the set of nontrivial subspaces (subspaces which have nonvanishing dimension and codimension). I can

1991 *Mathematics Subject Classification.* Primary 46C05.

Partially supported by NSF grant DMS-9222491.

only hope that the results of this paper can be ε-helpful in the future for solving of the invariant subspace problem.

§1. Norm topology

Denote by $\mathcal{H}$ a separable complex Hilbert space, $\mathcal{B}(\mathcal{H})$ the Banach algebra of all bounded linear operators in $\mathcal{H}$. We shall denote $\|\cdot\|$ the standard operator norm in $\mathcal{B}(\mathcal{H})$.

Denote by $G_{n,k}(\mathcal{H})$ the set of all closed linear subspaces $L \subset \mathcal{H}$ such that $\dim L = n$ and $\dim L^\perp = k$, where n, k are nonnegative integers or infinity and $n + k = \infty$. In particular $G_{0,\infty}(\mathcal{H}) = \{0\}$ and $G_{\infty,0}(\mathcal{H}) = \{\mathcal{H}\}$ consist each of one element only. Denote $G(\mathcal{H}) = \bigcup\{G_{n,k}(\mathcal{H}) : n + k = \infty\}$.

Let us identify $L \in G_{n,k}(\mathcal{H})$ with the selfadjoint projection in $\mathcal{H}$ with the image L; denote this projection by P_L. Denote $\mathcal{P}_{n,k}$ the set of all selfadjoint projections corresponding to the subspaces $L \in G_{n,k}(\mathcal{H})$. So $\mathcal{P}_{n,k} \subset \mathcal{B}(\mathcal{H})$. Denote $\mathcal{P} = \bigcup\{\mathcal{P}_{n,k} : n + k = \infty\}$.

LEMMA 1.1. *$\mathcal{P}_{n,k}$ is closed in $\mathcal{B}(\mathcal{H})$ in the norm topology.*

It follows that $G_{n,k}$, identified with $\mathcal{P}_{n,k}$, is a complete metric space with the metric $d(L, L') = \|P_L - P_{L'}\|$.

PROOF OF LEMMA 1.1. Clearly $\mathcal{P}$ is closed in $\mathcal{B}(\mathcal{H})$ because its elements are distinguished in $\mathcal{B}(\mathcal{H})$ by the conditions $P^2 = P = P^*$. To check that $\mathcal{P}_{n,k}$ is closed, it is sufficient to prove that each $\mathcal{P}_{n.k}$ is open in $\mathcal{P}$. This follows if we use the following well-known fact: if $P_1, P_2 \in \mathcal{P}$ and $\|P_1 - P_2\| < 1$, then $\dim P_1\mathcal{H} = \dim P_2\mathcal{H}$ (see, e.g., [**G-K**]). □

Denote by $U(\mathcal{H})$ the group of all unitary operators in $\mathcal{H}$ (with the norm topology). This is a complete metric space. Obviously $U(\mathcal{H})$ acts transitively on each of the Grassmannians $G_{n,k}(\mathcal{H})$ with the isotropy subgroup of a point $L \in G_{n,k}$ equal to $U(L) \times U(L^\perp)$. Therefore for any fixed $L \in G_{n.k}(\mathcal{H})$ we have an isomorphism (of sets)

$$G_{n,k}(\mathcal{H}) = U(\mathcal{H})/[U(L) \times U(L^\perp)], \tag{1.1}$$

where the quotient on the right means the set of the left cosets. We shall see that this is true with the topology as well.

PROPOSITION 1.2. *Let us fix a subspace $L_0 \in G_{n,k}(\mathcal{H})$. Then the map*

$$\pi : U(\mathcal{H}) \to G_{n,k}(\mathcal{H}), \qquad U \mapsto UL_0,$$

is a locally trivial fiber bundle with fiber $U(L_0) \times U(L_0^\perp)$.

PROOF. 1°. First note that the continuity of π follows from the fact that for any $L = UL_0$, $U \in U(\mathcal{H})$, we have $P_L = UP_{L_0}U^{-1} = UP_{L_0}U^*$.

2°. Suppose that for any $M \in G_{n,k}(\mathcal{H})$ we can construct a local section given in a neighborhood $\mathcal{U}$ of M, i.e., a continuous map $s : \mathcal{U} \to \pi^{-1}(\mathcal{U}) \subset U(\mathcal{H})$ such that $\pi \circ s = \mathrm{Id}_{\mathcal{U}}$. A trivialization of $\pi^{-1}(\mathcal{U})$ can then be constructed by

$$\mathcal{U} \times [U(L_0) \times U(L_0^\perp)] \to \pi^{-1}(\mathcal{U}), \qquad (L, u) \mapsto s(L)u.$$

This map is obviously one-to-one and continuous. The inverse map

$$\pi^{-1}(\mathcal{U}) \to \mathcal{U} \times [U(L_0) \times U(L_0^\perp)], \qquad V \mapsto (\pi(V), [s(\pi(V))]^{-1}V)$$

is obviously continuous too. Therefore to prove the proposition it is sufficient to construct local sections.

3°. Let us construct a local section near L_0. We shall do it on the set

$$\mathcal{U} = \{L : d(L, L_0) < 1\} = \{L : \|P_L - P_{L_0}\| < 1\}.$$

Denoting $s(L) = U_L$, we see that U_L must be a unitary operator U_L that depends continuously on L and $U_L L_0 = L$. Let us take

$$A_L = P_L P_{L_0} + (I - P_L)(I - P_{L_0}).$$

Let us check that A_L defines linear topological isomorphisms $L_0 \to L$ and $L_0^\perp \to L^\perp$. Since A_L maps L_0 to L and $L_0^\perp$ to $L^\perp$, it is sufficient to prove that $P_L P_{L_0}$ defines a linear topological isomorphism $L_0 \to L$ and then replace P_L and P_{L_0} by $I - P_L$ and $I - P_{L_0}$ respectively. Now $P_L P_{L_0} = P_L$ on L_0 and $P_{L_0} P_L = P_{L_0}$ on L, so it is sufficient to prove that the operator $P_L P_{L_0} \colon L \to L$ is invertible. Obviously $P_L P_{L_0} = I + P_L(P_{L_0} - P_L)$ on L, so $\|P_L(P_{L_0} - P_L)\| < 1$ implies the desired invertibility.

Let us construct the polar decomposition of A_L. We have

$$A_L^* = P_{L_0} P_L + (I - P_{L_0})(I - P_L),$$
$$A_L^* A_L = P_{L_0} P_L P_{L_0} + (I - P_{L_0})(I - P_L)(I - P_{L_0}),$$

so $A_L^* A_L$ is an invertible selfadjoint operator which maps L_0 to L_0 and $L_0^\perp$ to $L_0^\perp$ isomorphically. Besides A_L and $A_L^* A_L$ depend continuously on L. Denote as usual $|A_L| = \sqrt{A_L^* A_L}$. Then $|A_L|$ is also invertible, depends continuously on L, and leaves both spaces L_0 and $L_0^\perp$ invariant. Now we can take $U_L = A_L |A_L|^{-1}$. Then U_L is a unitary operator that maps L_0 to L and depends continuously on L, so it gives us the desired local section.

4°. To construct a section near an arbitrary point $M \in G_{n,k}(\mathcal{H})$, it is sufficient to use the homogeneity of the situation. □

COROLLARY 1.3. *$G_{n,k}$ is a connected component of $G(\mathcal{H})$.*

COROLLARY 1.4. *The topology of $G_{n.k}(\mathcal{H})$ is the quotient-topology of $U(\mathcal{H})$.*

PROOF. We must check that a set $\mathcal{U} \subset G_{n.k}(\mathcal{H})$ is open if and only if $\pi^{-1}(\mathcal{U})$ is open in $U(\mathcal{H})$. If $\mathcal{U}$ is open, then $\pi^{-1}(\mathcal{U})$ is open because π is continuous. Vice versa, suppose that $\pi^{-1}(\mathcal{U})$ is open. Then together with each point u_0, it contains its neighborhood which has the form $\mathcal{U}_0 \times V_0$ in the trivialization from Proposition 1.2. Here $\mathcal{U}_0$ is open in $G_{n,k}(\mathcal{H})$ and V_0 is open in $U(\pi(u_0)) \times U(\pi(u_0)^\perp)$. But then $\mathcal{U} = \pi(\pi^{-1}(\mathcal{U}))$ contains the neighborhood $\mathcal{U}_0$ of $\pi(u_0)$ as required. □

§2. Strong and weak topology

A. The strong and weak topologies on $G(\mathcal{H})$ and on its subsets are induced by the corresponding operator topologies on projections. We shall denote ${}^sG$ and ${}^wG$ the topological spaces obtained from $G(\mathcal{H})$ by supplying this set with strong and weak topology respectively. The notations ${}^sG_{n,k}$ and ${}^wG_{n,k}$ have a similar meaning.

Let us describe the strong and weak topologies more explicitly. We shall first describe the strong topology.

Neighborhoods in ${}^s G_{n,k}$: a base of neighborhoods of $L_0 \in {}^s G_{n,k}$ can be chosen as follows:

$$ {}^s\mathcal{U}_{x_1,\dots,x_N,\varepsilon}(L_0) = \{L : \|(P_L - P_{L_0})x_j\| < \varepsilon\}, \qquad j = 1, \dots, N. $$

Here $x_j \in \mathcal{H}$, $\varepsilon > 0$.

It is clear that we can restrict ourselves to x_j with $\|x_j\| \leqslant 1$ (or even with $\|x_j\| = 1$). Moreover, we can restrict ourselves to the vectors from any dense subset $X \subset \mathcal{H}$ (or a subset X which is dense in a unit ball or unit sphere of $\mathcal{H}$). Indeed, it is sufficient to prove that any neighborhood of L_0 contains a neighborhood of the form

$$ {}^s\mathcal{U}_{\tilde{x}_1,\dots,\tilde{x}_N,\tilde{\varepsilon}}(L_0) \quad \text{with } \tilde{x}_j \in X. $$

But for the inclusion

$$ {}^s\mathcal{U}_{\tilde{x}_1,\dots,\tilde{x}_N,\tilde{\varepsilon}}(L_0) \subset {}^s\mathcal{U}_{x_1,\dots,x_N,\varepsilon}(L_0) $$

to be true it is sufficient, for every $L \in {}^s\mathcal{U}_{\tilde{x}_1,\dots,\tilde{x}_N,\tilde{\varepsilon}}(L_0)$, to have

$$ \|(P_L - P_{L_0})x_j\| \leqslant \|(P_L - P_{L_0})\tilde{x}_j\| + \|(P_L - P_{L_0})(x_j - \tilde{x}_j)\| < \tilde{\varepsilon} + 2\|x_j - \tilde{x}_j\| \leqslant \varepsilon. $$

Hence we can choose $\tilde{\varepsilon} = \varepsilon/2$ and take $\tilde{x}_j$ so that $\|x_j - \tilde{x}_j\| < \varepsilon/4$. It follows that

$$ {}^s\mathcal{U}_{\tilde{x}_1,\dots,\tilde{x}_N,\varepsilon/2}(L_0) \subset {}^s\mathcal{U}_{x_1,\dots,x_N,\varepsilon}(L_0) \quad \text{if } \|x_j - \tilde{x}_j\| < \varepsilon/4,\ j = 1, \dots, N. $$

Furthermore, linearity implies that instead of a dense subset we can take X to be the set $\{e_j,\ j = 1, 2, \dots\}$ constituting a fixed orthonormal basis of $\mathcal{H}$. Hence we get the following base of neighborhoods of L_0:

$$ {}^s\mathcal{U}_{N,\varepsilon}(L_0) = \{L : \|(P_L - P_{L_0})e_j\| < \varepsilon,\ j = 1, \dots, N\}. $$

Indeed, suppose that $x_j = \sum_{1 \leqslant k \leqslant M} c_{jk} e_k$, $c_{jk} \in \mathbb{C}$ and $\|x_j\| \leqslant 1$, $j = 1, \dots, N$. Then $|c_{jk}| \leqslant 1$ and

$$ \|(P_L - P_{L_0})x_j\| \leqslant \sum_{1 \leqslant k \leqslant M} \|(P_L - P_{L_0})e_k\|. $$

It follows that

$$ {}^s\mathcal{U}_{M,\varepsilon/M}(L_0) \subset {}^s\mathcal{U}_{x_1,\dots,x_N,\varepsilon}(L_0). $$

Summarizing, we get the following

LEMMA 2.1. *In* ${}^s G_{n,k}$ *every point* L_0 *has a countable base of neighborhoods of the form* ${}^s\mathcal{U}_{N,1/N}(L_0)$, $N = 1, 2, \dots$. *The strong convergence* $\text{s-lim}_\alpha L_\alpha = L$ *is equivalent to the convergence* $\lim_\alpha P_{L_\alpha} e_j = P_L e_j$, $j = 1, 2, \dots$, *where* $\{e_j,\ j = 1, 2, \dots\}$ *is a fixed orthonormal basis in* $\mathcal{H}$.

COROLLARY 2.2. ${}^s G_{n,k}$ *is metrizable.*

PROOF. Identifying L with the sequence

$$ \{P_L e_j,\ j = 1, 2, \dots\} \in \mathcal{H}^\infty = \mathcal{H} \times \mathcal{H} \times \cdots, $$

we get an embedding $G_{n,k}(\mathcal{H}) \subset \mathcal{H}^\infty$. There is a metric on $\mathcal{H}^\infty$ given by

$$ d(\{f_j\}, \{h_j\}) = \sum_{j=1}^{\infty} \frac{1}{2^j} \frac{\|f_j - h_j\|}{1 + \|f_j - h_j\|}. $$

It is easy to see that this metric induces a metric on $G_{n,k}(\mathcal{H})$ such that the corresponding topology coincides with the strong topology. □

REMARK. Unlike the topology on ${}^sG_{n,k}$, the metric is not unitary invariant.

COROLLARY 2.3. *${}^sG_{n,k}$ is separable, i.e., it has a countable dense subset.*

PROOF. It is sufficient to notice that $\mathcal{H}^\infty$ is separable. □

B. Now let us describe the weak topology in $G_{n,k}(\mathcal{H})$.

Neighborhoods in ${}^wG_{n,k}$: for the base of neighborhoods of $L_0 \in {}^wG_{n,k}$ we can take

$$ {}^w\mathcal{U}_{x_1,\dots,x_N,\varepsilon}(L_0) = \{L : |((P_L - P_{L_0})x_j, x_k)| < \varepsilon\}, \qquad j,k = 1,\dots,N, $$

where $x_j \in \mathcal{H}$, $\varepsilon > 0$. Again we can restrict ourselves to the vectors x_j from a dense subset X in $\mathcal{H}$ or from the unit ball of $\mathcal{H}$. Moreover if is sufficient to take $x_j = e_j$, where $\{e_j,\ j = 1,2,\dots\}$ is an orthonormal basis in $\mathcal{H}$. Therefore we get a base of neighborhoods of each point L_0 of the form

$$ {}^w\mathcal{U}_{N,\varepsilon}(L_0) = \{L : |((P_L - P_{L_0})e_j, e_k)| < \varepsilon\}, \qquad j,k = 1,\dots,N. $$

Therefore we obtain the following

LEMMA 2.4. *At ${}^wG_{n,k}$ every point L_0 has a countable base of neighborhoods of the form ${}^w\mathcal{U}_{N,1/N}(L_0)$, $N = 1,2,\dots$. The weak convergence* w-$\lim_\alpha L_\alpha = L$ *is equivalent to the convergence of the matrix elements of the operators P_{L_α} to the matrix elements of P_L in the basis e_j, $j = 1,2,\dots$.*

COROLLARY 2.5. *${}^wG_{n,k}$ is metrizable and separable.*

Now note that we can describe exactly in the same way the strong and weak topologies on $G(\mathcal{H})$. The arguments do not change and we come to

LEMMA 2.6. *In ${}^sG$ and in ${}^wG$ every point L_0 has a countable base of neighborhoods of the form ${}^s\mathcal{U}_{N,1/N}(L_0)$ and ${}^w\mathcal{U}_{N,1/N}(L_0)$ respectively, $N = 1,2,\dots$. The convergence* s-$\lim_\alpha L_\alpha = L$ *and* w-$\lim_\alpha L_\alpha = L$ *in ${}^sG$ and ${}^wG$ is described exactly the same way as in ${}^sG_{n,k}$ and ${}^wG_{n,k}$.*

COROLLARY 2.7. *${}^sG$ and ${}^wG$ are metrizable and separable.*

§3. Comparison of the strong and weak topologies

A. In this section we shall compare the strong and weak topologies on $G_{n,k}(\mathcal{H})$ and $G(\mathcal{H})$ and describe the closures of these sets in $\mathcal{B}(\mathcal{H})$ in both topologies.

PROPOSITION 3.1. *Strong and weak topologies coincide on $G_{n,k}(\mathcal{H})$ and $G(\mathcal{H})$.*

PROOF. Due to Corollaries 2.2, 2.3, 2.5, and 2.7, it is sufficient to check that for any sequence $P_n \in \mathcal{P}$, $n = 1,2,\dots$, we have

$$ \operatorname*{w-lim}_{n\to\infty} P_n = P \in \mathcal{P} \implies \operatorname*{s-lim}_{n\to\infty} P_n = P. $$

This is well known (see [**H,** Problem 115]) and follows from the fact that $f_n \to f$ (weak) and $\|f_n\| \to \|f\|$ imply $f_n \to f$ (strong) ([**H,** Problem 20]). □

In this proof we assumed that the weak limit of the projections P_n is again a projection. This is not always true (see e.g. [**H,** Problem 115]). But this is true for strong convergence, even if we replace sequences by nets:

LEMMA 3.2. *Suppose $\{P_\alpha\}$ is a net of projections $P_\alpha \in \mathcal{P}$ and there exists a strong limit* $\text{s-lim}_\alpha P_\alpha = P$. *Then $P \in \mathcal{P}$, i.e., P is a selfadjoint projection too.*

PROOF. Taking limits in the equality $(P_\alpha u, v) = (u, P_\alpha v)$, $u, v \in \mathcal{H}$, we get $(Pu, v) = (u, Pv)$, i.e., P is selfadjoint. We also have

$$(Pu, u) = \lim_\alpha (P_\alpha u, u) \in [0, 1].$$

Therefore $0 \leqslant P \leqslant I$ in the operator sense. It follows that $\|P\| \leqslant 1$. Now writing

$$P_\alpha^2 u - P^2 u = P_\alpha (P_\alpha - P)u + (P_\alpha - P)Pu, \qquad u \in \mathcal{H},$$

we immediately see that $\|P_\alpha^2 u - P^2 u\| \to 0$. Hence $P_\alpha^2 = P_\alpha$ implies $P^2 = P$. □

PROPOSITION 3.3. *The weak closure of the set $\mathcal{P}$ of all selfadjoint projections in $\mathcal{B}(\mathcal{H})$ coincides with the set $[\mathbf{0}, \mathbf{I}]$ of all selfadjoint operators $A \in \mathcal{B}(\mathcal{H})$ such that $0 \leqslant A \leqslant I$.*

PROOF. Let us denote the strong and weak closures of any subset $M \subset \mathcal{B}(\mathcal{H})$ in $\mathcal{B}(\mathcal{H})$ by $\mathrm{Cl}_s(M)$ and $\mathrm{Cl}_w(M)$ respectively. Suppose that $A \in \mathrm{Cl}_w(\mathcal{P})$, i.e., $A \in \mathcal{B}(\mathcal{H})$ is a weak limit of selfadjoint projections. The first part of the proof of Lemma 3.2 shows that $A \in [\mathbf{0}, \mathbf{I}]$. This means that $\mathrm{Cl}_w(\mathcal{P}) \subset [\mathbf{0}, \mathbf{I}]$.

Let us prove the inverse inclusion. We must show that every $A \in [\mathbf{0}, \mathbf{I}]$ is a weak limit of selfadjoint projections. Let us choose an orthonormal basis $e_1, e_2, \dots$ in $\mathcal{H}$ and denote by P_n the selfadjoint projection of rank n on the subspace spanned by $e_1, \dots, e_n$. Then $P_n A P_n \to A$ strongly, and $P_n A P_n \in [\mathbf{0}, \mathbf{I}]$ for all n. Therefore it is sufficient to check the inclusion $A \in \mathrm{Cl}_w(\mathcal{P})$ for operators of finite rank from $[\mathbf{0}, \mathbf{I}]$. But by the spectral theorem every such operator is a finite sum of operators of the form $\alpha^2 P_1$, where P_1 is a selfadjoint projection of rank 1 and $\alpha \in [0, 1]$. Therefore it remains to prove that $\alpha^2 P_1 \in \mathrm{Cl}_w(\mathcal{P})$. To do this we shall practically follow the solution of Problem 115 in [**H**].

Let us choose an orthonormal basis $e_1, e_2, \dots$ of $\mathcal{H}$ so that the image of P_1 is spanned by e_1. Now let us consider the selfadjoint projections $P_{(n)}$ given by

$$P_{(n)} u = (u, \alpha e_1 + \beta e_n)(\alpha e_1 + \beta e_n), \qquad u \in \mathcal{H},$$

where $\beta \in [0, 1]$ is chosen so that $\alpha^2 + \beta^2 = 1$. Then we have for any $u, v \in \mathcal{H}$

$$\lim_{n \to \infty} (P_{(n)} u, v) = \lim_{n \to \infty} (u, \alpha e_1 + \beta e_n)(\alpha e_1 + \beta e_n, v) = \alpha^2 (u, e_1)(e_1, v) = \alpha^2 (P_1 u, v),$$

which means precisely that $P_{(n)} \to \alpha^2 P_1$ weakly as $n \to \infty$. □

REMARK. In the weak operator topology the set $[\mathbf{0}, \mathbf{I}]$ is a compact topological space.

Indeed, we can identify elements $A \in [\mathbf{0}, \mathbf{I}]$ with their matrices $[a_{ij}]$ in a fixed orthonormal basis $e_1, e_2, \dots$. (Here $a_{ij} = (Ae_i, e_j)$.) Then obviously $a_{ii} \in [0, 1]$ and also $|a_{ij}| \leqslant 1$ by the Cauchy–Schwarz inequality $|a_{ij}|^2 \leqslant a_{ii} a_{jj}$. Let K be the unit disk in the complex plane $\mathbb{C}$. Then we obtain a natural inclusion of $[\mathbf{0}, \mathbf{I}]$ in a product of a countable number of copies K_{ij} of K, identifying every element $A \in [\mathbf{0}, \mathbf{I}]$ with the set of its matrix elements. If we introduce the Tikhonov topology in the product, then it becomes a compact topological space and $[\mathbf{0}, \mathbf{I}]$ becomes its closed subset, hence also a compact topological space.

B. Now we shall describe strong and weak closures of the sets $G_{n,k}(\mathcal{H})$ in $G(\mathcal{H})$ and also the weak closures of the corresponding sets of projections in $\mathcal{B}(\mathcal{H})$.

PROPOSITION 3.4. *The strong and weak closure of* $G_{n,k}(\mathcal{H})$ *in* $G(\mathcal{H})$ *coincides with*

$$\bigcup\{G_{n',k'} : n' \leqslant n,\ k' \leqslant k,\ n' + k' = \infty\}.$$

In other words

(i) $\mathrm{Cl}_s(G_{\infty,\infty}(\mathcal{H})) = \mathrm{Cl}_w(G_{\infty,\infty}(\mathcal{H})) = G(\mathcal{H})$;

(ii) $\mathrm{Cl}_s(G_{n,\infty}(\mathcal{H})) = \mathrm{Cl}_w(G_{n,\infty}(\mathcal{H})) = \bigcup_{p=0}^{n} G_{p,\infty}$ *if* n *is finite*;

(iii) $\mathrm{Cl}_s(G_{\infty,n}(\mathcal{H})) = \mathrm{Cl}_w(G_{\infty,n}(\mathcal{H})) = \bigcup_{q=0}^{n} G_{\infty,q}(\mathcal{H})$ *if* n *is finite.*

PROOF. We have already seen that $\mathrm{Cl}_s(G_{n,k}(\mathcal{H})) = \mathrm{Cl}_w(G_{n,k}(\mathcal{H}))$ (if the closures are taken in $\mathcal{P}$).

Let us check that for every $P \in \mathrm{Cl}_w(\mathcal{P}_{n,k}) \cap \mathcal{P}$ we have

$$\dim P\mathcal{H} \leqslant n \quad \text{and} \quad \dim (I - P)\mathcal{H} \leqslant k.$$

We can find a sequence $\{P_j \mid j = 1, 2, \dots\}$ such that $P_j \in \mathcal{P}_{n,k}$ for all j and $P_j \to P$ weakly as $j \to \infty$. Let us consider the diagonal elements of P_j. Applying the Fatou Lemma, we obtain

$$\dim P\mathcal{H} = \mathrm{Tr}\, P \leqslant \liminf_{j\to\infty} \mathrm{Tr}\, P_j = \mathrm{Tr}\, P_j = n.$$

Similarly we can check that $\dim(I - P)\mathcal{H} \leqslant k$.

Conversely, suppose that $\dim P\mathcal{H} \leqslant n$ and $\dim(I-P)\mathcal{H} \leqslant k$. Let us prove that $P \in \mathrm{Cl}_s(\mathcal{P}_{n,k})$. Suppose that $\dim P\mathcal{H} = n' < \infty$ (the other cases are considered similarly). Then $k = \infty$. If $n' = n$, then $P \in \mathcal{P}_{n,k}$, and we have nothing to prove. Therefore assume that $n' < n$. Let us choose an orthonormal basis $\{e_1, e_2, \dots\}$ in $\mathcal{H}$ so that $e_1, \dots, e_{n'}$ span $P\mathcal{H}$. Now let us take a subspace $L_N \subset \mathcal{H}$ with $\dim L_N = n$ spanned by $e_1, \dots, e_{n'}, e_{N+1}, \dots, e_{N+n-n'}$. Denote by P_N the selfadjoint projection in $\mathcal{H}$ with the image L_N. Then $P_N \in \mathcal{P}_{n,k}$ and $P_N \to P$ strongly as $N \to \infty$. □

Now we shall consider the weak closure of $\mathcal{P}_{n,k}$ in $\mathcal{B}(\mathcal{H})$.

PROPOSITION 3.5. *The weak closure* $\mathrm{Cl}_w(\mathcal{P}_{n,k})$ *of* $\mathcal{P}_{n,k}$ *in* $\mathcal{B}(\mathcal{H})$ *coincides with the set*

$$\mathcal{A}_{n,k} = \{A : A \in [\mathbf{0}, \mathbf{I}],\ \mathrm{rank}\, A \leqslant n,\ \mathrm{rank}(I - A) \leqslant k\},$$

where $\mathrm{rank}\, A = \dim \mathrm{Im}\, A$.

PROOF. (i) Let us first consider the case $n = k = \infty$. Then the result follows from Proposition 3.3 and Proposition 3.4(i).

(ii) Now let us assume that $n < \infty$ (the case $k < \infty$ is treated similarly), so

$$\mathcal{A}_{n,k} = \mathcal{A}_{n,\infty} = \{A : A \in [\mathbf{0}, \mathbf{I}],\ \mathrm{rank}\, A \leqslant n\}.$$

Let us prove that $\mathrm{Cl}_w(\mathcal{P}_{n,k}) \subset \mathcal{A}_{n,k}$. Suppose that $P_j \in \mathcal{P}_{n,k}$, $j = 1, 2, \dots$, $\mathrm{rank}\, P_j \leqslant n$ for all j and there exists a weak limit A of P_j as $j \to \infty$. The inclusion $A \in [\mathbf{0}, \mathbf{I}]$ follows from Proposition 3.3, so we only have to check that $\mathrm{rank}\, A \leqslant n$.

Note first that $\mathrm{Tr}\, A \leqslant n$ by the argument from the proof of Proposition 3.4. Therefore A is in the trace class, hence compact. We can then find an orthonormal basis of eigenvectors of A. Denote the vectors of this basis $e_1, e_2, \dots$. The eigenvalues of A are nonnegative and have 0 as their only accumulation point. Therefore we can rearrange them in nonincreasing order. The rank of A is equal to the number of nonvanishing eigenvalues, multiplicities counted.

Assume, arguing by contradiction, that $\operatorname{rank} A \geqslant n+1$. Then in the chosen basis A will be represented by a diagonal matrix with positive first $n+1$ diagonal elements. For large j, the left-upper-corner $(n+1)\times(n+1)$ minor in the matrix of P_j will be nonvanishing. But this obviously implies that $\operatorname{rank} P_j \geqslant n+1$ because then $P_j e_1, \ldots, P_j e_{n+1}$ will be linearly independent. This contradicts the assumption $P_j \in \mathcal{P}_{n,k}$ and proves the inequality $\operatorname{rank} A \leqslant n$.

(iii) It remains to prove that for any $A \in \mathcal{A}_{n,\infty}$ with $n < \infty$ we can find a $P_N \in \mathcal{P}$, $N = 1, 2, \ldots$ with $\operatorname{rank} P_N \leqslant n$ so that $P_N \to A$ weakly as $N \to \infty$. By the spectral theorem, there exists an orthonormal basis $e_1, e_2, \ldots$ in $\mathcal{H}$ such that $Ae_j = \alpha_j^2 e_j$ with $0 \leqslant \alpha_j \leqslant 1$ for all j. This means that A can be written in the form

$$A = \sum_{j=1}^{n} \alpha_j^2 P^{(j)},$$

where $P^{(j)}$ is a selfadjoint projection of rank 1 acting by the formula $P^{(j)}u = (u, e_j)e_j$, $u \in \mathcal{H}$. We have seen in the proof of Proposition 3.3 that $\alpha_j^2 P^{(j)}$ is a weak limit of rank 1 selfadjoint projections, i.e., projections from $\mathcal{P}_{1,\infty}$. Making a specific choice of these projections, we can write

$$P_{j,N} \to \alpha_j^2 P^{(j)} \quad \text{weakly as } N \to \infty,$$

where $P_{j,N}$ is given by the formula

$$P_{j,N}u = (u, \alpha_j e_j + \beta_j e_{N+j})(\alpha_j e_j + \beta_j e_{N+j}),$$

with $\beta_j \in [0,1]$, $\alpha_j^2 + \beta_j^2 = 1$. Now for $N > n$ define

$$P_N = \sum_{j=1}^{n} P_{j,N}.$$

It is easy to check that $P_N \in \mathcal{P}_{n,\infty}$ and $P_N \to A$ weakly as $N \to \infty$. □

References

[G-K] I. Gokhberg and M. Krein, *Introduction to the theory of nonselfadjoint linear operators*, "Nauka", Moscow; English transl., Amer. Math. Soc., Providence, 1969.

[H] P. Halmos, *A Hilbert space problem book*, Second Edition, Springer-Verlag, New York, 1982.

Department of Mathematics Northeastern University Boston, MA 02115, USA

Translated by THE AUTHOR

Amer. Math. Soc. Transl.
(2) Vol. **175**, 1996

Asymptotic Completeness

I. M. Sigal

Dedicated to the memory of F. A. Berezin

Apology. A little more than 20 years ago my Ph.D. advisor, F. A. Berezin, suggested that I look into scattering theory. In his own words, I did not have time to do anything on my own, but it would keep me off the streets. I would get myself an education. For better or worse, it did keep me off the streets, but 20 years later I am still an amateur. In the first 10 years of my preoccupation with scattering theory, progress was very limited and extremely painful. In the last decade it was rather remarkable, but too fast too keep track of, unless one devoted all one's time to the subject. I did not. So this loosely written review reflects my bumpy ride. It sketches some general notions which I find fundamental, not only for scattering, but also outside of it. Unfortunately, I understand them only partially. I omit details, because I either find them boring or do not understand them well enough to find their proper place in the system of things. I dedicate this review to Felix Berezin. To his launching me on this tough subject, I owe many exhilarating moments of my scientific journey.

Acknowledgement. With the exception of the last section, this review keeps close to parts of the talks on mathematical problems in Quantum Mechanics I have given at MIT, Princeton, ETH, and Jerusalem. I am grateful to the audiences for remarks and encouragement. I was lucky to work on scattering theory in collaboration with Avy Soffer. My understanding of that theory, which I will try to convey below, is formed by this collaboration.

Problem. Scattering theory studies the asymptotic behavior of the time-dependent Schrödinger equation

$$i\frac{\partial \psi_t}{\partial t} = H\psi_t \tag{1}$$

as $t \to \pm\infty$. Here ψ_t is a differentiable path or orbit in the state space $L^2(X)$, where X is the configuration space of the system in question. Usually, it is $\mathbb{R}^{3N}$ or

1991 *Mathematics Subject Classification.* Primary 81U18.

a subspace thereof corresponding to the center-of-mass frame, and

$$H = -\Delta + V(x)$$

is the Schrödinger operator on $L^2(X)$, where Δ is the Laplace–Beltrami operator on X (the kinetic energy operator) and $V(x)$ is a real function on X (the potential). We assume that $V(x)$ is not too singular, so that H is self-adjoint (see Kato, 1953).

The self-adjointness of H is equivalent to the existence of unitary dynamics, i.e., the existence of global solutions to the Cauchy problem for (1) satisfying $\|\psi_t\| =$ const (see [**31**]). Once this is established, the next problem is the classification of solutions (orbits) according to their asymptotic behavior as $t \to \pm\infty$. This is called *asymptotic* completeness. It is the main mathematical problem of scattering theory. It was first attacked just soon after the birth of quantum mechanics. Although during the next 60 years the progress on the one-two body problem was rather satisfactory, the many-body problem eluded researchers and only limited progress was made. However, in the last decades, a remarkable development took place in this area. In this article I sketch the main highlights of this process.

I concentrate on many-body asymptotic completeness. To this end, consider a system consisting of N ν-dimensional particles. The configuration space of such a system in the center-of-mass frame is

$$X = \left\{ x \in \mathbb{R}^{\nu N} \,\middle|\, \sum m_i x_i = 0 \right\}.$$

Here $x = (x_1, \dots, x_N)$. We equip X with the inner product

$$\langle x, y\rangle = \sum_{i=1}^{N} m_i x_i y_i .$$

The kinematics of many particles in $\mathbb{R}^\nu$ translates to the geometry of X. This was understood by many people (see e.g. [**39, 6, 32, 24, 25**]). The passage from kinematic to geometric language, which is crucial in the latest constructions in the many-body scattering theory, was suggested by Agmon [**3**]. A brief kinematic-geometric dictionary is given below.

Kinematics	**Geometry**
N particles	System, $\mathfrak{X}$, of subspaces of X
$V(x) = \sum V_{ij}(x_i - x_j)$	$V(x) \to 0$ as $\lvert x\rvert \to \infty$, except along the subspaces from $\mathfrak{X}$
$N = 2$	$\mathfrak{X} = \phi$
$N = 3$	Subspaces from $\mathfrak{X}$ intersect only at the origin
$N \geqslant 4$	No restrictions (only on generations)
Break-up	
Cone in X	Subspaces from $\mathfrak{X}$

In the kinematic language, asymptotic completeness states that as $|t| \to \infty$, the system in question breaks up into independent, freely moving, stable subsystems. In the geometrical language one says that the system's motion is a superposition of a free motion along a ray starting at the origin (= free motion of the centers-of-

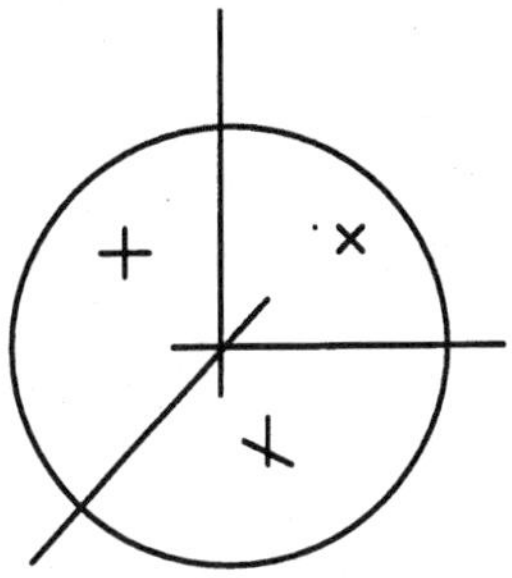

FIGURE 1. 3-body geometry

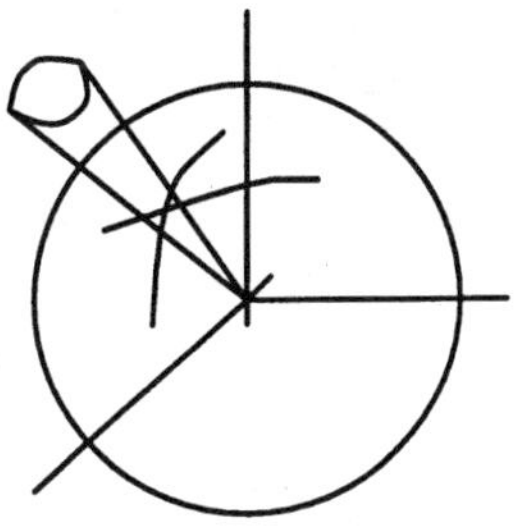

FIGURE 2. Break-up: Geometric representation

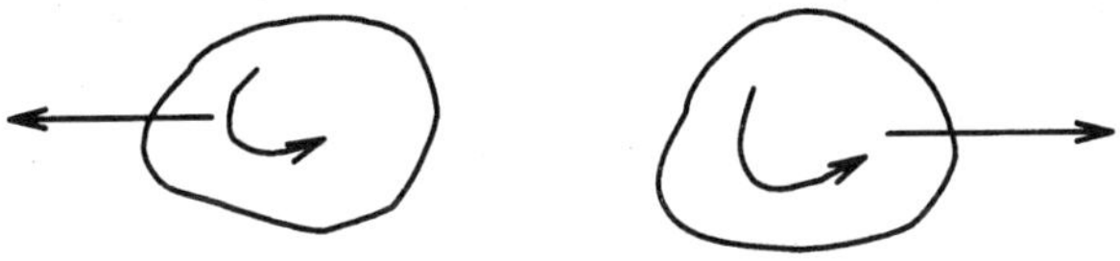

FIGURE 3. Break-up: Physical picture

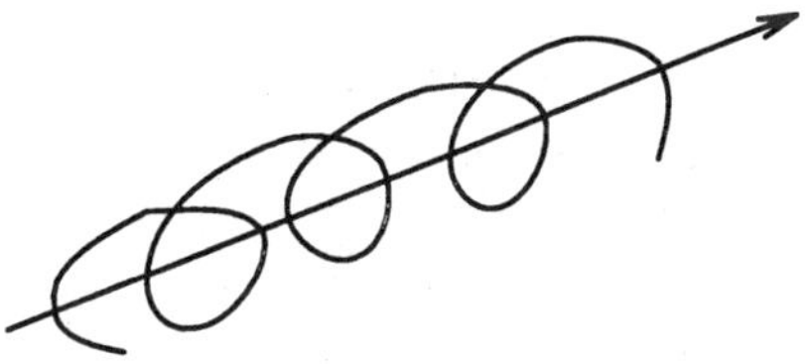

FIGURE 4. Free motion of the center of mass and bounded internal motion of clusters

mass of the subsystems) and a bounded motion in the transversal direction (stable internal motion of the subsystems):

$Y \in \mathfrak{X}$

(free motion of the center-of-mass)

In rigorous terms, asymptotic completeness states that for any orbit ψ_t (i.e., a solution to (1) with an L^2 initial condition) there are $\varphi_Y \in L^2(Y)$ for all $Y \in \mathfrak{X}$, such that

$$\left\| \psi_t - \sum_{(Y,\psi_{Y^\perp})} \psi_{Y^\perp} \otimes e^{-i(\varepsilon_{Y^\perp} - \Delta_Y)t} \varphi_Y \right\| \to 0$$

as $t \to +\infty$ (and similarly for $t \to -\infty$). Here Δ_Y is the Laplace–Beltrami operator on $Y \in \mathfrak{X}$, while $\psi_{Y^\perp}$ is an eigenfunction (labeled by ψ!) of $H_{Y^\perp}$ (the "trace of H on the subspace $Y^\perp$", to be defined later) with an eigenvalue $\varepsilon_{Y^\perp}$ and the sum runs over all pairs $(Y, \psi_{Y^\perp})$, where $Y \in \mathfrak{X}$ and $\psi_{Y^\perp}$ is an eigenfunction of $H_{Y^\perp}$. The collection of such pairs can be called the *boundary, ∂_∞, of X at ∞*. Thus this boundary is determined by a certain type of geometry of X, but also by dynamical data of our system. Now,

$$H_{Y^\perp} = -\Delta_{Y^\perp} + V_{Y^\perp}$$

on $L^2(Y^\perp)$, where, as before, $\Delta_{Y^\perp}$ is the Laplace–Beltrami operator on $Y^\perp$ and $V_{Y^\perp}(x_{Y^\perp})$ is the limit of $V(x)$ as $|x_Y| \to \infty$ while $x_{Y^\perp}$ is kept fixed. Here x_Y is the projection of x onto Y, etc.

Conditions on the potential. In the kinematic language, one imposes restrictions on the pair potentials, say, that for $y \in \mathbb{R}^\nu$ in a neighborhood of infinity

$$V_{ij}(y) = O(|y|^{-\mu})$$

for some $\mu > 0$. The behavior of V_{ij} in a bounded part of $\mathbb{R}^\nu$ is not important as long as H is self-adjoint. In geometric language, one way to formulate conditions on $V(x)$ is to assume that for each $Y \in \mathfrak{X}$

$$\nabla_Y V(x) = O(|x|^{-\mu}) \quad \text{on any closed cone in } Y \setminus \Big(\bigcup_{Z \subsetneq Y} Z \Big)$$

again for some $\mu > 0$.

Result and brief history. The main result I would like to report is the following

THEOREM. *Let $\mu > \sqrt{3} - 1$. Then asymptotic completeness holds.*

Careful investigation of scattering in quantum systems began with the appearance of quantum mechanics. The mathematical methods used date back to works of Lord Rayleigh and A. Sommerfeld on the wave equation in the 19th and the beginning of the 20th century. Since then scattering theory counted among the most active areas among physicists and mathematicians with early important works by W. Heisenberg, J. Schwinger, H. Eckstein, S. Weinberg, W. Hunziker, T. Kato, S. Birman, Ya. Povzner, T. Ikebe, and K. Friedrichs, among others. A major breakthrough was made by L. D. Faddeev in 1963, who solved the three-body problem. While the one-body problem has undergone significant development, especially in works of S. Agmon and L. Hörmander, the events in the many-body case took a slow turn and concentrated, with the exception of two papers, [**15**] and [**21**], on the improvement of the three-body result.

The situation changed with the works of V. Enss [**7**], who introduced phase-space (or micro-local in the terminology of PDE's) methods in scattering theory and of E. Mourre [**22**], who introduced the method of local positive commutators.

Merging these two sets of ideas and developing the method which can be loosely called that of microlocally positive commutators, I. M. Sigal and A. Soffer [**27**]

proved asymptotic completeness for N-body *short-range* (i.e., with $\mu > 1$) systems. Finally, extending microlocal analysis to also include so-called asymptotic projections J. Derezinski [**5**] and I. M. Sigal and A. Soffer [**30**] proved asymptotic completeness for *long-range* systems with $\mu > \sqrt{3} - 1$ and $\mu \geqslant 1$, respectively (see also [**38**]). Powerful ideas of G. M. Graf [**14**] and, in the former case, of D. Yafaev [**36**], played important roles in these proofs.

The propagation set. Now I shall describe the key notion which guides the proof of asymptotic completeness and which, I believe, can also be useful outside the problem at hand. For the Schrödinger equation it plays a role similar to that of the wave front set for the wave equation.

Consider a class of symbols on the space-time $X \times \mathbb{R}$ (endowed into the usual symplectic form $dx \wedge dp - dE \wedge dt$) satisfying the estimates

$$|\partial^{\alpha}_{(x,t)} \partial^{\beta}_{(p,E)} \varphi| \leqslant C_{\alpha,\beta} \langle (x,t) \rangle^{-|\alpha|}$$

on any set which is compact in p and E. These are classical symbols in which the role of the coordinate and momentum are interchanged. Now for any $\psi \in L^{\infty}(\mathbb{R}, L^2(X))$, we define its propagation set by

$$PS(\psi) = \bigcap_{\|\phi\psi\|_{L^2(\langle t \rangle^{-1} dx\, dt)} < \infty} \operatorname{char} \phi .$$

Here the intersection is taken over pseudodifferential operators ϕ with symbols described above and satisfying the estimate indicated and $\operatorname{char} \phi$, the characteristic set of ϕ, is the null set of its symbol. The inequality under the intersection sign means that $\|\phi\psi\|^2_{L^2(dx)}$, the probability that ψ is localized in $\operatorname{supp} \varphi$, vanishes in some probabilistic sense as $|t| \to \infty$. In other words, as $|t| \to \infty$, the state ψ concentrates with rarer and rarer exceptions on $PS(\psi)$.

For each point at ∞, $\omega = (Y, \psi_{Y^{\perp}})$, we introduce the classical Hamiltonian function

$$h_{\omega}(z) = \varepsilon_{Y^{\perp}} + h_Y(z),$$

where $z \in T^*Y$, while $\varepsilon_{Y^{\perp}}$ is the eigenvalue of $H_{Y^{\perp}}$ corresponding to the eigenfunction $\psi_{Y^{\perp}}$ and h_Y, the "trace of h along Y", is given by

$$h_Y(z) = \frac{1}{2} |p_Y|^2 + V_Y(x_Y),$$

where p_Y and x_Y are projections of p on Y' and x on Y, respectively, and $V_Y(x) = V(x) - V_{Y^{\perp}}(x)$. Recall that the bicharacteristics of h_{ω} are just its classical trajectories, i.e., the solutions of the Hamiltonian equations

$$\dot{x} = \frac{\partial h_{\omega}}{\partial p}, \qquad \dot{p} = -\frac{\partial h_{\omega}}{\partial x}, \qquad h_{\omega} = E.$$

THEOREM. *Let $\mu > \sqrt{3} - 1$ and let ψ solve (1). Then $PS(\psi)$ consists of asymptotics to the bicharacteristics of h_{ω} for $\omega \in \partial_{\infty}$.*

Discussion. The Hamiltonians h_{ω}, $\omega \in \partial_{\infty}$, play the role of the principal symbol of H. These Hamiltonians, as well as their bicharacteristics, are partially quantized. A bicharacteristic for h_{ω} with $\omega = (Y, \psi_{Y^{\perp}})$ describes the free (classical) motion along Y and the bounded (quantum) motion in $Y^{\perp}$ (see Figure 1).

The motion along partially quantized bicharacteristics is unstable. At random moments of time the system jumps from one bicharacteristic to another. That is why the convergence of $\|\phi\psi\|_{L^2(dx)}$ to zero as $|t| \to \infty$, for the symbol, φ, of ϕ supported outside $PS(\psi)$, is in certain average. We expect that if φ is supported on those parts of the phase space outside of $PS(\psi)$ through which tunneling between different bicharacteristics is not ruled out by energy or similar considerations, then uniform decay of $\|\phi\psi\|_{L^2(dx)}$ as $|t| \to \infty$ does not occur, i.e., the estimate $\|\phi\psi\|_{L^2(\langle t\rangle^{-1}dx\,dt)} < \infty$ cannot be sharpened.

The theorem above was first proved in **[27]** for $\mu > 0$, but only for the part of the phase space away from the critical set

$$(2) \qquad \bigcup_{\omega\in\partial_\infty,\omega>Y} \{z \mid \nabla h_\omega^Y(z) = 0\},$$

where, for $\omega = (Z, \psi_{Z^\perp})$, $\omega > Y$ means $Z \supset Y$ and h_ω^Y stands for

$$\varepsilon_{Z^\perp} + \frac{1}{2}|p_Z^Y|^2 + I_Z(x_Z^Y)$$

with p_Z^Y and x_Z^Y denoting the projections of p and x on $Z' \ominus Y'$ and on $Z \ominus Y$, respectively. This result sufficed to prove asymptotic completeness for short-range forces, i.e., for $\mu > 1$. In the long-range case, i.e., for $\mu \leqslant 1$, one must control $PS(\psi)$ near the critical set (which corresponds to the case of multiple characteristics in the problem of propagation of singularities). Moreover, near the critical set not only the bicharacteristic motion becomes rather complicated (in particular, not ballistic), but the microlocal nature of the problem breaks down, the quantum phenomena begin playing the crucial role.

Asymptotic observables. To deal with these problems, one introduces fractional time scales and uses, in addition, asymptotic projection operators (or asymptotic observables) which allow one to localize on the space $T^*X \otimes L^2(X)$, rather than on T^*X or on $T^*(X \times \mathbb{R})$ alone.

I shall explain briefly what is going on here. To begin with, in order to stay in the same class of Hamiltonians in the process of induction in the number of particles, we consider more general Hamiltonians of the form $H(t) = H + W(x,t)$, where H is the N-particle Schrödinger operator described above and $W(x,t)$ is a real smooth function satisfying

$$|\partial_{(x,t)}^\alpha W(x,t)| \leqslant C_\alpha \langle t\rangle^{-\mu-|\alpha|},$$

μ being the same as in the conditions on $V(x)$, although this is not necessary. Let $U(t)$ be the evolution group (from time 0 to time t) generated by $H(t)$. For a family B of self-adjoint operators (= observables) B_t, we introduce the asymptotic cut-off functions

$$f^\pm(B) = \lim_{t\to\pm\infty} U(t) f(B_t) U(t)^*,$$

if the limits exist. In particular, we shall be interested in the asymptotic cut-off functions for the velocity operator $v = \{|x|/t\}$. Originally, such limits are defined for $f \in C_0^\infty$ and then extended by continuity to more general f's, which include in particular the characteristic functions, E_Δ, of intervals. In this way one can show that $E_\Delta^\pm(v)$ exists for any interval whose boundary does not contain zero. The point now is that the critical set of $H(t)$ (see (2)) can be characterized, adequately for our purposes, by the spectral projection $E^{\pm}_{\{0\}}(v)$. More precisely, one shows that if $\psi \in (\operatorname{Ran} E^{\pm}_{\{0\}}(v))^{\perp}$, then for any $\varepsilon > 0$ there is a ψ_ε such that $\|\psi - \psi_\varepsilon\| \leqslant \varepsilon$ and $PS(\psi_\varepsilon)$ is a closed subset of the complement of the set (2) in the set of asymptotics of bicharacteristics of h_ω for all ω. Moreover, if $\psi \in \operatorname{Ran} E^{\pm}_{\{0\}}(v)$, then $U(t)\psi$ is localized, in a rather strong sense, in the ball $\{|x| \leqslant t^{\alpha}\}$ with $\alpha > 2(2+\mu)^{-1}$. This information is sufficient to prove the asymptotic completeness of the decay rate, μ, of potentials greater than $\sqrt{3}-1$.

Asymptotic observables were used for the quantum mechanical scattering theory in [**8, 20**]. The characterization of the critical set in terms of asymptotic projections was introduced in [**29**] for the observable of energy, i.e., H (in this case the relevant asymptotic projection is associated with the threshold set of H) and implicitly for A/t, where A is the dilation generator, and in [**5**] for the velocity observable x/t. A review of the results up to June 1991 can be found in [**34**].

Recent results. In some parts of the phase-space the notion of propagation set can be considerably sharpened, as was shown in [**33, 10**]. This leads to a sharper notion of propagation set in the three-body case. Ch. Gérard and I. Łaba [**11, 12**] have proved asymptotic completeness for quantum many-particle systems placed in homogeneous magnetic fields whose potential decays at the rate $\mu > \sqrt{3}-1$. Asymptotic completeness for particles interacting via hard core potentials was proven by A. Iftimovici [**19**]. The latter result is based on a generalization of the Mourre theory to rough systems due to W. Amrein, A. M. Boutet de Monvel and G. Georgescu [**4**]. Some of the ideas of the many-body scattering theory were used and further developed in the existence problem for the nonlinear Schrödinger equation in [**35, 13**]. For other results and a more detailed discussion, see [**34**] and also [**26**].

References

1. S. Agmon, *Spectral properties of Schrödinger operators and scattering theory*, Ann. Scuola Norm. Sup. Pisa (4) **2** (1975), 151–218.
2. ———, *Some new results in spectral and scattering theory of differential operators on* $\mathbb{R}^n$, Sém. Goulaouic–Schwartz 1978–1979, Exp. II (1979), 1–11.
3. ———, *Lectures on exponential decay of solutions of second order elliptic equations*, Princeton Univ. Press, Princeton, NJ, 1982.
4. W. Amrein, A. M. Boutet de Monvel, and V. Georgescu, *Notes on the N-body problem, Part* I, UGVA-DPT, No. 11-598A (1988), 1–156.
5. J. Dereziński, *Asymptotic completeness for N-particle long-range quantum systems*, Ann. Math. **138** (1993), 427–476.
6. V. Enss, *A note on Hunziker's theorem*, Comm. Math. Phys. **52** (1977), 233.
7. ———, *Asymptotic completeness for quantum mechanical potential scattering*, Comm. Math. Phys. **61** (1978), 285–291.
8. ———, *Asymptotic observables on scattering states*, Comm. Math. Phys. **89** (1983), 245–268.
9. L. Faddeev, *Mathematical aspects of the three-body problem in quantum scattering theory*, Steklov Institute (1963).
10. C. Gérard, *Sharp propagation estimates for N-particle systems*, Duke Math. J. **67** (1992), no. 3, 483–515.
11. C. Gérard and I. Łaba, *Scattering theory for N-particle systems in constant magnetic fields*, Duke Math. J. **76** (1994), 433-465.
12. ———, *Scattering theory for N-particle systems in constant magnetic fields,* II. *Long-range interactions*, Comm. Partial Differential Equations **20** (1995), 1791–1830.

13. J. Ginibre, A. Soffer, and G. Velo, *The global Cauchy problem for the critical nonlinear wave equation*, J. Funct. Anal. **110** (1992), 96–130.
14. G. M. Graf 1990, *Asymptotic completeness for N-body short range quantum systems: A new proof*, Comm. Math. Phys. **132**, 73–101.
15. K. Hepp, *On the quantum mechanical N-body problem.*, Helv. Phys. Acta **42** (1969), 425–458.
16. L. Hörmander, *The existence of wave operators in scattering theory*, Math. Z. **146** (1976), 69–91.
17. ———, *The analysis of linear partial differential operators*, vol. II, IV, Springer-Verlag, Berlin–Heidelberg, 1985.
18. W. Hunziker and I. M. Sigal, *The general theory of N-body quantum systems*, Mathematical Quantum Theory II. Schrödinger Operators (J. Feldman et al, eds.), CRM Proc. and Lecture Notes, vol. 8, Amer. Math. Soc., Providence, RI, 1995.
19. A. Iftimovici, *Hard core scattering for N-body systems*, Preprint (1993), Bielefeld.
20. R. Lavine, *Commutators and scattering theory*, Comm. Math. Phys. **20** (1971), 301–323.
21. ———, *Completeness of the wave operators in the repulsive N-body problem*, J. Math. Phys. **14** (1973), 376–379.
22. E. Mourre, *Absence of singular continuous spectrum for certain self-adjoint operators*, Comm. Math. Phys. **78** (1981), 391–408.
23. D. Ruelle, *A remark on bound states in potential scattering theory*, Nuovo Cimento A **61** (1969), 655–662.
24. I. M. Sigal, *Geometric methods in the quantum many-body problem ...*, Comm. Math. Phys. **85** (1982), 309–324.
25. ———, *Mathematical questions of quantum many-body problem*, Seminare Equations aux Derivees Partielles 1986–1987, Ecole Polytechnique, Expose No. XXIII, 26 May 1987 (1987).
26. ———, *Quantum mechanics of many-particle systems*, Proceedings of the International Congress of Mathematicians, Kyoto, 1990, 1991.
27. I. M. Sigal and A. Soffer, *The N-particle scattering problem: Asymptotic completeness for short-range quantum systems*, Ann. Math. **125** (1987), 35–108.
28. ———, *Asymptotic completeness of three-body long-range scattering*, Preprint (1989), Princeton.
29. ———, *Asymptotic completeness for $N \leqslant 4$ particle systems with the Coulomb-type interactions*, Duke Math. J. **71** (1993), no. 1, 243–298.
30. ———, *Asymptotic completeness of N-particle long-range scattering*, J. Amer. Math. Soc. **7** (1994), 307–334.
31. B. Simon, *Quantum dynamics: From automorphism to Hamiltonian*, Studies in Mathematical Physics, Essays to honor Valentine Bargmann, Princeton Press, 1976.
32. ———, *Geometrical methods in multiparticle quantum systems*, Comm. Math. Phys. **55** (1977), 259–274.
33. E. Skibsted, *Propagation estimates for N-body Schrödinger operators*, Comm. Math. Phys. **142** (1991), 67–98.
34. A. Soffer, *On the many-body problem in quantum mechanics*, Astérisque **207** (1992), 109–152.
35. ———, *Phase-space analysis of nonlinear waves and global existence*, Ann. of Math. (1993) (to appear).
36. D. R. Yafaev, *Radiation conditions and scattering theory for N-particle Hamiltonians*, Comm. Math. Phys. **154** (1993), 523–554.
37. K. Yajima, *An abstract stationary approach to three-body scattering*, J. Fac. Sci. Univ. Tokyo Sect. IA Math. **25** (1978), 109–132.
38. L. Zelinski, *Une estimation de propagation avec applications en théorie de Schrödinger des systèmes quantiques*, C. R. Acad. Sci. Paris Sér. I Math. **315** (1992), 357–362; *A proof of asymptotic completeness for N-body Schrödinger operators*, Comm. Partial Differential Equations **19** (1994), 455–522.
39. G. M. Zhislin, *Discussion of the spectrum of the Schrödinger operator for systems of many particles*, Trudy Moscov. Mat. Obshch. **9** (1960), 81–128. (Russian)

Department of Mathematics University of Toronto, Toronto, CANADA M5S 1A1
E-mail address: sigal@math.toronto.edu

Amer. Math. Soc. Transl.
(2) Vol. **175**, 1996

Constructive Modules and the Reductivity Problem in the Category $\mathcal{O}$

D. P. Zhelobenko

ABSTRACT. The axiomatics for a new class of contragredient associative algebras and, in particular, for Cartan-type algebras are given. Special extensions of Cartan-type algebras are considered. The reductivity problem is considered in a certain category of "constructive" modules and in the standard category $\mathcal{O}$. In the latter case a partial solution of the reductivity problem is given in terms of projective modules and in terms of homomorphisms of the Verma modules.

In the paper a new approach is proposed for the investigation of general cohomology problems in the theory of representations of contragredient (associative) Cartan-type algebras. The notion of contragrediency was first introduced by V. G. Kac (1968) in connection with the theory of infinite-dimensional Lie algebras [**K1**]. We shall use this notion in a much wider sense with reference to associative algebras (or rings). A preliminary version of this definition is given in [**Z1**]. In this paper we adhere to a more general concept from [**Z7**].

The notion of contragrediency is associated with the following three conditions: the availability of a $\mathbb{Z}$-grading (with the possible replacement of the group $\mathbb{Z}$ by an additive ordered group), the existence of an involutive symmetry connected with the involution in the group $\mathbb{Z}$, and the presence of a more specific property, namely that of "local triangularity" (possibly strengthened, which makes sense for a wide class of meaningful examples).

The category of contragredient associative algebras (CAA) contains the somewhat more special subcategory of Cartan-type algebras (CTA). The latter include, in particular, the classical and quantum hulls of contragredient Lie algebras as well as certain related constructions (Weyl algebras, Yangians, various superanalogs, quantum analogs, etc.).

For example, compact connected Lie groups are "inherently contragredient", which manifests itself in the case of Lie algebras (their complex hulls, to be more precise). The category of contragredient Lie algebras is included in the CAA by means of the classical functor $U\colon \mathfrak{g} \to U(\mathfrak{g})$.

1991 *Mathematics Subject Classification.* Primary 81U10; Secondary 47A40.

A unified approach to the theory of representations of Cartan-type algebras is connected with the development of the "extremal projection" method, which originated in the representation theory of reductive Lie algebras [**Z2, Z3, Z4**]. The development of the general axiomatics of CAA is motivated by the desire to find the natural area of application of this method. The possibility of such an approach was noted in [**Z3**]. The approach was largely stimulated by recent studies in quantum groups (Hopf algebras), many of which are also contained in the category of CAA.

The development of the extremal projection method (see Theorem 2.6) is possible because natural extensions in the category of CTA (localization, formal series) exist. The use of extremal projections enables a rather easy solution of the reductivity problem in the special category of constructive A-modules (see Theorem 3.5).

Use of the extremal projection method in the standard category $\mathcal{O}$ meets with certain difficulties due to the singularities of the extremal projection (lying in the spectrum of an associative Cartan subalgebra). A partial solution of the reductivity problem is provided by Theorems 4.5 and 4.7, which connect the reductivity problem with the study of projective A-modules and the description of homomorphisms of the appropriate Verma modules.

As is known, the further development of this method is connected with the investigation of a number of specific problems of representation theory [**Z2–Z6**].

§1. Contragredient algebras

1.1. Let A be an associative algebra with unit over a field k which is $\mathbb{Z}$-graded by A_n ($n \in \mathbb{Z}$). The subspace $B \subset A$ is said to be *graded* if it coincides with the sum of its homogeneous components $B_n = B \cap A_n$.

DEFINITION. An algebra A is said to be *locally triangular* if it is generated by a graded subspace T whose components satisfy the following conditions:

(T_0) Let H be the subalgebra with unit generated by the component T_0, then we have $HT_i = T_iH$ for all $i \in \mathbb{Z}$.

(T_1) Let A_+ (A_-) be the subalgebra with unit generated by the component T_i for $i > 0$ ($i < 0$, respectively). Then we have

$$A = A_- H A_+, \tag{1}$$

where T is called the *base subspace* (or the *base triple*) of the algebra A. The decomposition (1) is called the *triangular decomposition* of the algebra A. The algebra A is said to be *locally finite* if $\dim T < \infty$.

From now on we write the base subspace T in the form $T = F \oplus T_0 \oplus E$, where E (F) is the linear hull of the components T_i for $i > 0$ ($i < 0$, respectively).

PROPOSITION. *Let the components T_i satisfy condition* (T_0) and the condition (T_{ij}) $T_iT_j \equiv T_jT_i \mod T_{i+j}$ for $i, j < 0$.
Then the algebra A is locally triangular.

PROOF. Set $E_i = T_i$, $F_i = T_{-i}$ for $i \geqslant 0$ (so that $E_0 = F_0 = T_0$). We have

$$E_iF_j \subset F_jE_i + E_{i-j} \quad \text{for } i > j,$$
$$E_iF_j \subset F_jE_i + F_{j-i} \quad \text{for } i < j.$$

Using these relations inductively, for the subalgebra $X = A_-H$ we find:

$$E_i X \subset \sum_{k=0}^{i} X E_k .$$

Hence each monomial in the components T_i ($i \in \mathbb{Z}$) is contained in A_-HA_+. □

EXAMPLE. For any $\mathbb{Z}$-graded Lie algebra $\mathfrak{g}$, its universal enveloping algebra $U(\mathfrak{g})$ is a locally triangular algebra with base subspace $\mathfrak{g}$.

REMARK. The definition of local triangularity admits natural generalizations. First, we can drop the condition that the algebra A be graded (preserving the grading of the subspace T). Second, instead of the group $\mathbb{Z}$, we can consider other additive ordered groups (see, for example, 2.1). Finally, the definition of local triangularity can be extended in a natural way to other algebraic structures (rings, groups, etc.).

The triangular decomposition (1) can also be written in the form

$$A = B_-A_+ = A_-B_+,$$

where $B_\pm = HA_\pm$ is a subalgebra of the algebra A. Hence

$$A = B_- + AE = FA + B_+ = H + N,$$

where $N = FA + AE$.

1.2. DEFINITION. An algebra A is said to be *weakly* (*strongly*) *triangular* if the following condition (Wt) ((St), respectively) is fulfilled:
(St) $A \approx A_- \otimes H \otimes A_+$.
(Wt) $A \approx (A_- \otimes H) \oplus AE$.
Here the symbol $\approx$ denotes the natural isomorphism defined for each pair of subspaces $X, Y \subset A$ according to the rule $x \otimes y \mapsto xy$ for $x \in X$, $y \in Y$.

EXAMPLE. It follows from the Poincaré–Birkhoff–Witt theorem that the algebra $U(\mathfrak{g})$ defined in 1.1 is strongly triangular.

Let us regard the algebra A as an A-bimodule (with respect to the multiplication defined in A), and observe that AE is an $(A \times H)$-bimodule. Therefore, the quotient space

$$M = A/AE$$

is also equipped with an $(A \times H)$-bimodule structure. Identifying the unit element $1 \in A$ with its image in M, we find by condition (Wt) that

$$M = B_- \cdot 1 \approx B_- \approx A_- \otimes H \tag{2}$$

with respect to the map $x \mapsto x \cdot 1$ ($x \in B_-$). Clearly, the space M inherits the $\mathbb{Z}$-grading $M_n = A_n \cdot 1 \approx (B_-)_n$, so that $M_n = 0$ for $n > 0$.

The left A-module M is called (by analogy to the case of a Lie algebra) the *universal Verma module* over the algebra A. According to (2), the module M is a free B-module with one generator 1.

1.3. Definition. A weakly triangular algebra A is said to be *contragredient* if it possesses an involution $a \mapsto a'$ such that $h' = h$ for all $h \in T_0$ and $T_i' = T_{-i}$ for all $i \in \mathbb{Z}$.

Here the term "involution" denotes an involutive antiautomorphism of the algebra A. Accordingly, $A_n' = A_{-n}$ for all $n \in \mathbb{Z}$.

By condition (Wt), we have $A = H \oplus N$, where $N = FA + AE$. Let π denote the projection of A on H with respect to this decomposition. For each pair of elements $x, y \in A$, we set

$$\varphi(x, y) = \pi(x'y).$$

Obviously, $\varphi\colon A \times A \to H$ is an H-bilinear form with respect to the right action of the algebra H in the space A. It follows from the relation $\pi(x') = \pi(x)$ for all $x \in A$ that the form φ is symmetric, i.e., $\varphi(x, y) = \varphi(y, x)$ for all $x, y \in A$.

It is also clear that AE is contained in the kernel $\ker\varphi$ of the form φ. Accordingly, we can regard φ as an H-bilinear symmetric form on the space M. The form φ on the algebra A (on the space M) satisfies the following contravarience relation:

$$\varphi(ax, y) = \varphi(x, a'y) \tag{3}$$

for all $a \in A$ and all $x, y \in A$ (hence $x, y \in M$). From the inclusion $A_m' A_n \subset AE$, for $n > m$, it follows that the weighted grading A_n (hence M_n) is orthogonal with respect to the form φ, i.e., $A_n \perp A_m$ for $n \neq m$.

The form φ is called the *canonical bilinear form* on the algebra A (on the space M).

Remark. The form φ on the space M is uniquely determined by the contravarience condition (3) and by the normalization condition $\varphi(1, 1) = 1$.

1.4. An algebra A is said to be *nonsingular* if $\ker\varphi = AE$, i.e., the form φ is nonsingular on the space M.

Proposition. *Let the algebra A be nonsingular. Then we have:*

(i) *the algebra A is strongly triangular*;

(ii) *the A-module M is exact (i.e., the equality $aM = 0$ for $a \in A$ is possible only if $a = 0$).*

Proof. (i) It is sufficient to verify that for each homogeneous basis $a_i \in A_+$ and for each collection of elements $h_{ij} \in H$ the relation

$$\sum_{i,j} a_i' h_{ij} a_j = 0 \tag{4}$$

implies $h_{ij} = 0$ for all i, j. Setting $n_i = \deg a_i$, we may suppose that $h_{ij} = 0$ for $n_i < n$, $n_j < m$ (where $n, m \geqslant 0$). If we multiply relation (4) from the right by a_k' for $n_k = m$, we find that $a_j a_k' \equiv c_{jk} \mod AE$, where $c_{jk} = 0$ for $n_j > m$, $c_{jk} = \pi(a_j, a_k')$ being a nondegenerate matrix for $n_j = n_k = m$. Therefore $h_{ij} = 0$ for $n_j = m$. In a similar way, $h_{ij} = 0$ for $n_i = n$. Thus the assertion is proved.

(ii) Applying the same arguments to the operator $a \in \operatorname{End} M$ written in the form (4), we obtain $h_{ij} = 0$ for all i, j, i.e., $a = 0$ in the algebra A. □

1.5. Suppose the algebra A is strongly triangular, i.e., each element $x \in A$ may be represented uniquely in the form

$$x = \sum_{i,j} a_i x_{ij} b_j, \tag{5}$$

where a_i (b_j) is the homogeneous basis of the vector space A_- (A_+, respectively).

DEFINITION. Let A^σ be a set of all formal series (5) satisfying the following conditions:

(F_1) Summation in (5) is subject to the constraint $|n_i + m_j| \leqslant$ constant (with the constant depending on x), where $n_i = \deg a_i$, $m_j = \deg b_j$.

(F_2) For each $m \in \mathbb{Z}_+$, the series (5) contains only a finite number of elements of fixed degree $m_j = m$.

Obviously, A^σ is a vector space graded by the subspaces A^σ_p $(p \in \mathbb{Z})$, where A^σ_p contains the elements (5) only for $n_i + m_j = p$. It is also clear that A^σ does not depend on the choice of bases a_i, b_j. If the algebra A is locally finite, then condition (F_2) is automatically fulfilled.

PROPOSITION. *The space A^σ is an algebra with respect to multiplication of formal series. In addition, the algebra A^σ is graded by the subspaces A^σ_p $(p \in \mathbb{Z})$.*

PROOF. Let $x, y \in A^\sigma$ be homogenous. Writing these elements as formal series (5) and calculating the product xy, we must consider the corresponding partial products of the form

$$a_i x_{ij} b_j a_s y_{st} b_t, \tag{6}$$

where $n_i + m_j = \deg x$, $n_s + m_t = \deg y$. Let us fix $p, q \in \mathbb{Z}_+$, and let N_{pq} be the set of all terms of (6) giving nonzero contributions to the component $(A_-)_{-p} H (A_+)_q$. It is clear this condition implies $|n_i| \leqslant p$, $m_t \leqslant q$. Hence, we conclude that $\operatorname{card} N_{pq} < \infty$, i.e., the product xy is well defined. □

REMARK. Conditions (F_1) and (F_2) are essential. For example, let A be the Weyl algebra with generators e, f and fundamental relation $[e, f] = 1$. Then the product of the formal series

$$x = \sum_{n=0}^{\infty} e^n, \qquad y = \sum_{n=0}^{\infty} f^n \tag{7}$$

is not defined. Indeed, $e^n f^n \equiv n! \mod Ae$, i.e., the multiplication of the elements (7) makes an infinite contribution to the zero component of the product xy.

1.6. The algebra A^σ is said to be the *σ-extension* (or the *σ-completion*) of the algebra A. If the algebra A is contragredient, then A^σ inherits the contragredient structure of the algebra A.

REMARK. All the constructions of this section are well known for the case $A = U(\mathfrak{g})$. For example, the canonical form φ in this case is called the Shapovalov form [**K2**]. The extension $U(\mathfrak{g})^\sigma$ was constructed in 1983 (see [**Z3**]) in connection with the development of projective methods for the algebra $U(\mathfrak{g})$.

§2. Cartan-type algebras

2.1. Let A be a locally triangular algebra with base subspace $T = F \oplus T_0 \oplus E$. Suppose $\operatorname{Aut} H$ is the group of automorphisms of the algebra H. For each $\alpha \in \operatorname{Aut} H$, let

$$A_\alpha = \{a \in A \mid ha = ah_\alpha \text{ for all } h \in H\}, \tag{1}$$

where $h \mapsto h_\alpha$ denotes the right action of the automorphism $\alpha \in \operatorname{Aut} H$ on an element $h \in H$. It follows from this definition that $A_\alpha A_\beta \subset A_{\alpha\beta}$ for all $\alpha, \beta \in \operatorname{Aut} H$.

DEFINITION. An algebra A is said to be *Q-equipped* (or a *Cartan-type algebra*) if the following conditions are fulfilled:

(CT_0) The subalgebra H is commutative and has no divisors of zero.

(CT_1) The algebra A is a direct sum of the subspaces (1) for $\alpha \in Q$, where Q is a commutative subgroup in $\operatorname{Aut} H$.

(CT_2) Each homogeneous component T_n ($n \in \mathbb{Z}$) inherits the Q-grading $T_{n\alpha} = T_n \cap A_\alpha$.

In this case the subalgebra H is called the *Cartan subalgebra* of the algebra A.

It is convenient to use an additive notation for multiplication in the group Q (with neutral element 0). According to (CT_1), the family A_α ($\alpha \in Q$) constitutes the Q-grading of the algebra A. We hope that the notations A_n ($n \in \mathbb{Z}$) and A_α ($\alpha \in Q$) can be distinguished from the context.

The subalgebra $A_\pm$ inherits the Q-grading from the algebra A for $\alpha \in Q_\pm$, where Q_+ (Q_-) is the semigroup generated by the elements $\alpha \in Q$ for which $T_{n\alpha} \neq 0$ for some $n > 0$ ($n < 0$, respectively). The algebra A is said to be *strongly equipped* if the following condition is fulfilled:

(SQ) The zero component of the Q-grading in $A_\pm$ coincides with $k \cdot 1$.

In particular, let A be contragredient, and let $Q_- \cap Q_+ = \{0\}$. In this case we shall regard Q as an ordered group by the rule $\alpha \geqslant \beta$ for $\alpha - \beta \in Q_+$.

PROPOSITION. *Let A be a contragredient Cartan-type algebra. Then the grading A_α ($\alpha \in Q$) is orthogonal with respect to the canonical form φ.*

PROOF. Observe that $A'_\alpha = A_{-\alpha}$, in which case $A'_\beta A_\alpha \subset A_{\alpha-\beta}$ for all $\alpha, \beta \in Q$. Therefore,

$$h\varphi(x, y) = \varphi(x, y)\, h_{\alpha-\beta} \tag{2}$$

for all $h \in H$ and all $x \in A_\beta$, $y \in A_\alpha$. If $\alpha \neq \beta$, then there is an $h \in H$ such that $h \neq h_{\alpha-\beta}$. Using (2) and (CT_0), we obtain $\varphi(x, y) = 0$, i.e., $A_\alpha \perp A_\beta$ for $\alpha \neq \beta$. □

2.2. Let A be a Cartan-type algebra, and let K be a commutative extension of the algebra H such that $Q \subset \operatorname{Aut} K$ (i.e., the action Q in the algebra H is extended to the action Q in the algebra K). We regard the space

$$A(K) = A \otimes_H K \tag{3}$$

as an algebra over the field k with respect to the multiplication given by $ax = a \otimes x$ for $a \in A$, $x \in K$ and the commutation law $xa = ax_\alpha$ for $a \in A_\alpha$. Thus, $A(K)$ is an extension of the algebra A. Obviously, $A(K)$ is a Cartan-type algebra with Cartan subalgebra K and triangular decomposition $A(K) = A_- K A_+$. Moreover, the universal Verma module over the algebra $A(K)$ is naturally identified with the space

$$M(K) = M \otimes_H K. \tag{4}$$

In particular, let the algebra A be contragredient. The relation $x' = x$ for $x \in K$ allows us to extend the involution $a \mapsto a'$ to an involution of the algebra $A(K)$. Consequently, $A(K)$ becomes a contragredient Cartan-type algebra. We also set $A^\sigma(K) = A(K)^\sigma$.

The algebra $A^\sigma(K)$ for $K = \operatorname{Fract} H$ (the field of quotients of the algebra H) is called the *canonical extension of the algebra* A (see [**Z2–Z4**]) for the particular case $A = U(\mathfrak{g})$.

2.3. We shall consider the action of the algebra A ($A(K)$) in the space M ($M(K)$, respectively) defined by the operators $l(a)x = ax$ for $a \in A$, $x \in M$.

Let $\operatorname{End} M$ be the algebra of endomorphisms of the right H-module of M. Set

$$L_p = \{a \in \operatorname{End} M \mid aM_n \subset M_{n+p} \text{ for all } n \in \mathbb{Z}\}, \tag{5}$$

where $p \in \mathbb{Z}$. Set

$$L = \bigoplus_p L_p, \tag{6}$$

so that L is a $\mathbb{Z}$-graded algebra (subalgebra in $\operatorname{End} M$).

It is readily seen that $l\colon A \to L$ is a homomorphism of $\mathbb{Z}$-graded algebras, in which case $l(A_p) \subset L_p$ for all $p \in \mathbb{Z}$. Observe that $E^n M_k = 0$ for $n + k = 0$, i.e., the action of the subalgebra A_+ in the space M is locally nilpotent. It is thus seen that each formal series $x \in A^\sigma$ can be regarded as an operator in the space M. In other words, the map $l\colon A \to L$ can be extended to a homomorphism of graded algebras $l\colon A^\sigma \to L$ so that $l(A^\sigma_p) \subset L_p$ for all $p \in \mathbb{Z}$.

PROPOSITION. *Suppose the algebra A is nonsingular. Then the map $l\colon A^\sigma \to L$ is injective. In other words, M is an exact A^σ-module.*

PROOF. It is sufficient to repeat literally the arguments of 1.4 for the elements $a \in A^\sigma$. □

COROLLARY. *Let $L(K)$ be defined similarly to the algebra L with M replaced by $M(K)$. Then the map $l\colon A^\sigma(K) \to L(K)$ is injective.*

2.4. THEOREM. *Let A be a nonsingular locally finite Cartan-type algebra, and let* $\operatorname{Fract} H \subset K$. *Then the map $l\colon A^\sigma(K) \to L(K)$ is an isomorphism.*

PROOF. Suppose $z \in L(K)_p$. Let us prove the existence of an element $x \in A^\sigma(K)_p$ such that $l(x) = z$. The required element will be written as a series $x = x_0 + x_1 + \cdots + x_n + \cdots$, where x_n is of the form (5) from 1.5 with coefficients $x_{ij} \in K$, $x_{ij} = 0$ for $n_j \neq n$. It follows that

$$l(x_n)a_k = \sum_i a_i x_{ij} c_{jk} \tag{7}$$

in the notation of (1.5), where $n_k \leqslant n$. In particular, $l(x_n)M_i = 0$ for $|i| < n$.

Suppose the elements x_i for $i < n$ have already been constructed. Subtracting the elements $l(x_i)$, we can assume that $zM_i = 0$ for $|i| < n$. We also have

$$za_k = \sum_i a_i z_{ik} \tag{8}$$

for some $z_{ik} \in K$. Using the fact that the finite matrix c_{jk} is nondegenerate (see 1.5), we can select elements x_{ij} so that relation (7) coincides with (8) on M_i for $|i| \leqslant n$.

Thus, the map $l\colon A^\sigma(K) \to L(K)$ is surjective. According to 2.3, it is injective. Consequently, $A^\sigma(K) \approx L(K)$. □

2.5. For each A-module V the subspace $V^E = \ker E$ is said to be *extremal* (with respect to E).

PROPOSITION. *Let the algebra A be nonsingular. Then the extremal subspace M E coincides with the zero component of the $\mathbb{Z}$-grading $M_0 = H \cdot 1$.*

PROOF. From the relation $EH = HE$ it follows that $EH \cdot 1 = 0$ in the space M, i.e., $M_0 \subset M^E$. On the other hand, the contravarience condition (3) from 1.3 implies that $M^E \perp \operatorname{im} F$, i.e., $M^E \subset M_0$ because the form φ is nondegenerate in the space M. Thus, $M^E = M_0$. □

2.6. THEOREM. *Under the conditions of Theorem 2.4 we have*

(i) *There exists a unique element $p \in A^\sigma(K)$ satisfying the system of equations*

$$Ep = pF = 0 \tag{9}$$

and the normalization condition $\pi(p) = 1$. Moreover,

$$\mathbb{Z} - \deg p = Q - \deg p = 0, \quad p^2 = p = p'.$$

(ii) *The operator p projects each $A^\sigma(K)$ =module V on its extremal subspace V^E parallel to $\operatorname{im} F$. In particular, we have*

$$V = \ker E \oplus \operatorname{im} F. \tag{10}$$

PROOF. (i) The required element $l(p) \in \operatorname{End} M(K)$ is uniquely defined as the projection onto $M(K)^E$ parallel to $\operatorname{im} F$. It follows that $\mathbb{Z} - \deg p = 0$ and also $hp = ph$ for all $h \in H$, i.e., $Q - \deg p = 0$. The element p' satisfies equations (9) and also the normalization condition $\pi(p') = 1$; therefore, $p' = p$.

(ii) Since $p^2 = p$, we have $V = \operatorname{im} p \oplus \ker p$; in this case equations (9) imply that $\operatorname{im} p \subset \ker E$ and $\operatorname{im} F \subset \ker p$. On the other hand, it follows from the equality $px = x$ for $x \in \ker E$ that $\ker E \subset \operatorname{im} p$. It is also clear that $\ker p = \operatorname{im}(1 - p) \subset \operatorname{im} F$. □

DEFINITION. The element $p \in A^\sigma(K)$ defined in the theorem is called the *extremal projection* of the algebra A.

REMARK. If we stipulate that $\mathbb{Z} - \deg p = 0$ or $Q - \deg p = 0$, then each group of equations $Ep = 0$ or $pF = 0$ together with the normalization condition $\pi(p) = 1$ suffices to define the element p. Indeed, in this case the decomposition $M(K) = \ker E \oplus \operatorname{im} F$ is invariant with respect to the operator $l(p)$, and the condition $Ep = 0$ is equivalent to $pF = 0$.

In what follows, we identify the elements x and $l(x)$, i.e., we preserve the module notation for the action A^σ in M.

2.7. The algebra A is said to be *regular* [**Z1**] if its Cartan subalgebra H satisfies the following condition:

(Re) The set $X(H)$ of all characters of the algebra H is total in H, i.e., if we have $\lambda(h) = 0$ for all $\lambda \in X(H)$, then $h = 0$.

Condition (Re) can be reformulated as follows. The relation

$$h(\lambda) = \lambda(h) \quad \text{for } h \in H,\ \lambda \in X(H) \tag{11}$$

defines an isomorphism of the algebra H and a certain algebra of functions on a complete subset $P \subset X(H)$.

EXAMPLES. The following regular Cartan-type algebras were examined in [**Z1**]:

- *Weyl algebras* (or algebras of type W);
- *Chevalley algebras* $A = U(\mathfrak{g})$ (or algebras of type CH), where $\mathfrak{g} = \mathfrak{g}(a)$ is a contragredient Lie algebra of the Kac type [**K1**], including the symmetrized Chevalley algebras (or algebras of type SCH), where a is a symmetrized matrix in the sense of [**K1**];
- *Drinfeld–Jimbo algebras* $A = U_q(\mathfrak{g})$ (or algebras of type DJ), where $\mathfrak{g} = \mathfrak{g}(a)$ is a symmetrized Kac–Moody algebra ($U_q(\mathfrak{g})$ is the quantum hull of the algebra $\mathfrak{g}$);
- *Kashiwara algebras* $A = D_q(\mathfrak{g})$ (or algebras of type K) associated with DJ-type algebras (also see [**Z2, Z5**]).

Note that $H = k$ for the types W and K; $H = \mathbb{C}[x_1, \dots, x_n]$ for the type CH. As is known, algebras of type W, K, SCH, and DJ are nonsingular [**Z2**].

§3. Constructive modules

3.1. From now on we assume that A is a nonsingular locally finite Cartan-type algebra.

DEFINITION. A vector space V over a field k is called a *constructive A-module*, if V carries the structure of $(A \times H)$-bimodule satisfying the following conditions:

($\mathcal{C}_1$) The module V is locally nilpotent with respect to the action of E, i.e., $E^n v = 0$ for $n \geqslant n_0(v)$ for each $v \in V$.

($\mathcal{C}_2$) The module V is Q-graded with homogeneous components of the following type:

$$V_\alpha = \{v \in V \mid hv = vh_\alpha \text{ for all } h \in H\}, \tag{1}$$

where $\alpha \in Q$.

Here the term "$(A \times H)$-bimodule" denotes a "left A-module and right H-module" with commuting operations of the algebras A and H, i.e., $(av)h = a(vh)$ for all $a \in A$, $v \in V$, and $h \in H$.

Obviously, it follows from condition $\mathcal{C}_1$ that the A_+-module V is locally finite, i.e., $\dim Av < \infty$ for all $v \in V$. The converse also holds for the class of $\mathbb{Z}$-graded A-modules (i.e., $A_n V_m \subset V_{n+m}$ for all $n, m \in \mathbb{Z}$).

The constructive module V is said to be *dominant* if each component V_α ($\alpha \in Q$) is free as a right H-module. We shall use the notation $\mathcal{C}$ ($\mathcal{D}$) for the category of all constructive (respectively dominant) A-modules.

Note that the category $\mathcal{C}$ is closed under the passage to submodules and to quotient modules (in the category of $(A \times H)$-bimodules).

If $V \neq 0$ is a module from the category $\mathcal{C}$, then $V^E \neq 0$. In fact, for $0 \neq v \in V$ we have $0 \neq E^n v \subset V^E$ (for sufficiently large values of n).

3.2. Associate with each constructive module V the following family of left A-modules:

$$V * N = V \otimes_H N, \tag{2}$$

where N is an arbitrary left H-module. In particular, we shall consider the following special cases:

$$V(K) = V * K, \qquad S_\lambda V = V * k_\lambda, \tag{3}$$

where K is the commutative extension of the algebra H defined in 2.2, and k_λ is a one-dimensional H-module specified by the character $\lambda \in X(H)$.

The module $V(K)$ for $K = \operatorname{Fract} H$ is called the *rational hull of the module* V. The module $S_\lambda V$ is called the *specialization of the module* V *at the point* $\lambda \in X(H)$.

Using the rule $xv = vx_\alpha$ for $x \in K$ and $v \in V_\alpha$, we equip $V(K)$ with the structure of a left K-module. It is readily seen that this structure is consistent with the left action of the algebra A, i.e., their joint action defines the structure of a left $A(K)$-module in the space $V(K)$.

Moreover, condition ($\mathcal{C}_1$) allows us to extend this action to the formal series $a \in A^\sigma(K)$, i.e., it allows us to regard $V(K)$ as a left $A^\sigma(K)$-module.

3.3. EXAMPLES. According to 1.2, the universal Verma module M over the algebra A is constructive (with respect to the Q-grading determined by the Q-grading of the algebra A).

In the case under consideration, $M(K)$ coincides with the universal Verma module M over the algebra A (see 2.2). The module $M(\lambda) = S_\lambda M$ is called the *Verma module* over the algebra A with highest weight $\lambda \in X(H)$. Note that $M(\lambda)$ admits the following definition:

$$M(\lambda) = A/I(\lambda) \tag{4}$$

(up to a standard isomorphism), where $I(\lambda)$ is the left ideal of the algebra A generated by the elements $e \in E$ and $h - \lambda(h) \cdot 1$, where $h \in H$. According to 1.2, $M(\lambda)$ is a cyclic A-module generated by the vector 1_λ, where 1_λ is the image of the unit element $1 \in A$ in the space (4). Moreover, $M(\lambda)$ is a free A_--module with the unique generator 1_λ.

Note also that the module M is dominant with $hv = vh_\mu$, where $h \in H$ and $v \in M_\mu$ (here M_μ denotes the Q-grading of the module M). If we replace this equality by the relation $hv = vh_{\alpha+\mu}$ for some fixed $\alpha \in Q$, then in the space M we obtain a new dominant module structure, which will be denoted by the symbol $D(\alpha)$.

In other words, let H_α be the space H, in which the algebra H acts on the left via the multiplication defined in H and acts on the right according to the rule $hv = vh_\alpha$ for all $h, v \in H$. Then we have $D(\alpha) = M * H_\alpha$ for all $\alpha \in Q$. Clearly, we also have

$$S_\lambda D(\alpha) = M(\alpha\lambda) \tag{5}$$

for all $\alpha \in Q$ and all $\lambda \in X(H)$.

Let us agree to use the notation $1_\alpha = 1 \otimes 1$ for the cyclic vector in $D(\alpha)$. According to Proposition 2.5, we have $D(\alpha)^E = H \cdot 1_\alpha$ for all $\alpha \in Q$.

3.4. Let V be a module from the category $\mathcal{C}$. A nonzero vector $v \in V$ is said to be *extremal* with weight $\alpha \in Q$ if it is contained in the subspace $V_\alpha^E = V_\alpha \cap V^E$.

The module V is said to be *extremal of type* $\alpha \in Q$ if it is generated by an extremal vector V_0 with weight α, i.e., $V = Av_0$, where $0 \neq v_0 \in V_\lambda^E$.

PROPOSITION. (i) *The module $D(\alpha)$ for $\alpha \in Q$ is universal in the class of extremal modules V of type α, i.e., for each such module V we have*

$$V \approx D(\alpha)/N, \tag{6}$$

where N is an $(A \times H)$-submodule of the module $D(\alpha)$ that does not contain 1_α.

(ii) *Each dominant extremal module of type α is isomorphic to the module $D(\alpha)$.*

(iii) *If H is a field, then the A-module $D(\alpha)$ is simple for each $\alpha \in Q$.*

PROOF. (i) The map $\varphi\colon x \cdot 1_\alpha \mapsto xv_0$, where $x \in A\_H$, is well defined (in view of the fact that A_H-module $D(\alpha)$ is free) and defines the isomorphism (6) for $N = \ker\varphi$.

(ii) According to 3.3, the extremal subspace N^E consists of elements $h \cdot 1_\alpha$, where $h \in H$. In this case $hv_0 = v_0 h_\alpha = 0$ in the space (6); hence $h = 0$, since the left (right) H-module V_α is free. Therefore, $N = 0$, i.e., $V \approx D(\alpha)$.

(iii) In this case the inclusion $h \cdot 1_\alpha \in N$ for $h \neq 0$ implies $1_\alpha \in N$, i.e., $N = V$ for each submodule $N \neq 0$. □

3.5. THEOREM. *Suppose H is a field. Then the category $\mathcal{C} = \mathcal{D}$ is semisimple, i.e., each module V from the category $\mathcal{C}$ is the direct sum of simple modules $D(\alpha)$ with $\alpha \in Q$.*

PROOF. According to Proposition 3.4, the submodule $N = AV^E$ is a direct sum of simple modules $D(\alpha)$, where $\alpha \in Q$. If $N \neq V$, then there exists a vector v_0 such that $v_0 \notin N$, $ev_0 \in N$ for all $e \in E$ (in which case v_0 defines an extremal vector in V/N). Applying the extremal projection $p \in A^\sigma$ to this vector, we obtain the extremal vector $pv_0 \equiv v_0 \bmod N$; hence $pv_0 \notin N$, which contradicts the definition of N. Thus, $V = N$, i.e., the module V is semisimple. □

COROLLARY. *In the general case, the theorem still holds if we replace the algebra A by its rational hull $A(K)$, where $K = \operatorname{Fract} H$.*

3.6. Let us consider the following family of induced A-modules:

$$\operatorname{Ind} N = A \otimes_{A_+} N, \tag{7}$$

where N is an arbitrary (left) A_+-module.

Note that $\operatorname{Ind} N \approx A\_H \otimes N$ (isomorphism of vector spaces). In particular, $D(\alpha) = \operatorname{Ind} H_\alpha$ in the notation of 3.3.

Suppose $\mathcal{F}$ is the category of $(B \times H)$-modules N, where $B = HA^+$, satisfying the following conditions:

($\mathcal{F}_1$) The module N is locally nilpotent with respect to E (see 3.1).

($\mathcal{F}_2$) The module N has the Q-grading n_α ($\alpha \in Q$) defined similarly to 3.1 (replacing V by N).

It is readily seen that the map Ind defines a covariant functor from the category $\mathcal{F}$ to the category $\mathcal{C}$ with respect to the natural extension of the Q-grading from the subspace $N \approx 1 \otimes N$ to the subspace $V = \operatorname{Ind} N$.

PROPOSITION. *Suppose H is a field. Then for each module N from the category $\mathcal{F}$ we have*

$$\operatorname{Ind} N = \bigoplus_\alpha n_\alpha D(\alpha), \tag{8}$$

where $n_\alpha = \dim_H N_\alpha$ and the isotypical component $\operatorname{Ind}_\alpha N = n_\alpha D(\alpha)$ is of the form

$$\operatorname{Ind}_\alpha N = Ap(1 \otimes N_\alpha). \tag{9}$$

Here $p \in A^\sigma$ is an extremal projection of the algebra A.

PROOF. It is easily verified that $(\operatorname{Ind} N)^E = p(1 \otimes N)$; hence we have (8) and (9) by virtue of Theorem 3.5. □

COROLLARY. *In the general case the decomposition* (8) *is valid for the rational hull* $\operatorname{Ind} N(K)$ *over the algebra* $A(K)$, *where* $K = \operatorname{Fract} H$.

3.7. The induced modules (7) enable us to construct an analog of partial sums of the formal series $a \in A^\sigma$.

Indeed, each element $a \in A^\sigma$ is uniquely determined by its action on the weight elements $v_0 \in M_\alpha$, where $\alpha \in Q$. According to Proposition 3.4, this action lifts to an action on the cyclic vector v_0 in the induced module $\operatorname{Ind} N$, where $N = HA_+v_0$. Thus, the operator $a_N = a$ in $\operatorname{Ind} N$ (or even the vector $a_n v_0$) may be interpreted as a partial sum of the formal series $a \in A^\sigma$ (the series a_N is finite, because the A_+-module M is locally nilpotent).

Suppose the algebra A is strictly equipped (see 2.1), and also let α_i ($i = 0, 1, \dots, n$) be a nondecreasing sequence of weights of the module $N = HA_+v_0$, where $\alpha_0 = \alpha$ is the weight of the vector v_0. Thus, $\alpha_i < \alpha_j$ implies $i < j$ and $\alpha_0 \neq \alpha_i$ for $i \neq 0$ (in view of the fact that the algebra A is strictly equipped).

Suppose N_i is the direct sum of the weight components N_α for $\alpha \geqslant \alpha_i$, so that $N = N_0 \supset N_1 \supset \cdots \supset N_n = 0$ is the filtration of the module N by the submodules N_i ($i = 0, 1, \dots, n$). Consequently, the module $\operatorname{Ind} N$ possesses a decreasing filtration

$$\operatorname{Ind} N = \operatorname{Ind} N_0 \supset \operatorname{Ind} N_1 \supset \cdots \supset \operatorname{Ind} N_n = 0,$$

where

$$\operatorname{Ind} N_i / \operatorname{Ind} N_{i+1} \approx n_i D(\alpha_i). \tag{10}$$

Here $n_i = n_\alpha$ for $\alpha = \alpha_i$. It also follows that the module $V = \operatorname{Ind} N$ possesses an increasing family of quotient modules $V_i = V/\operatorname{Ind} N_{i+1}$, so that $V_0 = D(\alpha_0)$ and $V_n = V$. In this case $V_i \approx \operatorname{Ind}(N/N_{i+1})$.

Thus, with each partial sum a_N we can associate a sequence of (subordinate) partial sums $a_i = a_{V_i}$, where $i = 0, 1, \dots, n$.

3.8. The definition given in 3.7 yields a natural (cohomological) interpretation of the extremal projection $p \in A^\sigma(K)$ in terms of the partial sums defined in 3.7.

For simplicity, assume that H is a field (replacing H by $\operatorname{Fract} H$ in the general case), so that $p \in A^\sigma$. According to Proposition 3.6, the operator p_i projects V_i onto $D(\alpha_0)$ parallel to the subspace $W_i = \operatorname{Ind}(N_1/N_{i+1})$. In particular, $p_0 = 1$ and $p_n = p_N$ in the notation of 3.7. Moreover, with each operator p_i we associate a homomorphic embedding

$$(11) \qquad \gamma_i \colon D(\alpha_0) \to V_i \qquad (i = 0, 1, \ldots, n)$$

defined by the rule $\gamma_i(x \cdot 1_\lambda) = x p_i v_0$, where $x \in A$. Thus, the passage from p_i to p_{i+1} is associated with the lifting of the homomorphism γ_i to the homomorphism γ_{i+1}. In particular, the embedding $D(\alpha_0)$ to the module $V = \operatorname{Ind} N$ is determined by γ_n.

In the general case (H arbitrary), with the module N we associate the following family of commutative diagrams:

$$(12) \qquad \begin{array}{ccc} & & V_i \\ & \overset{\psi}{\nearrow} & \downarrow \\ D(\alpha_0) & & \\ & \underset{\varphi}{\searrow} & \downarrow \\ & & V_{i-1} \end{array}$$

Here $\psi = \pi\gamma_i$ and $\phi = \pi\gamma_{i-1}$, where $\pi \in H$ is the left denominator of the rational function p_i (replacing γ_i by $\pi\gamma_i$ with $\pi \in H$ corresponds to replacing p_i by πp_i).

Thus, the lifting of homomorphisms for the module $\operatorname{Ind} N$ is determined by the diagram (12).

§4. The reductivity problem

4.1. Definition. Let us fix a subspace $P \subset X(H)$ invariant with respect to the action of the group Q and let $\mathcal{O} = \mathcal{O}(P)$ be the category of A-modules V satisfying the following conditions:

($\mathcal{O}_1$) The module V is locally nilpotent with respect to E (see 3.1).

($\mathcal{O}_2$) The module V possesses a P-grading V_λ ($\lambda \in P$) with homogeneous components of the form

$$(1) \qquad V_\lambda = \{v \in V \mid h_v = \lambda(h) v \text{ for all } h \in H\}.$$

It is easily verified that the Q-grading of the algebra A is consistent with the weighted grading (1) by relations of the form

$$(2) \qquad A_\alpha V_\lambda \subset V_{\alpha\lambda} \quad \text{for all } \alpha \in Q,\ \lambda \in P.$$

It is also easily verified that the category $\mathcal{O}$ is closed under the operations of passing to submodules and quotient modules. For each module $V \neq 0$ from the category $\mathcal{O}$ we have $V^E \neq 0$.

The module $V \neq 0$ is said to be *extremal of type* $\lambda \in P$ if it is generated by an extremal vector $v_0 \in V_\lambda^E$, i.e., $V = A v_0$. It is readily verified that the Verma module $M(\lambda)$ is universal in the class of extremal modules of type λ (see 3.4).

In particular, each simple module of category $\mathcal{O}$ is isomorphic to one of the modules of the form

$$E(\lambda) = M(\lambda)/N(\lambda), \tag{3}$$

where $N(\lambda)$ is the greatest submodule of the module $M(\lambda)$ not containing the cyclic vector 1_λ.

For simplicity, let us agree that the semigroup Q acts effectively on the elements $\lambda \in P$, i.e., the equality $\alpha\lambda = \lambda$ for $\alpha \in Q_-$ and $\lambda \in P$ implies $\alpha = 1$. In this case the simple module (3) is defined for each $\lambda \in P$.

REMARK. In the general case (see [**Z1**]) we can define $N(\lambda)$ as the greatest $\mathbb{Z}$-graded submodule in $N(\lambda)$ not containing the vector 1_λ. The corresponding module (3) is quasisimple, i.e., it does not contain any $\mathbb{Z}$-graded submodules that are different from 0 and $E(\lambda)$.

The reductivity problem for the category $\mathcal{O}$ consists in establishing reductivity criteria for modules V from category $\mathcal{O}$, including singling out some subcategories possessing the reductivity (semisimplicity) property.

4.2. Suppose the algebra A is regular, i.e., its Cartan subalgebra H is isomorphic to the algebra of functions on a complete subspace $P \subset X(H)$. We agree to write the elements $x \in A^\sigma$ in the form

$$x = \sum_{i,j} a_i b_j x_{ij} \tag{4}$$

with coefficients $x_{ij} \in H$. Accordingly, with each series (4) we associate a formal series

$$x(\lambda) = \sum_{i,j} a_i b_j x_{ij}(\lambda) \tag{5}$$

that depends on the parameter $\lambda \in P$.

The symbol (5) can also be used for elements (4) with $x \in A^\sigma(K)$, where $K = \operatorname{Fract} H$. More precisely, we define the singular set $\sigma(x)$ of the element x as the union of the countable family of sets $\sigma(x_{ij})$, where $\sigma(x)$ with $x \in K$ denotes the set of singular points of the numerical function $x(\lambda)$. Obviously, this definition does not depend on the choice of bases a_i, b_j. Accordingly, the symbol (5) is defined for $\lambda \notin \sigma(x)$.

The elements (5) can be interpreted as operators in the category $\mathcal{O}$ defined on the components V_λ for $\lambda \notin \sigma(x)$.

In particular, let $p \in A^\sigma(K)$ be an extremal projection. In this case for $\lambda \notin \sigma(p)$ we have

$$p(\lambda) V_\lambda = V_\lambda^E, \qquad p(\lambda)(V_\lambda \cap \operatorname{im} F) = 0. \tag{6}$$

Below (in 4.3) we shall see that the singular set $\sigma(p)$ plays an essential role in the investigation of the reductivity problem for the category $\mathcal{O}$. Therefore, we introduce the notation

$$P_{\text{reg}} = P \setminus \sigma(p).$$

4.3. For each module V from the category $\mathcal{O}$ let $P(V)$ be the set of all weights of the module V, i.e., the set of all $\lambda \in P$ for which $V_\lambda \neq 0$.

The weight $\lambda \in P(V)$ is said to be *primitive* if there exists a vector $v_0 \in V_\lambda$ that is mod N-*extremal* for some submodule N (i.e., $ev_0 = 0$ for all $e \in E$).

Let $P'(V)$ be the set of all primitive weights of the module V, and let $P''(V)$ be the subset of all $\lambda \in P'(V)$ having a strict majorant in $P'(V)$ (i.e., $\lambda < \mu$ for some $\mu \in P'(V)$).

The module V is said to be *regular* (*quasiregular*) if $P'(V) \subset P_{\text{reg}}$ ($P''(V) \subset P_{\text{reg}}$, respectively).

Let $\mathcal{O}_{qr}$ be the subcategory in $\mathcal{O}$ consisting of quasiregular modules. Similarly to 3.5 (see also [**Z2**]), it is easy to show that the category $\mathcal{O}_{\text{gr}}$ is semisimple.

Actually, as will be shown in 4.7, the quasiregularity condition is, in a certain sense, necessary for a certain subcategory in the category $\mathcal{O}$ to be semisimple.

4.4. DEFINITION. An algebra A will be called τ-*regular* if the set P is equipped with topology τ satisfying the following conditions:

(τ_1) Each function $h \in H$ is continuous and its null set $N(h)$ for $h \neq 0$ has a complement everywhere dense in the set P.

(τ_2) Each countable intersection of open subsets everywhere dense of the set P is everywhere dense in P.

For example, condition (τ_2) is fulfilled for the Zariski topology or for the complete metric topology of the space P.

It follows from conditions (τ_1) and (τ_2) that for each $x \in A^\sigma(K)$ the set $P \backslash \sigma(x)$ is everywhere dense in P.

4.5. Recall that the module M is said to be *projective* (in some category $\mathcal{X}$) if for each morphism $\varphi\colon M \to V$ (in the category $\mathcal{X}$) and for each covering $\pi\colon W \to V$ there exists a morphism $\psi\colon M \to W$ such that $\phi = \pi \circ \psi$. In other words, this means that the following diagram is commutative:

(7)

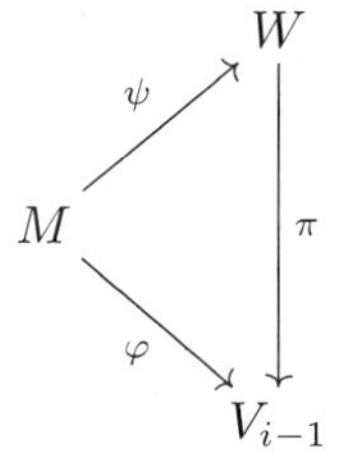

THEOREM. *Suppose the algebra A is τ-regular. In this case the set P_{reg} coincides with the set Π of all $\lambda \in P$ for which the module $M(\lambda)$ is projective.*

PROOF. If in the diagram (7) we have $M = M(\lambda)$ and $\lambda \in P_{\text{reg}}$, then it suffices to apply the operator $p(\lambda)$ to the extremal vector $v_0 = \varphi(1_\lambda)$. The corresponding vector $w_0 = pv_0$ defines the desired homomorphism ψ according to the rule $\psi(x1_\lambda) = xw_0$, where $x \in A$. Thus, $P_{\text{reg}} \subset \Pi$.

On the other hand, if $\alpha_0\lambda \in \Pi$, then for each sequence of morphisms (11) from 3.8 the following commutative diagrams are defined:

(8)

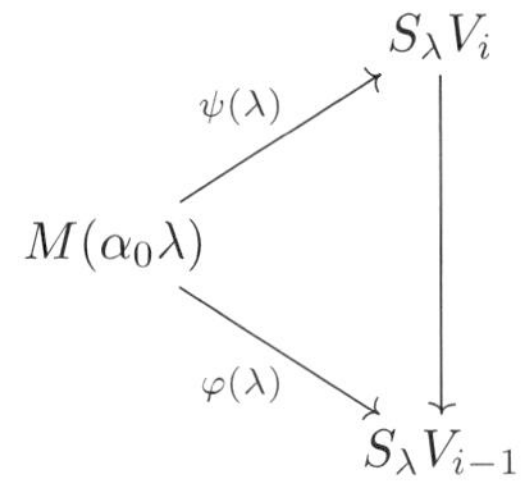

Here $\varphi(\lambda)$ ($\psi(\lambda)$) is a specialization of the morphism φ (ψ) defined by substituting $p_i(\alpha_0\lambda)$ for p_i.

Consider the following system of "extremal" equations in the space $S_\lambda V_i$:

$$Ex = 0, \qquad x \equiv v_0 \mod S_\lambda W_i \tag{9}$$

(in the notation 3.8). Choosing a suitable homogeneous basis of the free H-module V_i, we can identify each vector $x \in V_i$ with the vector-valued function $x(\lambda) = (x_0(\lambda), \ldots, x_m(\lambda))$, where $x_0(\lambda) = 1$, under the normalization condition (9).

Thus, for each $\lambda \in \alpha_0^{-1}\Pi$, there exists a solution $x(\lambda)$ of the system (9). For $\alpha_0\lambda \in P_{\text{reg}}$ this solution is unique and determined by the equation $x(\lambda) = p_i(\alpha_0\lambda)v_0$.

Since P_{reg} is everywhere dense in Π, it follows that the function $p_i(\alpha_0\lambda)v_0$ has no singularities at points $\lambda \in \alpha_0^{-1}\Pi$, i.e., $\Pi \subset P_{\text{reg}}$. Consequently, we obtain $\Pi \subset P_{\text{reg}}$. □

REMARK. The τ-regularity condition was not used in the proof of the inclusion $P_{\text{reg}} \subset \Pi$.

4.6. DEFINITION. An algebra A is said to be *σ-regular* if the functions $h \in H$ are continuous (with respect to some topology of the set P) and satisfy the following conditions:

(σ_1) The algebra H is factorial, i.e., each element $h \in H$, $h \neq 0$ can be represented in the form $h = h_1 \cdots h_n$ with prime (nondecomposable) factors $h_i \in H$ ($i = 1, \ldots, n$).

(σ_2) For each $h \in H$ and each prime $h_0 \in H$ the divisibility condition $h = gh_0$ (for some $g \in H$) is equivalent to the vanishing of the function h in a nonempty neighborhood of the null set $N(h_0)$.

In particular, these conditions are satisfied for the algebra of polynomials $\mathbb{C}[x_1, \ldots, x_n]$ with respect to the Zariski topology in $\mathbb{C}^n$.

4.7. THEOREM. *Suppose the algebra A is σ-regular. Then the singular set $\sigma(p)$ coincides with the set Γ of all $\lambda \in P$ for which there exists a nonzero homomorphism $M(\lambda) \to M(\alpha\lambda)$ for some $\alpha \neq 1$.*

PROOF. If $\lambda \in P_{\text{reg}}$, and $\gamma\colon M(\lambda) \to M(\alpha\lambda)$, $\alpha \neq 1$, is a homomorphism, then $\gamma(1_\lambda) \in \operatorname{im} F$; hence $\gamma(1_\lambda) = p\gamma(1_\lambda) = 0$, i.e., $\gamma = 0$. Thus, $\Gamma \subset \sigma(p)$.

On the other hand, let us consider the diagrams (8) generated by the commutative diagrams 3.8(12). Here $\psi(\lambda)$, $\varphi(\lambda)$ correspond to the operators $\pi(\mu)p_i(\mu)$, $\pi(\mu)p_{i-1}(\mu)$ for $\mu = \alpha_0\lambda$.

According to the definition of the denominator π, there exists a point $\mu \in P$ such that $\pi(\mu) = 0$, $\varphi(\mu) = 0$, and $\psi(\mu) \neq 0$. Hence $h_0(\mu) = 0$ for some prime factor of the element $\pi \in H$. The condition $\psi(\mu) \neq 0$ is satisfied in some neighborhood $\Omega \subset N(h_0)$. In view of diagram (8), the condition $\varphi(\mu) = 0$ in the neighborhood Ω indicates that for all $\mu \in \Omega$ there exists a nonzero homomorphism

$$\gamma(\mu)\colon M(\mu) \to M(\alpha\mu), \tag{10}$$

where $\alpha = \alpha_i \alpha_0^{-1}$. It should be noted that the set $\Gamma(\alpha)$ for all $\mu \in P$ for which (10) holds is the intersection of the null sets $N(h_i)$ for some $h_i \in H$ $(i = 1, \dots, m)$. Consequently, $h_i = 0$ on Ω $(i = 1, \dots, m)$; hence, according to condition (σ_2), we have $h_i = g_i h_0$ for some $g_i \in H$ $(i = 1, \dots, m)$. Now we have $\sigma(p_i) \subset \Gamma$, i.e., $\sigma(p) \subset \Gamma$. □

REMARK. The σ-regularity condition was not used in the proof of the inclusion $\Gamma \subset \sigma(\pi)$.

COROLLARY. *Suppose the algebra A is both τ-regular and σ-regular. Then we have $\Pi = P \setminus \Gamma$. In other words, the module $M(\lambda)$ is projective if and only if there are no nonzero homomorphisms $M(\lambda) \to M(\alpha\lambda)$ for $\alpha \neq 1$.*

EXAMPLE. The module $M(\lambda)$ over the Lie algebra $sl(2)$ is projective if and only if $\lambda \neq -2, -3, \dots$.

4.8. Theorem 4.7 reduces the description of the set P_{reg} to the investigation of the zeros of the canonical bilinear form

$$\varphi(\lambda)(x, y) = \varphi(x, y)(\lambda) \tag{11}$$

defined in $M(\lambda)$ for $\lambda \neq P$.

Indeed, $\sigma(p)$ (under the conditions of Theorem 4.7) is contained in the union of the sets αP_0 $(\alpha < 0)$, where P_0 is a set of all $\lambda \in P$ for which the module $M(\lambda)$ is not simple. On the other hand (see, for example, [**Z1, Z2**]), $\ker \varphi(\lambda) = N(\lambda)$. Consequently, P_0 coincides with the set of all $\lambda \in P$ for which the form (11) is degenerate in the space $M(\lambda)$.

For the types W and K the problem is trivial ($P_0 = \varnothing$). For the types SCH and DJ the description of the set P_0 is one of the main results of representation theory [**K2, CK**]. Accordingly, in all these cases the complete description of the set P_{reg} is known [**Z4, Z6**].

References

[CK] C. De Concini and V. G. Kac, *Representations of quantum groups at roots of* 1, Progr. Math. **92** (1990), 471–506.

[K1] V. G. Kac, *Infinite-dimensional Lie algebras*, Cambridge University Press, Cambridge, 1985.

[K2] V. G. Kac and D. A. Kazhdan, *Structure of representations with highest weight of infinite-dimensional Lie algebras*, Adv. Math. **34** (1979), 97–108.

[Z1] D. P. Zhelobenko, *Contragredient algebras*, J. Group Theory in Physics **1** (1990), no. 1.

[Z2] ———, *Representations of reductive Lie algebras*, "Nauka", Moscow, 1970. (Russian)

[Z3] ———, *S-algebras and Verma modules over reductive Lie algebras*, Dokl. Akad. Nauk SSSR **273** (1983), no. 4, 785–788; English transl. in Soviet Math. Dokl.

[Z4] ———, *Constructive modules and extremal projections over Chevalley algebras*, Funktsional. Anal. i Prilozhen. **27** (1993), no. 3, 5–14; English transl. in Functional Anal. Appl **27** (1993), no. 3.

[Z5] ——, *The algebra of quantum bosons, Schubert filtering and Lustzig bases*, Izv. Russ. Akad. Nauk Ser. Mat. **57** (1993), no. 6, 3–32; English transl. in Math. Russia-Izv.

[Z6] ——, *Extremal projections and reductivity problem for classical and quantum modules*, Vestnik Ross. Univ. Druzhby Narodov **1** (1993), no. 1, 22–33. (Russian)

[Z7] ——, *Cartan-type algebras*, Dokl. Russ. Akad. Nauk **339** (1994), no. 4; English transl. in Russian Math. Dokl.

Translated by N. K. KULMAN

Amer. Math. Soc. Transl.
(2) Vol. **175**, 1996

Alik Berezin
in the Recollections of Friends

V. P. Maslov's recollections

When I was a third-year student at the Physics Department of Moscow University, I made Alik[1] Berezin's acquaintance through my close friend N. Korst, who lived in the same house with him and had known him from childhood. Alik's nickname among the boys in the courtyard was "Beribes" (*beri, bes* in Russian means "take it, devil" and sounds like his last name). He lived in a two-room apartment with his mother, who adored him and was very supportive of his passion for physics and mathematics. She was a good cook and enjoyed playing the role of hostess. She was a regular listener of Radio Israel and showed keen interest in and felt emotionally involved with what was going on there.

Alik used to study with his window flung wide open both in summer and winter alike, the table moved up close up to the window-sill. When he was thinking about a problem, he used to walk back and forth in his room. He would often think while strolling in the woods. Once, when he was visiting me at my dacha, we took a walk in the nearby wood, talking as we went along the paths. After a short pause I asked him about something. He answered: "You are interfering with my thinking. Don't disturb me for a while, I am working" (on some problem in mathematics, no doubt).

We often talked about politics and politicians. I remember that he surprised me by saying that Khrushchev is the only guarantee of our security. That was all the more surprising since both his mother and I had a feeling of disgust and revulsion towards Khrushchev. Alik seemed to have strong misgivings about the future.

Many years later I gave him my first book as a present. He read the book and later returned it with a dedication of his own, which was a funny parody on my own dedication, which I had thought rather touching. The book returned also contained numerous reasonable remarks and some witty ones. One passage in the book was commented on in the following facetious way: "this is a piece of voluntarism, which is censured by our beloved Party!"

We had lengthy discussions on mathematical craft, on the major trends in mathematics, and talked about the leading mathematicians. He was of the opinion

[1] *Translator's note.* Felix Aleksandrovich Berezin was known to his friends as Alik.

that one should take up two widely differing branches of mathematics and do research in both with the utmost vigor and professionalism. According to him, this would benefit both branches.

He "awarded" military ranks to our "great" mathematicians. Thus, in those days he "promoted" L. Pontryagin to the rank of colonel, and I. Gelfand and M. Krein to that of general. The reason was that, as he explained, unlike I. Gelfand and M. Krein, Pontryagin had no army (of disciples and followers) of his own.

In general, he divided mathematicians into "essentially creators" and "essentially teachers". His own appraisal of himself was rather modest. He used to say that his own destiny was to teach and not to create. Of course, this was not true, but this indicates how devout he was in his calling as a teacher. Moreover, in his heart of hearts he realized that his research work was very important. He was much aggrieved that some of his work was not immediately accepted for publication. He felt that it was a shame not being able to put his ideas across to the reading public, since he realized that he understood the overall picture better than his more famous competitors.

Generally, his attitude had always been absolutely independent and ahead of its time. Even in 1979, a year before his tragic death, he was irritated at the mathematics community for not being able to grasp where the frontiers of science were. In a way, he resembled Nabokov's Pnin; in particular, he always looked somehow unprotected. He was not a fighter. People who live before their time die early.

M. A. Shubin's recollections

These are personal notes concerning my acquaintance and relationship with F. A. Berezin.

I believe that the fact that I met him and was able to talk with him and learn from him for a considerable period of my life was one of the best gifts of fortune to me.

We first met in 1967 during a Summer School on Representation Theory and Spectral Theory in a small summer resort called Zagulba on the Caspian sea near Baku. I was then 22 and had just completed the first year of my Ph.D. studies (my advisor was Professor M. I. Vishik). I guess that there were not very many single rooms in the place were we lived and certainly I did not expect the privilege of being given one. So there were three participants of the school in my room, the two others being F. A. Berezin and another graduate student of my age, Grisha Litvinov. Berezin was then 37, and from our (Grisha's and mine) point of view, very famous and deserving the highest respect. But for the administration of the school, he was still not up to the level of those who rate a separate single room.

Anyway all three of us lived peacefully in this room and, as far as I understand, Berezin first attentively observed the behavior of his roommates and our occasional mathematical conversations. At some point he asked us the following question:

"Do you want to learn quantum mechanics?"

"Of course we do," we answered. What followed was totally unexpected for Grisha and me.

"Let us write a book about it then," said Berezin. I do not know precisely what Grisha felt about this, but for me this proposal was astonishing. Authors of all books seemed persons on the level of demigods, not less. To see my name on the cover page of a book seemed something totally out of reach. But of course it was an attractive proposal. I could not believe that Berezin was serious, inviting such inexperienced people to participate in an act of creation. But he definitely was serious. He explained to us what he wanted us to do. We were expected to transform into a book some existing lecture notes of his and add something that he would explain to us. This is how my relationship with Berezin started.

Unfortunately, progress on the book was very slow. I was virtually illiterate in most aspects of physics, spectral theory and all the other things that were necessary. The lecture notes contained almost no proofs, so that we had to invent them (this was much easier than to find them in books and papers by other authors because of the poor situation with the libraries, not to mention the complete absence of copying facilities). Berezin was ready to help when we asked him, but his proofs usually contained new gaps to be filled and we did not dare to ask him for too many clarifying explanations. Grisha quit after learning the basics of quantum mechanics and writing a chapter about them. I continued, and as I see now this was one of the most profitable learning enterprises in my life. I acquired an active knowledge not only of quantum mechanics, but also of functional analysis, spectral theory, and representation theory. Moreover, my fear of physics and physicists gradually dispersed. I was reluctant to come to the Berezin seminar, where many talks in physics took place, because I was afraid that I would not understand anything. But once, before a physics talk, I shared my reluctance with Berezin and he said:

"Do you know already that a physical quantity is a self-adjoint operator?"

"Yes, I do," I answered.

"Then you don't need anything else," Berezin assured me. Listening to this talk I found that this was almost true. This situation was repeated again, and after a while I was not scared by physics talks anymore, realizing that actually not so many things are needed to understand them.

Many times I was a guest in Berezin's apartment, to talk about mathematics (there were no offices in Moscow State University, so people worked mostly at home). He also often shared his nonmathematical views about different aspects of life with me. He was a man of courage, never stepping back from his moral principles, although he was never an overt dissident. Sometimes we went skiing together. But this was too difficult for me, because he was much stronger and his level of fitness was incomparably better than mine.

Unfortunately the book was never finished in the form which Berezin had planned. The reason was that I was lazy, and besides I felt that I could do good things which might be more interesting. A preliminary version of the book was published in Russian in 1973 as "Lectures in Quantum Mechanics." But this was a very small part of what was planned. Then the work slowly advanced. Amazingly, Berezin never became angry at me about it (although I definitely deserved severe reproach). Only after his tragic death did I realize how criminally lighthearted I had been about the book. It became suddenly clear to me that I had lost a unique opportunity to learn a lot more from Berezin (being sure that I still had plenty of time) and that the book would never be completed at the desired level of quality. Then at last, strongly urged by A. S. Schwarz, I made an effort to do what I could, and the book under the title "The Schrödinger Equation" appeared in 1983 in Russian. The translation into English took another 8 years, although work on it began immediately (the English version was published by Kluwer in 1991).

Another period in my life in which Berezin played a very important role came around 1974, when I decided to try to submit a Doctor of Science thesis. I had done some things about spectral theory of almost periodic differential operators and in particular found a way to apply von Neumann algebras there. Everyone around me said that I had gathered enough material for this thesis. All except Berezin. He explained a few things about it to me. He said that my results lack applications to mathematical physics and suggested a direction for getting these applications. He called my attention to solid state physics and physics of disordered systems and conjectured that the integrated density of states should coincide with the spectrum distribution function that I had defined with the help of von Neumann algebras. Eventually I proved this conjecture and it became one of the best parts of my thesis. Berezin also insisted that the main results should be published in the Russian Mathematical Surveys and reported at the Moscow Mathematical Society. I did all this, although at the moment I was reluctant about it because this caused a considerable delay (and I knew perfectly well that theses of much lower quality were defended without such additional precautions). But later on I had a chance to see how wise Berezin was in giving me his advice. Shortly after I submitted my thesis (already satisfying all Berezin's requirements), certain people started trouble, trying to topple me. For me it was a fight for survival and I do not know what could have happened if I had lost it. But I won in the end, because of support of such mathematicians as V. Arnold, L. Faddeev, A. Kolmogorov, V. Marchenko, V. Maslov, S. Novikov, S. Sobolev, support that they could hardly have provided if my thesis had been as I had planned it before talking with Berezin.

Now it is obvious that the name of Berezin will be never forgotten in mathematics and mathematical physics (if only because of the Berezinian or of the Berezin integral over anticommuting variables). But it is important to remember that he was also a man of the highest moral standards, and, besides, was very kind and warmhearted. This helped people who learned from him, worked with him, and just surrounded him. Many Moscow students who were simply in his classes remember him as one of the teachers who regarded students as their equals (and to my understanding, such teachers were not a majority in Moscow then). I personally know that my meeting with him in 1967 changed my life and am infinitely grateful to Berezin for this.

A. M. Vershik's recollections

Berezin's departure from this world was unexpected and mysterious. There is an element of mystery in any death, but Alik's death, far away from Moscow, in a geological party, where there were no friends, under circumstances that remain unclear, was a strange kind of diappearance. I remember the moment when I was told about it, and the overwhelming feeling of disbelief...

Our last long conversation occured half a year before that, at a summer workshop near Minsk. This conversation made a strong impression on me: I recognized Alik's acute pessimism. The year was 1979, we were deep in the "stagnation period" with its oppressive atmosphere, the absence of any hope for the liberalization of our society, vicious attacks against any form of dissent, active emigration, dismal days at the university. We strolled for a long time in a quiet forest. Alik touched upon the usual topics of our discussions—the situation at *mekhmat*[2], on the hopelessness of trying to improve it, on the problem of mutual friends who were emigrating or weren't, on the impossibility of real contacts with Western mathematicians, and

[2] *Translator's note.* The Mechanics and Mathematics Department of Moscow University.

how this fact is exploited by some both here and abroad. But the main topic of our talk that evening was the Jewish problem, which we had rarely discussed before. I recall that it was this part of the conversation that struck me most of all; such gloomy forecasts of events to come I had rarely heard from anyone before: Alik was saying that he was afraid of pogroms and open persecution, that communo-fascist ideas were in the air, and so on. To the workshop I had taken with me some *samizdat* and *tamizdat* materials[3] and our journal *Summa* (a Leningrad *samizdat* publication, surveying a wide spectrum of political, social, and literary questions), where the Jewish problem was also discussed; I had shown all this to Alik. Like some authors from the dissidence movement, I viewed the future dangers under a different angle, and tried to convince Alik that pogroms were hardly an imminent reality. Today, after all these years, one can say that Berezin's predictions were, several times, on the verge of coming true, his intuition did not entirely deceive him. However, later, coming back to that conversation, I always felt that Alik's apocalyptic vision, in some mystic sense, was not an accident.

I first met Berezin in the early sixties here in Leningrad; he was at the height of his popularity and was often invited to our city. My first conversation with him disconcerted me somewhat: he said that my results on Gaussian dynamical systems had been known for a long time, but then agreed with my objections; I appreciated his ability to immediately get at the core of any subject and, of course, his wide erudition. Since then we met many times and talked about mathematics, about friends and acquaintances. These encounters did not occur too often, maybe two or three times a year, and I mostly listened and asked questions, at least in the first years. I think that it was progressively that an inner contact developed between us, the inner contact commonly known as friendship. It was easy to understand why close interaction with him was not easy to achieve: Alik often spoke with an aplomb that allowed no questions, was sometimes superficial, but these were minor traits in his overall intellectual image.

We in Leningrad often invited him for talks at sessions of the Leningrad Mathematical Society, at the V. I. Smirnov–O. A. Ladyzhenskaya seminar, and elsewhere. It seems that he enjoyed coming to visit us and engaging in discussions with colleagues whose work was within his sphere of interest (L. D. Faddeev, O. A. Ladyzhenskaya, M. S. Birman, V. S. Buslaev, and others). I participated in many workshops together with him (Katsiveli 1966, Kazan 1971, Tashkent 1975, Minsk 1979, among others). Always his participation was active and significant. His desire and aptitude to adapt to new interesting developments in mathematics that were presented at conferences and workshops was remarkable: he always strived to find a place for them in his continuously reconstructed physico-mathematical universum. I remember how early he had pointed out the role of topology in mathematical physics. On the other hand, he did not hide his attitude to pseudoscientific or superficial publications or reports, especially in areas adjoining physics.

Berezin occupied a special position in the Moscow and All-Union mathematical scene. He began as one of the most successful and cherished pupils of I. M. Gelfand in representation theory. The huge number of facts, inventions, unexpected relationships obtained at the time (the end of the fifties) constituted the core of this new theory: the role of Gelfand himself in this process is difficult to overestimate.

---

[3] *Translator's note.* Illegal literature, edited in the Soviet Union or abroad, respectively.

However, many of these achievements were then only sketched, others required corrections and additions. This is also true of Berezin's first papers, written together with or under the guidance of his teacher. The shortcomings of these papers, or perhaps other circumstances, resulted, unfortunately, in the end of the cooperation between teacher and pupil, which occured rather quickly, and this cooperation was never resumed again. Working with Gelfand in the seventies, and being friends with Alik, I tried to convince both of them in the usefulness and importance of their reunification, but was not very successful.

In attempting to assess his role in our mathematics, I should begin by saying that, in my opinion, it is precisely Berezin who was at the origin of the essential turning point in the work of many mathematicians and their rapprochement with physics. He was the first in his generation who, following his understanding of science, decided to carry out the huge efforts necessary to go into theoretical physics as a physicist rather than only as a mathematician, and succeeded in doing that. One can discuss at length to what extent this can or should be done, while still remaining a mathematician; there are examples when mathematicians became physicists, but Alik found his own proportion and became an active mathematical physicist and a propagandist of physical problems. He is the one who introduced many mathematicians to this circle of ideas, and some of them became outstanding experts in the theory of mathematical models of contemporary physics. Several participate in the present collection. The attitude of the leading physicists in the thirties, forties, and fifties to mathematics, although some of them were able to use the mathematical techniques of the times, was, to say the least, quite reserved. The fact that this has changed radically in the seventies, eighties, and nineties should to a great extent be credited to Berezin.

Berezin's favorite topics were quantization and Grassmanian analysis. In the creation of Grassmanian analysis and supermathematics, he played the leading role. Essentially the general program for the construction of this theory was sketched in his first book (incidentally, this book, first written in the form of an article, remained an unsolicited manuscript for several years with the editors of *Uspekhi*, and was finally published in the series *Biblioteka Uspekhov*). For many years, he presented these ideas with great enthusiasm and tried to convince many people to work on them, but the real explosion of interest came much later.

Unfortunately, Alik only lived to see the very beginning of the unquestioned acceptance of his ideas. Today supermathematics (the term is accidental and imperfect, but that is no fault of Berezin's) has become a kind of parallel mathematics: any result must have its "superanalog".

Berezin's contributions to the mathematical theory of quantization are so varied that it is hardly possible to sketch them briefly here. I will mention only a few fragments. The popular notion of a quantum group is the development of the idea of deformation of the universal enveloping algebra put forward by Berezin, although in a form somewhat different from the one considered in the eighties. The history of the Lie–Berezin–Kirillov–Kostant bracket, which also appeared in connection with quantization, is well known. My impression is that Berezin himself regarded his cycle of papers on quantization as his main and favorite theme.

I recollect many conversations with him on various topics. Most of them, in one way or another, had to do with different outgrowths of the main theme and can serve as examples of the application of physical ideas to purely mathematical problems. One of them (approximation in dynamical systems) was widely developed, although Berezin's role remained hidden in the background. He was interested in path integrals, where he justly considered himself to be one of the initiators, in the theory of von Neumann factors, C^*-algebras, asymptotic problems in algebra, and other topics. All the variations of the spectral theory of operators, scattering theory, the theory of matrix spectra, all this was always in the center of his attention. I remember discussions about the calculus of variations, on nonholonomic mechanics, on the algebraic aspects ot the theory of integrable systems.

The heritage of a mathematician is never limited to his published work and even to his manuscripts; a part of it, usually difficult to perceive for future generations, is transmitted through reports, talks, and conversations, in the ideas talked about with colleagues, and finally by the influence on others. All these components were strongly represented in Berezin's scientific life: by his talent and enthusiasm, his work, seminars, numerous reports and contacts, he succeeded in getting mathematicians, both well-known and young, interested in new problems.

His life in research, rich in scientific events, came to an early end. The hardships of existence in the Soviet Union for a talented scientist, who was not very loyal to the establishment and a Jew to boot, left a deep trace in Alik. I would not like to list here all the injustices and blows that he suffered from those in power, from his illwishers. Courageous by nature, Berezin always found the strength to rise above everyday annoyances and work, work. Was his outstanding talent fully realized? Can such a question be answered? Be that as it may, I feel he succeeded in telling us a great deal.

N. D. Vvedenskaya's recollections

These notes make no claim to completeness and no attempts at any generalizations. They are simply the reminiscenses of a close friend.

I met Felix Aleksandrovich Berezin in 1948, when we were both admitted to the Mechanics and Mathematics Department (*mekhmat*) of Moscow State University and were placed in the same study group[4]. We all called him Alik (this was the name used by his family and his close aquaintances), and I will use this name here. By the time we graduated, we had became close friends, but we got along quite well from the first year. I learned about his family and home later from his words.

Alik was brought up mainly by his mother's parents, because his mother herself, Esther Abramovna Rabinovich, was too preoccupied with her career (she was divorced from Alik's father). Alik recalled that in her youth she had envisioned the career of a concert pianist, but decided to become a doctor instead for ethical reasons. When he obtained his passport, Alik chose his mother's "nationality", officially becoming Jewish, and this choice was to play a crucial role in his life.[5] His mother's family came from Moldavia. Her brother as a very young man had enlisted in Yakir's bolshevik troops, becoming a soldier of the Red Army. Later he went on to be an important industrial administrator and as such perished during the Great

[4] *Translator's note.* Students of universities in the Soviet Union and present day Russia normally have the same prescribed curriculum and are organized into "study groups" of 20–30 people; all the groups attend the lectures together but separate for the exercise classes. Each group is often a closeknit organizational unit, not only in the administrative sense.

[5] *Translator's note.* Each Soviet citizen, upon receiving a passport (at the age of sixteen) had to indicate his so-called "nationality" (ethnic group); if both parents had the same nationality, this would be the one of the parents, if not, he/she could choose between the two nationalities of his parents or write "Russian".

Terror in the late thirties. (Incidentally, Yakir's name was always pronounced with respect in the family.) Alik's mother was not arrested because she left Moscow at the time (a timely departure would sometimes save people from arrest, especially if they were not the "principal candidates for arrest").

I write about this because the dread of terror always hung as a dark shadow over the family.

In the beginning of 1953, during the height of the campaign against "rootless cosmopolites" and the "doctors' case", rumors began to circulate in Moscow about imminent arrests of Jews and their expulsion (the rumors were founded, Stalin did have such intentions). Alik and his mother were also expecting arrest, they had already prepared little suitcases with underwear and medical supplies (the experience of 1937!).

Unfortunately, this complex system of fears remained with Alik for his whole life. He believed that all of us were under constant surveillance and was inclined toward gloomy predictions. For example, before the Moscow Olympiad (in 1980, the last year of his life) there was a rumor that during the games people would be allowed to come to Moscow only with special passes. Alik surmised that this restriction would remain for all time.

We, people from Alik's circle of friends, often made fun of his fears, but he was never able to overcome his fright and the anxiety acquired in childhood. (I am writing this in a completely different time; it is difficult, fortunately, for younger people to imagine the atmosphere of the Stalin and post-Stalin years. I hope that Alik's daughter Natasha will never experience such feelings.)

Having completed his high school education, Alik wanted to study at the Physics Department of Moscow University. He had graduated from school with a gold medal, which according to the existing rules gave him the right of having the entrance exams replaced by an interview. But in 1948 practically no ethnic Jews were accepted to the Physics Department; after an unsuccessful interview there, Alik succeeded in entering the Mechanics and Mathematics Department.

From our first year we became part of a circle of friends, among whose other members I should mention M. Agranovich. A. Yushkevich, S. Kamenomostskaya, O. Ziza, V. Ryzhik. Usually we sat together at lectures. Later all of us would go ice-skating. Outwardly Alik was not particularly striking. He was always dressed neatly but modestly, not to say poorly (although, of course, most people dressed in that way then). Actually he had a snug jacket with warm lining. At lectures (which were obligatory) we would sit near an open window. I would feel cold, and while I heard most of the lectures in that jacket, Alik remained in shirtsleeves without any problems. I later had the opportunity of noting his immunity to cold during our long walks, which are described below. I recall how one very cold spring we were canoeing near Moscow and Alik's canoe companion paddled in such a way that Alik was wet from head to toe, but during the whole day he never said a word.

Before the begining of studies in our first year, all the freshmen students were sent to work on the construction of the new Moscow University building on Lenin Hills (we studied in the old building on Mokhovaya Street). Working at the construction site, students became acquainted, got to know each other. After that Alik was elected *komsorg* of our study group[6], and this meant that he was popular

[6] *Translator's note.* Leader of the Komsomol (Young Communist League) organization of the group.

among the students of his group; one should not forget that at the time almost all students were *komsomol* members. Of course Alik never became a person making a political career, although he was a very noticeable figure in *mekhmat*. Incidentally, at the end of our fifth year (our last year of studies), he edited a very nonstandard issue, with "purely imaginary" number i, of the wall newspaper of our graduating class.

Alik studied extremely well and was one of the best, if not the very best, student of our class. From the freshman year he participated in E. B. Dynkin's seminar, then started going to I. M. Gelfand's. Nevertheless, he was not even recommended for graduate studies (this was 1953, and he was Jewish), but his "diploma thesis" was highly commended. The defense of the thesis took place at the Chair of Algebra, then headed by A. G. Kurosh.

After graduating from university, Alik began working as a school teacher: no other job was available for him.

At the university he was on friendly terms with many people, but had few close friends. He himself regarded some two to three people as his close friends in his youth and four to five people in his mature years. In them he appreciated originality of mind, sense of humor, sense of purpose. Professionalism was also very important, but not understood in the narrrow sense of professional success. Thus, in his friend Valery Nikolskii, a physicist, he mainly appreciated erudition, Valery's rebellious nature and even his devil-may-care attitude and relish of boisterous festivities (Alik himself was not inclined to the latter). In himself Alik mainly appreciated his professionalism as a mathematician.

Alik was an introverted person, he was easily hurt. He suffered considerably from the cool reaction to his mathematical ideas and achievements. For this reason, I believe, he parted with several people for whom he felt a strong sympathy in his youth.

His faith in mathematics and his studies was the main trait of his character. I recall that in some context (I suppose I was trying to convince him to go somewhere to relax) he said sharply: "I have no private life other than mathematics" (this is when he had a wife and daughter, and his mother was still living with him). Many times he refused to go on vacation, declined other entertainment, so as not to be distracted from work. Here he was quite capable of not keeping a promise, letting down his friends with whom he had intended to spend some time.

In Moscow, living in the city, Alik subjected everything to his working schedule and was very serious about keeping himself in good mathematical shape, which included taking care of his health and making sure he had enough sleep. (He even claimed that he became seriously ill as the result of one sleepless night.) One of his favorite sayings was "I love having a good time, especially having some sleep". He was always very anxious about his health (which was excellent), worried a lot about it, and, as the pessimist that he was, predicted his own early death (alas, this prediction came true).

Also very important for Alik were recreation and physical activity—he liked to move. Recreation was ordinarily "to take a walk". In the evening after work he would stroll in the streets around the house, on weekends would leave the city. These weekly hikes played an important role in his life. They were in fact the source of our friendship.

Alik knew the outskirts of Moscow quite well, and visited them often. On weekends he would hike by himself or, more often, with a few friends. Beginning from our two last years at *mekhmat* (1952–53), I started participating in these walks regularly, on foot in summer, on cross country skis in the winter. At different times various people from our graduating class took part: we often hiked with A. Yushkevich and V. Nikolskii, less often with D. Kazhdan, A. Schwartz, E. Fradkin. The women were usually S. Kamenomostskaya, I. Karpova (Alik's first wife), R. Kallosh. We would take the suburban train, and walk for 20–30 kilometers, sometimes more. There were some records, say to hike from Serpukhov to Kashira in one day, that is 60 kilometers. We would walk practically without stopping, eat sandwiches, stopping to swim if there was water on our way (Alik was fond of swimming and diving). Occasionally we would make campfires. In the first years we usually hiked around the Savyelovskaya railroad line (or walked from it to the Yaroslavlskaya or Leningradskaya line), because Alik lived near the Savyelovkii terminal. The fact that I, say, lived in the opposite end of the city was never taken into consideration, especially since the Northern outskirts of Moscow are reputed to be the most beautiful. When Alik moved nearer to the university, we started hiking mostly near the Kievskaya rail line. We walked through forests and fields, often made new tracks in virgin snow in winter. During the spring thaw we usually walked along the *betonkis*[7] encircling Moscow, at that time there would be practically no traffic on them. During his youth, in the spring and fall Alik would wear cheap army-type boots, which he considered very comfortable (for me these boots were always associated with soldiers on march). On skis he always wore a flannel sports outfit of the type that was worn by everyone in the early fifties. Only at the end of his life did he acquire a "modern" training suit and lovely ski cap. On the other hand, his skis were always of high quality. During our trips Alik always chose the itinerary. When the two of us went for walks, somehow there was no incentive to argue, but when there were many of us, arguments did arise. But usually Alik would listen to no one, and walked where he saw fit, and the others obediently followed suit. He exhibited the same stubborness in longer trips as well.

Camping trips for several days, summer or winter "wild" excursions, were a very special part of our existence. Describing the lives of our friends and ourselves, it is impossible to bypass them. We loved to travel, we were fond of remaining alone with nature, and at that time travel within the country was relatively cheap. Usually in the summer we would not go to vacation resort boarding houses, nor to somebody's dacha: we would go camping, to the mountains and the *taïga* by foot, to waterways by canoe (actually in the summer Alik would usually rent a dacha in order to be able to live and work there and have the opportunity to take little walks and go swimming). Several times Alik went on winter skiing trips and, in the summer, on canoeing trips on northern and Siberian rivers and lakes. He went only to very few mountain trips, and, I think, did not like or understand mountains. In the mountains also he prefered to "go his own·way": avoiding paths progressively curling uphill, he would go straight up the steepest incline, and move downhill in whatever direction he saw fit. In a similar way, on a lake he might choose to go "straight out to the open sea" even in stormy weather. In other words, he was not an easy companion on such trips.

In the spring, during flood time, we often went canoeing for several days, usually during the May Day holidays. Alik was a strong paddler, was never afraid of the cold (as I mentioned previously), was an expert campfire maker at our stops, and could

[7] *Translator's note.* Concrete highways made for the military.

blow up a soccer ball with his mouth. In general he was quite strong, especially in his hands and fingers.

He was also fond of working manually, say to repair a damaged canoe or even to fix broken skis. Once, using parts of a broken kayak and ski poles, he contrived a kitchen table of which he was very proud. In his house and mine he drove nails into concrete walls and was indeed a jack-of-all-trades.

As I have already mentioned, Alik's general appearance was not especially striking. He dressed very modestly, to a suit and tie he preferred informal checkered shirts and sweaters. At social gatherings he tended to remain quiet, although he was an excellent conversationalist. Many thought that he was conceited, because he often failed to say hello to persons he met at the department, but this was due to his poor eyesight—he simply did not recognize people (he only wore glasses at lectures and classes). He was unpretentious in his everyday needs, in particular concerning food. I think this was due to family traditions (in his youth his family lived rather poorly). When he began working at the university, Alik always had leftover money from his salary, which he readily loaned to people, not reacting if the money was not returned ("I understand that he has children and no money", he once said about one of his debtors).

On our walks together, or telephoning each other several times a week, we spoke of many things: about life and people around us, about politics, about books we had read. Conversing, we enjoyed evoking events from our student life or camping trips. Together with A. Yushkevich, the three of us sometimes decided in advance who would be the storyteller of one of our favorite episodes. In a more serious vein, I should mention that Alik's forecasts for the future of our country were not very cheerful (but who could then predict the approaching disintegration of the Soviet system!). Nevertheless, he loved Moscow, the university, and his way of life. At different times he had different thoughts about emigration (they depended on his mood and the level of unpleasantness at the university). But once, during the peak of emigration, to my direct question about leaving the country he answered: "No. I love the countryside near Moscow. I love the Russian language. I have several friends here, you among them."

Like all of us, Alik actively desired to see the world, to visit various countries. Even more ardently, he wanted to travel for contacts with mathematicians and physicists. But he was deemed "unsuited for travel"[8] by the administration. He was, however, allowed out to Poland and Hungary (iron curtain countries). In Hungary he first lived at Karoly Majus' place and later simply in an office of the Mathematics Institute (an example of his unpretentiousness). He also visited Mongolia. He gave interesting descriptions of all these trips. He also told me about his first trip to Kolyma (I vividly recall the picture of ecological disaster due to the effects of gold prospecting dredges, which had literally turned a river bed inside out, that he had drawn). During one of his vacations he had gotten himself hired as a manual worker in a geological party. We sometimes took such trips instead of our traditional camping excursions; one of Alik's school friends, a geologist, worked in Magadan. The second trip to Kolyma was to cost Alik his life.

---

[8] *Translator's note.* An ordinary Soviet citizen was not allowed to travel abroad; at the time this was a special privilege granted to those who were on especially good terms with their superiors and not blacklisted by the KGB.

As I mentioned before, Alik wanted to see the world, especially so because in the seventies he had numerous invitations from abroad. But trips abroad were then impossible, and Alik wrote a letter to the rector of Moscow University, R. V. Khokhlov, describing the appaling status of all creative mathematicians and their pupils at *mekhmat.* After Khokhlov's accident and death, this letter came into the possession of the department's party and administrative leaders, and they did all they could to "punish" Alik and not to allow him to travel anywhere. Even after the new rector, Logunov, had authorized his trip to Poland, the department let the affair drag on for over a year, shuffling the official papers back and forth between bureaucrats' offices. Alik died before the final permission was given. I would like to stress that despite his pervailing interest in mathematics, Alik was not indifferent to questions of social justice. The letter to Khokhlov is an example.

Alik had a very poor command of foreign languages. In the last years of his life, when physicists desiring to talk to him would come to Moscow, he asked me several times to participate in walks with them in order to help translate from and into English.

But in Russian Alik spoke beautifully, in a rich, sometimes somewhat bookish language. He read a great deal and had an excellent literary library (besides the mathematical one). He was a widely educated person, was fond of classical literature, history, and read with pleasure such things as Chinese chronicles. His favorite books were *The Adventures of Brave Soldier Schweick, Penguin Island,* and the verse of Heine. Alik also loved music. He did not go to concerts often, but his mother always played the piano at home.

In the summer of 1980 Alik went on his second trip to Kolyma, again to work in a geological party. According to the account of the other workers in his party, they were going downstream in an inflatable boat on the day Alik was to leave camp to return to Moscow. These boats are always difficult to maneuver. They were in a strong current heading for a heap of logs. Such accumulations of logs in Siberian rivers can be extremely dangerous: the stream plunges under the logs that stick out like the needles of a huge hedgehog. Everyone jumped out of the boat, pulled it out (thus the water was not deep), and when they looked around, Alik was not there. No one saw him getting carried under the logs. He drowned. His wife, Elena Grigorievna Karpel' and V. P. Palomodov flew to Kolyma and brought back his remains to Moscow. Alik was survived by a four year old daughter.

As any actively working mathematician, Alik was in need of recognition of his work by his peers. It came to him at the very end of his life, but to a greater extent only after he was no more.

I have not written about Alik as a mathematician since we did not intersect professionally. But we met very often, he sometimes came to my parties, would usually come to my birthday celebrations and give me roses. At my housewarming he presented me with a hammer with an inscription engraved on it. Very rarely, only a few times in all these years, did he organize parties at his home, inviting his pupils and friends. When I broke my leg, he visited me at the hospital. He was my close friend, a person I could always rely on.

Translated by A. B. SOSSINSKY

Selected Titles in This Series

(Continued from the front of this publication)

144 **S. D. Berman et al.,** Thirteen Papers Translated from the Russian
143 **V. A. Belonogov et al.,** Eight Papers Translated from the Russian
142 **M. B. Abalovich et al.,** Ten Papers Translated from the Russian
141 **H. Draškovičová et al.,** Ordered Sets and Lattices
140 **V. I. Bernik et al.,** Eleven Papers Translated from the Russian
139 **A. Ya. Aĭzenshtat et al.,** Nineteen Papers on Algebraic Semigroups
138 **I. V. Kovalishina and V. P. Potapov,** Seven Papers Translated from the Russian
137 **V. I. Arnol′d et al.,** Fourteen Papers Translated from the Russian
136 **L. A. Aksent′ev et al.,** Fourteen Papers Translated from the Russian
135 **S. N. Artemov et al.,** Six Papers in Logic
134 **A. Ya. Aĭzenshtat et al.,** Fourteen Papers Translated from the Russian
133 **R. R. Suncheleev et al.,** Thirteen Papers in Analysis
132 **I. G. Dmitriev et al.,** Thirteen Papers in Algebra
131 **V. A. Zmorovich et al.,** Ten Papers in Analysis
130 **M. M. Lavrent′ev, K. G. Reznitskaya, and V. G. Yakhno,** One-dimensional Inverse Problems of Mathematical Physics
129 **S. Ya. Khavinson,** Two Papers on Extremal Problems in Complex Analysis
128 **I. K. Zhuk et al.,** Thirteen Papers in Algebra and Number Theory
127 **P. L. Shabalin et al.,** Eleven Papers in Analysis
126 **S. A. Akhmedov et al.,** Eleven Papers on Differential Equations
125 **D. V. Anosov et al.,** Seven Papers in Applied Mathematics
124 **B. P. Allakhverdiev et al.,** Fifteen Papers on Functional Analysis
123 **V. G. Maz′ya et al.,** Elliptic Boundary Value Problems
122 **N. U. Arakelyan et al.,** Ten Papers on Complex Analysis
121 **V. D. Mazurov, Yu. I. Merzlyakov, and V. A. Churkin, Editors,** The Kourovka Notebook: Unsolved Problems in Group Theory
120 **M. G. Kreĭn and V. A. Jakubovič,** Four Papers on Ordinary Differential Equations
119 **V. A. Dem′janenko et al.,** Twelve Papers in Algebra
118 **Ju. V. Egorov et al.,** Sixteen Papers on Differential Equations
117 **S. V. Bočkarev et al.,** Eight Lectures Delivered at the International Congress of Mathematicians in Helsinki, 1978
116 **A. G. Kušnirenko, A. B. Katok, and V. M. Alekseev,** Three Papers on Dynamical Systems
115 **I. S. Belov et al.,** Twelve Papers in Analysis
114 **M. Š. Birman and M. Z. Solomjak,** Quantitative Analysis in Sobolev Imbedding Theorems and Applications to Spectral Theory
113 **A. F. Lavrik et al.,** Twelve Papers in Logic and Algebra
112 **D. A. Gudkov and G. A. Utkin,** Nine Papers on Hilbert's 16th Problem
111 **V. M. Adamjan et al.,** Nine Papers on Analysis
110 **M. S. Budjanu et al.,** Nine Papers on Analysis
109 **D. V. Anosov et al.,** Twenty Lectures Delivered at the International Congress of Mathematicians in Vancouver, 1974
108 **Ja. L. Geronimus and Gábor Szegő,** Two Papers on Special Functions
107 **A. P. Mišina and L. A. Skornjakov,** Abelian Groups and Modules
106 **M. Ja. Antonovskiĭ, V. G. Boltjanskiĭ, and T. A. Sarymsakov,** Topological Semifields and Their Applications to General Topology

(See the AMS catalog for earlier titles)